Lecture Notes in Mathematics 2054

Editors:
J.-M. Morel, Cachan
B. Teissier, Paris

T0207360

For further volumes:
http://www.springer.com/series/304

Jakob Stix

Rational Points and Arithmetic of Fundamental Groups

Evidence for the Section Conjecture

 Springer

Jakob Stix
Mathematics Center Heidelberg (MATCH)
University of Heidelberg
Heidelberg, Germany

ISBN 978-3-642-30673-0   ISBN 978-3-642-30674-7 (eBook)
DOI 10.1007/978-3-642-30674-7
Springer Heidelberg New York Dordrecht London

Lecture Notes in Mathematics ISSN print edition: 0075-8434
                             ISSN electronic edition: 1617-9692

Library of Congress Control Number: 2012945519

Mathematics Subject Classification (2010): 14H30, 14G05, 14H25, 11G20, 14G32, 14F35

Printed on acid-free paper

Springer is part of Springer Science+Business Media (www.springer.com)

To Antonia, Jaden and Lucie
and to Sabine

Preface

The section conjecture, as stated by Grothendieck [Gr83] in a letter to Faltings in 1983, speculates about a representation of rational points in the realm of anabelian geometry. Every k-rational point of a geometrically connected variety X/k gives rise to a conjugacy class of sections

$$s : \pi_1(\mathrm{Spec}(k)) \to \pi_1(X)$$

of the natural projection $\pi_1(X) \to \pi_1(\mathrm{Spec}(k))$ of étale fundamental groups. The section conjecture suggests that the converse also holds for smooth, projective curves of genus at least 2 over fields k that are finitely generated over $\mathbb{Q}$.

If the section conjecture turns out to be true, then it would shed new Galois theoretic light on the old Diophantine problem of describing rational points.

This volume of Lecture Notes consists of the author's Habilitationsschrift and aims to develop the foundations of the anabelian geometry of sections and to present our knowledge about the section conjecture with a natural bias towards the work of the author. In addition, we discuss the section conjecture over number fields from a local to a global point of view and provide detailed discussions of various analogues of the section conjecture, which might serve as supporting evidence in favour of the conjecture.

Acknowledgments

As this present book builds on several years of work, it is quite natural that over this period I have been influenced by many people and institutions and owe to them a certain debt of gratitude that I will now try to balance.

I have worked at several places: the Mathematische Institut Bonn, the Institute for Advanced Study in Princeton for the academic year 2006/2007, the University of Pennsylvania in Philadelphia in 2007/2008, the Isaac Newton Institute in Cambridge during the summer of 2009, and most recently at the Mathematisches Institut Heidelberg and in particular the MAThematics Center Heidelberg (MATCH).

I thank all these institutes, particularly the people there, I thank for the excellent working conditions and the supportive environment that they provided, for the coffee and for the discussions on mathematics, life, the universe, and all the rest.

I am grateful to Florian Pop for his ongoing interest in my work, for the numerous discussions, and for becoming, beyond an academic teacher, a collaborator and friend. I thank Hiroaki Nakamura and Akio Tamagawa for my several stimulating visits to Okayama and Kyoto. To Kirsten Wickelgren and Gereon Quick, I owe thanks for the enlightening discussions on the point of view of homotopy fixed points towards the section conjecture. My thanks also go to Jochen Gärtner for the discussions on Galois groups of number fields with restricted ramification. I thank David Harari for the pleasant collaboration on the descent obstruction, and I am grateful to Hélène Esnault and Olivier Wittenberg for the wonderful, stimulating, and very intense meetings in Essen, Paris, and Heidelberg, and also for their enthusiastic approach towards mathematics.

There are certainly aspects of the section conjecture beyond what is covered in this book, and also some aspects that were originally intended to be included. But then, among other things, Antonia and Jaden would not sleep in their beloved bunk bed. So, by keeping me busy otherwise, my kids earned the honour of forcing me to focus on finding an end point for this piece of work. I am thankful to Antonia, Jaden, and Lucie for the joy that they are and give every day, and for bringing me coffee and breakfast while these final lines were being written in January 2011.

To my parents, Gisela and Michael Stix, I am extremely grateful for being there whenever a helping hand was needed.

Writing my Habilitationsschrift as a culmination of research conducted over several years while being part of a growing family would have been simply impossible without the warm, enthusiastic, and inexhaustible support of my beloved wife, Sabine. I thank her from the bottom of my heart with wholehearted love and admiration, although there are no words that capture the uncountable gratefulness she deserves.

Heidelberg, Germany   Jakob Stix

Contents

Introduction

This volume of Lecture Notes explores the anabelian geometry of the section conjecture that was disclosed by Grothendieck [Gr83] in a letter to Faltings in 1983. On page 6 of the letter the conjecture reads:

> ...Es sei $\Gamma(X/K)$ die Menge aller K-wertigen Punkte (also "Schnitte") von X über K, man betrachtet die Abbildung
>
> $$\Gamma(X/K) \rightarrow \operatorname{Hom ext}_{\pi_1(K)}\big(\pi_1(K), \pi_1(X)\big), \tag{7}$$
>
> wo die zweite Menge also die Menge aller "Spaltungen" der Gruppenerweiterung (3) ist, ..., oder vielmehr die Menge der üblichen Konjugationsklassen solcher Spaltungen via Aktion der Gruppe $\pi_1(\overline{X})$. Es ist bekannt, daß (7) injektiv ist, und die Hauptvermutung sagt aus, daß sie bijektiv ist. [siehe unten Berichtigung]

This quote leaves us with a question, a remark and an expectation. The question, besides that the quote is in German, posed as: *what is (3) and what is the map in (7)?*, will be answered when we properly state the conjecture below, while the remark follows after the conjecture has been properly stated. The expectation refers to the *Berichtigung* (an erratum) and will be discussed when it becomes necessary.

The Section Conjecture

Let k be a field, let $\bar{k}$ be a fixed separable closure, and let $\operatorname{Gal}_k = \operatorname{Gal}(\bar{k}/k)$ be the absolute Galois group of k. Let X/k be a geometrically connected variety, let $\overline{X} = X \times_k \bar{k}$ be the base change of X to $\bar{k}$, and let $\bar{x} \in \overline{X}$ be a geometric point. The étale fundamental group $\pi_1(X, \bar{x})$ with base point $\bar{x}$ is an extension, see [SGA1] IX Theorem 6.1,

$$1 \rightarrow \pi_1(\overline{X}, \bar{x}) \rightarrow \pi_1(X, \bar{x}) \rightarrow \operatorname{Gal}_k \rightarrow 1, \tag{3}$$

where $\pi_1(\overline{X}, \bar{x})$ is the geometric fundamental group of X with base point $\bar{x}$. In the sequel, we denote the extension (3) by

$$\pi_1(X/k),$$

with the notation indicating that (3) captures the relative π_1-datum of the projection
pr $: X \to \mathrm{Spec}(k)$. The choice of a base point $\bar{x}$ with $\mathrm{pr}_*(\bar{x})$ equivalent to the
canonical (due to fixing $\bar{k}$) base point

$$* : \mathrm{Spec}(\bar{k}) \to \mathrm{Spec}(k)$$

of $\mathrm{Spec}(k)$, such that $\mathrm{Gal}_k = \pi_1(\mathrm{Spec}(k), *)$, is silently understood.

To a rational point $a \in X(k)$ the functoriality of π_1 for pointed spaces gives rise
to a continuous homomorphism

$$s_a : \mathrm{Gal}_k \to \pi_1(X, \bar{a}),$$

where $\bar{a}$ is a geometric point above a compatible with $*$. An étale path γ from $\bar{a}$ to
$\bar{x}$ on $\overline{X}$ defines an isomorphism

$$\gamma(-)\gamma^{-1} : \pi_1(X, \bar{a}) \xrightarrow{\sim} \pi_1(X, \bar{x})$$

compatible with the projections pr_* to Gal_k. The composition $\gamma(-)\gamma^{-1} \circ s_a$ thus
defines a section of

$$\mathrm{pr}_* : \pi_1(X, \bar{x}) \to \mathrm{Gal}_k$$

or a splitting of (3). Changing the étale path γ on $\overline{X}$ varies the section over a
$\pi_1(\overline{X}, \bar{x})$-conjugacy class of splittings/sections, so that the conjugacy class itself,
denoted by $[s_a]$, depends only on the rational point $a \in X(k)$. We denote the set of
$\pi_1(\overline{X}, \bar{x})$-conjugacy classes of sections of $\pi_1(X/k)$ by

$$\mathscr{S}_{\pi_1(X/k)} = \{s : \mathrm{Gal}_k \to \pi_1(X, \bar{x}) ; \; \mathrm{pr}_* \circ s = \mathrm{id}_{\mathrm{Gal}_k}\} /_{\pi_1(\overline{X}, \bar{x})\text{-conjugacy}}.$$

Changing the base point $\bar{x}$ leads to another description of $\mathscr{S}_{\pi_1(X/k)}$ with a canonical
identification between the two descriptions, which moreover satisfies the cocycle
relation for composing the identifications between three choices of base points. The
set $\mathscr{S}_{\pi_1(X/k)}$ is therefore canonically associated to the variety X/k.

Definition 1. The *(profinite) Kummer map* is the well defined map $a \mapsto [s_a]$

$$\kappa : X(k) \to \mathscr{S}_{\pi_1(X/k)}$$

denoted by (7) in the quote from [Gr83] above, see Sect. 2.4 and in particular
Remark 21 for an origin of the terminology.

    The section conjecture of Grothendieck gives a conjectural description of the set
of all the sections in an arithmetic situation as follows.

Conjecture 2 (Grothendieck [Gr83]). *Let X be a smooth, projective and geometrically connected curve of genus ≥ 2 over a field k that is finitely generated over $\mathbb{Q}$.
Then the profinite Kummer map $a \mapsto [s_a]$ is a bijection from the set of rational points
$X(k)$ onto the set $\mathscr{S}_{\pi_1(X/k)}$ of $\pi_1(\overline{X}, \bar{x})$-conjugacy classes of sections of $\pi_1(X/k)$.*

Remark 3. The form of the section conjecture given in Conjecture 2 is one of the standard forms for the conjecture. Actually, the quote above shows that Grothendieck originally stated that injectivity is clear and referred to the surjectivity question as the "Hauptvermutung" (Main Conjecture). For more on the issue of injectivity also in cases beyond the claim of the section conjecture we refer to Chap. 7.

The original form of the section conjecture addressed anabelian curves which are not necessarily projective. The affine case however needs the erratum alluded to above, see Conjecture 6 below for a precise version of the conjecture.

We take the opportunity to distinguish the sections s_a by a name.

Definition 4. A *Diophantine section* is a section $s : \mathrm{Gal}_k \to \pi_1(X, \bar{x})$ of the extension (3) for a geometrically connected variety X/k that is of the form $s = s_a$ for a rational point $a \in X(k)$, more precisely: $s = \gamma(-)\gamma^{-1} \circ s_a$ for an étale path γ on $\overline{X}$.

The Erratum

Conjecture 2 should apply to all anabelian varieties, in particular, at least to hyperbolic curves, the only varieties that are known to definitely show anabelian behaviour.

Definition 5. A *hyperbolic curve* is a smooth, geometrically connected curve U/k with a smooth, projective completion $U \subseteq X$ of genus g such that the complement $Y = X \setminus U$ is étale over $\mathrm{Spec}(k)$ and the Euler characteristic

$$\chi_U = 2 - 2g - \deg(Y)$$

is negative. We will frequently use the term *curve* to include the property of being geometrically connected.

In short: a hyperbolic curve is a map $U \to \mathrm{Spec}(k)$ with completion $X \to \mathrm{Spec}(k)$ that is a good Artin neighbourhood of $\dim(U) = 1$.

An erratum to Conjecture 2 when *projective* gets replaced by *hyperbolic* becomes necessary because a rational point $y \in Y(k)$ gives rise to non-Diophantine sections, see Chap. 18. In the postscript of [Gr83] we can read the following erratum in the case of anabelian curves which are not necessarily projective.

> ...so läßt sich genau angeben, welches die Spaltungsklassen im zweiten Glied sind, die keinem "endlichen" Punkt also keinem Element des ersten Gliedes entsprechen; ...Es sei
>
> $$\pi_1(K)^0 = \text{Kern von } \pi_1(K) \to \hat{\mathbb{Z}}^* \text{ (der zyklotomische Character).}$$
>
> Gegeben sei eine Spaltung $\pi_1(K) \to \pi_1(X)$, also $\pi_1(K)$ und deshalb auch $\pi_1(K)^0$ operiert auf $\pi_1(\overline{X})$, der geometrischen Fundamentalgruppe. Die Bedingung ist nun, daß die Fixpunktgruppe dieser Aktion nur aus **1** bestehen soll!

Thus Grothendieck states the section conjecture more precisely as follows.

Conjecture 6 (Grothendieck [Gr83]). *Let k be a field that is finitely generated over $\mathbb{Q}$. For a hyperbolic curve U/k with geometric point $\bar{u} \in U$, the profinite Kummer map induces a bijection*

$$\kappa : U(k) \to \left\{ s \in \mathscr{S}_{\pi_1(U/k)} ; \begin{array}{l} \textbf{1} \text{ is the only fixed point of } \mathrm{Gal}_k^0 \\ \text{acting on } \pi_1(\overline{U}, \bar{u}) \text{ via } s \end{array} \right\}$$

where Gal_k^0 is the kernel of the cyclotomic character $\mathrm{Gal}_k \to \hat{\mathbb{Z}}^*$.

We explain in Proposition 104 that Conjecture 6 is well posed because the image of the profinite Kummer map satisfies the fixed point condition. Moreover, in Proposition 256 we infinitesimally generalize work of Nakamura to show that the sections of $\mathscr{S}_{\pi_1(X/k)}$ excluded by the fixed point condition all arise from rational points of the boundary Y. The latter type of sections are called *cuspidal sections*, see Chap. 18 for a proper definition.

The State of the Art in the Section Conjecture

The section conjecture suggests a fascinating way of representing rational points in the realm of anabelian geometry. A genuine arithmetic problem, the Diophantine question of rational solutions to certain polynomials, transforms to a description in the different habitat of profinite groups.

Let us now summarize what is known about the section conjecture prior to these notes or due to the work of the author which built up towards it.

(1) The group theory of cuspidal sections was characterised by Nakamura in a series of works [Na90a, Na90b, Na91], see Theorem 253.

(2) The profinite Kummer map

$$\kappa : X(k) \to \mathscr{S}_{\pi_1(X/k)}$$

is indeed injective in many cases, including those claimed in Conjecture 2 and Conjecture 6. Due to Mochizuki [Mo99] §19, we even know that the pro-p version of κ is injective in many cases. All of this is recalled in Chap. 7.

(3) It took a while until the first examples occurred of hyperbolic curves over algebraic number fields that satisfy the section conjecture. Examples with local obstructions at p-adic places are constructed in [Sx10b], see Sect. 10.1. Subsequently, Harari and Szamuely [HaSz09] with the help of Flynn for numerical data constructed curves of genus 2 over $\mathbb{Q}$ that are counter-examples to the Hasse principle and yet satisfy the section conjecture. One more example of this kind, Schinzel's curve, was shown by Wittenberg [Wi12] to satisfy Conjecture 2 making use of the Brauer–Manin obstruction for sections established in [Sx11], see Chap. 11.

Hain showed in [Ha11b], that the generic curve of genus $g \geq 5$ over a field such that the ℓ-adic cyclotomic character has infinite image does not admit sections and thus also satisfies the section conjecture.

All these examples share the following feature: none of them has a section and therefore also none of them has a rational point, which was known before. There is even no example known of a hyperbolic curve such that the space of sections is finite but non-empty.

Moreover, there is no example known, such that Conjecture 6 holds for the hyperbolic curve X/k together with all finite scalar extensions $X' = X \times_k k'$.

(4) A *weak form of the section conjecture*, Conjecture 100, claims that a projective hyperbolic curve X/k over a field k that is finitely generated over $\mathbb{Q}$ admits a rational point as soon as $\pi_1(X/k)$ splits. It is well known that this *weak form* is far from being weak because it is in fact equivalent to Conjecture 6, see Corollary 101. Consequently, it suffices to treat the case of curves with no rational points, see Corollary 102.

A result on Galois descent in [Sx11], see Sects. 3.3 and 9.4, shows that on the contrary we may as well restrict to the case of curves that have at least one rational point, see Corollary 108.

(5) In the arithmetic case, the space of sections $\mathscr{S}_{\pi_1(X/k)}$ is a profinite set, and in particular compact, see [Sx11]. In Chap. 4 and Sect. 9.1 we describe the topology of $\mathscr{S}_{\pi_1(X/k)}$ in detail.

(6) A series of counter-examples to the pro-p versions of the section conjecture was found by Hoshi [Ho10], see Sect. 14.6, even with an infinite space of sections thus dashing the hope that nilpotent methods alone would solve the conjecture or at least help bounding the set of rational points.

(7) Analogues of the section conjecture have served to prove other anabelian conjectures. Tamagawa's success in anabelian geometry for affine hyperbolic curves [Ta97] relies among other ideas on a group theoretic description of Diophantine sections for hyperbolic curves over finite fields, see Sect. 15.4.

In order to achieve anabelian results for hyperbolic curves over sub-p-adic fields, Mochizuki introduced Hodge–Tate sections in [Mo99] and showed that this property is preserved under open maps between fundamental groups of p-adic curves.

(8) The *real section conjecture*, the analogue of Conjecture 2 with $\mathbb{R}$ as its base field, Theorem 229, is also known due to Mochizuki [Mo03], although it is related and could in fact have been deduced earlier from work of Miller [Mi84], Dwyer, Miller and Neisendorfer [DMN89], Carlsson [Ca91], and Lannes [La92] on Sullivan's conjecture in homotopy theory. Plenty of alternative proofs are now known, for example by the author [Sx10b], Pál [Pa11], Wickelgren [Wg10], Esnault and Wittenberg [EsWi09], see Sect. 16.1 for a survey.

(9) The analogue of the section conjecture for hyperbolic curves over p-adic local fields has been addressed recently by Pop and the author [PoSx11]. The result yields a *valuative section conjecture* that is potentially the final statement, if the section conjecture in the narrow sense is wrong over a p-adic local field, see Sect. 16.2. Instead of a rational point a section would then also potentially originate from a very special Berkovich point or even an adic point.

(10) A birational analogue of the section conjecture was proven by Koenigsmann [Ko05] for curves over p-adic local fields. This has been extended recently to higher transcendence degree in [Sx12a]. A minimalistic pro-p version of Koenigsmann's theorem is due to Pop [Po10].

   A recent result put together in final form by Harari and the author building on the work of Koenigsmann and Stoll [St06] establishes the birational section conjecture, Conjecture 260, in a few cases for curves over number fields, see Sect. 18.4. This approach has been extended in [Sx12b], and independently by Hoshi [Ho12], to yield a group theoretic description of Diophantine sections in the birational case for curves over $\mathbb{Q}$. The extra condition imposes a restriction on ramification for certain 2-dimensional representations associated to a section, see Theorem 269.

   Esnault and Wittenberg [EsWi10] related the existence of a birational abelian section to the existence of a rational zero cycle of degree 1 under favourable arithmetic assumptions.

(11) In the sense of general theory for sections the following has been achieved. Esnault and Wittenberg [EsWi09] established a theory of the cycle class of a section which was in special cases implicitly used by Parshin [Pa90] and Mochizuki [Mo03], see Chap. 6 for a survey.

   An attempt to set up a deformation theory for outer Galois representations was explored by Rastegar [Ra11a, Ra11b].

A Guide Through the Lecture Notes

The present book aims to develop the foundations for the anabelian geometry of sections (Part I) not necessarily limited to sections for curves and to present the state of the art in our knowledge about the section conjecture (Part II and Part IV) with a natural bias towards the work of the author. For example, no effort is taken on working in a pro-$\sum$ context or with truncated versions of the extension $\pi_1(X/k)$, since later, if and when we know the conjecture is true, then we probably see easily what is essentially needed in order to state an optimal result in these directions.

   Moreover, we discuss the section conjecture over number fields from a local to global point of view (Part III) and provide detailed discussions of various analogues of the section conjecture (Part IV) which might serve as supporting evidence for the conjecture by showing the arithmetic content hidden in a section of the fundamental group extension.

   The focus in this work lies on the section conjecture over *algebraic number fields*[1], since we have a vague idea that the general case over fields that are finitely generated over $\mathbb{Q}$ follows from this case. In order to encourage future research we have included open questions and also cases where only partial results were obtained.

---

[1]We follow the agreement that a *number field* is an algebraic extension of $\mathbb{Q}$ while an *algebraic number field* is a finite algebraic extension of $\mathbb{Q}$.

A reader with some familiarity with the section conjecture should be able to read the various chapters independently and is encouraged to do so. Let us discuss the content in more detail and in order of appearance in view of new and notable results.

(12) In Sect. 9.5 we examine *going up* and *going down* for the section conjecture with respect to a finite étale map. The discussion relies on results of Chap. 3 on Galois descent and fibres.

(13) Section 10.3 contains some new examples of projective hyperbolic curves X/k without rational points constructed by a principle which works over many base fields k.

(14) Let k be an algebraic number field and let X/k be a geometrically connected variety over k. In Chap. 11, which reports on joint work with Harari in [HaSx12], we reformulate the main result of [HaSx12] in terms of a description of the image of the localisation map

$$\text{res} \; : \; \mathscr{S}_{\pi_1(X/k)} \to \prod_v \mathscr{S}_{\pi_1(X/k)}(k_v)$$

with respect to all places v of k and the corresponding completions k_v. The image consists of all tuples of local sections that survive all finite constant descent obstructions, see Theorem 144.

This point of view is pushed further in Chap. 12 towards at least fragments of a non-abelian Tate–Poitou sequence in Theorem 168.

(15) In Sect. 13.5 we show that each k-rational zero-cycle of degree 1 on a smooth projective variety X/k leads to a section of the abelianized fundamental group extension $\pi_1^{\text{ab}}(X/k)$, see Theorem 192. This result aims beyond curves, because it is only remarkable for a variety X with nontrivial torsion in its geometric Néron–Severi group $\text{NS}_{\overline{X}}$. For such X/k the natural surjection

$$\pi_1^{\text{ab}}(\overline{X}) \twoheadrightarrow \pi_1(\overline{\text{Alb}_X^1})$$

is not an isomorphism, recalled in Proposition 69, where $X \to \text{Alb}_X^1$ is the universal Albanese torsor, see [Wi08].

(16) In Sect. 14.7 we find refinements of the counter-examples for the pro-p version of the section conjecture due to Hoshi. Theorem 218 on the one hand relies on a finer use of the known structure of S-unramified pro-p Galois groups of number fields in the *degenerate case*, see [NSW08] Definition 10.9.3. On the other hand we make use of our discussion of the section conjecture over finite fields, see Chap. 15, and in particular the counting result Theorem 226. As a proof of concept we manage to give an explicit new example, namely the smooth projective curve $C/\mathbb{Q}(\zeta_3)$ given by

$$Y^3 = X(X-1)(X-3)(X-9).$$

The curve has genus 3, an injective pro-3 Kummer map

$$\kappa_3 \; : \; C(\mathbb{Q}(\zeta_3)) \to \mathscr{S}_{\pi_1^{\text{pro-}3}}(C/\mathbb{Q}(\zeta_3))$$

with finite image, and an uncountable space of pro-3 sections. It is amusing to note that the example makes use of the Catalan solution $3^2 - 2^3 = 1$.

(17) In Chap. 15 we investigate the profinite Kummer map in the case of projective hyperbolic curves over a finite field $\mathbb{F}_q$. At first this seems strange as $\mathrm{Gal}_{\mathbb{F}_q} = \hat{\mathbb{Z}}$ is profinite free and thus all extensions $\pi_1(X/\mathbb{F}_q)$ will split. But for example, for abelian varieties over $\mathbb{F}_q$, the analogue of the section conjecture holds as a corollary to a theorem of Lang–Tate. For a projective hyperbolic curve $X/\mathbb{F}_q$ however, Theorem 224 states that the profinite Kummer map is never surjective.

The naive approach to Theorem 224 relies on the estimate (15.5) of the Picard number $\#\mathrm{Pic}^0_X(\mathbb{F}_q)$ in terms of the genus g of X and the number of rational points $N = \#X(\mathbb{F}_q)$

$$\#\mathrm{Pic}^0_X(\mathbb{F}_q) \geq (q-1)^2 \cdot \frac{1 + q^{g-1} + N(g - 2 + \frac{q^{g-1}-1}{q-1})}{(g+1)(q+1) - N},$$

an infinitesimal improvement over the estimates in [LMD90] Theorem 2. We deduce that

$$\#\mathrm{Pic}^0_X(\mathbb{F}_q) > \#X(\mathbb{F}_q)$$

in all but few exceptional cases—later called *space filling curves in their Jacobians*. So then there are sections of $\pi_1(X/\mathbb{F}_q)$ that on the abelian level, as sections of $\pi_1^{\mathrm{ab}}(X/\mathbb{F}_q)$, do not come from a point in $X(\mathbb{F}_q)$. We also wrote a program in SAGE [S$^+$08] to compute (all) exceptional cases, see Sect. 15.2.

In the case of a projective hyperbolic curve, a more refined approach computes the cardinality of the space of pro-ℓ sections by cohomology computations for the associated graded of the descending central series, see Sect. 15.3. It turns out that for bounded order of nilpotency, even taking the product over all $\ell \neq p$, we have only finitely many conjugacy classes of sections, while in the limit for every $\ell \neq p$ the space of pro-ℓ sections is uncountable, see Theorem 226. The result relies on work of Labute [La67] used in a careful study of the associated Lie algebra

$$\mathfrak{g} = \mathrm{Lie}\left(\pi^{\mathrm{pro}\text{-}\ell}(\overline{X})\right)$$

and a general result on the space of invariants of a finite abelian group acting on $\mathfrak{g}$ being infinite dimensional, see Proposition 207.

(18) Finite fields again occur in Sect. 17.2 in order to help constructing a field $\mathbb{F}$, an infinite algebraic extension of $\mathbb{F}_p$, with the remarkable property that for every smooth, projective geometrically connected variety $X/\mathbb{F}$ of dimension $\dim(X) > 0$ that injects into an abelian variety, the profinite Kummer map

$$\kappa : X(\mathbb{F}) \to \mathscr{S}_{\pi_1(X/\mathbb{F})}$$

is injective with dense image with respect to the natural topology of Chap. 4. But unfortunately, the map κ is never surjective, see Theorem 243.

Part I
Foundations of Sections

Chapter 1
Continuous Non-abelian H^1 with Profinite Coefficients

We recall the theory of non-abelian first cohomology as presented in [Se97] I §5 and emphasize the role played by conjugacy classes of sections and by equivariant torsors. The difference between two sections can be described by either a cocycle or an equivariant torsor.

The analogue in the non-abelian setup of the familiar abelian long exact sequences for short exact sequences of coefficients or of low degree terms in the Hochschild–Serre spectral sequence turn out to exist but in a restricted form depending on the amount of commutativity that one is willing to spend as an assumption.

1.1 Torsors, Sections and Non-abelian H^1

Non-abelian first cohomology comes in three different flavours: a description in terms of cocycles, as conjugacy classes of sections, and as isomorphy classes of equivariant torsors.

Cocycles. Let Γ be a profinite group. A Γ-*group* is a profinite group N together with a continuous action of Γ by group automorphisms. A 1-*cocycle* of Γ with values in N is a continuous map $a : \Gamma \to N$ such that

$$a_{\sigma\tau} = a_\sigma \sigma(a_\tau)$$

for all $\sigma, \tau \in \Gamma$. Two 1-cocycles a, b are by definition equivalent if there is a $c \in N$ such that for all $\sigma \in \Gamma$ we have

$$b_\sigma = c^{-1} a_\sigma \sigma(c).$$

J. Stix, *Rational Points and Arithmetic of Fundamental Groups*, Lecture Notes in Mathematics 2054, DOI 10.1007/978-3-642-30674-7_1,
© Springer-Verlag Berlin Heidelberg 2013

The *first non-abelian cohomology* $H^1(\Gamma, N)$ of Γ with values in N is the pointed set of equivalence classes of continuous 1-cocycles with the trivial cocyle $a_\sigma \equiv 1$ as the special element.

Sections. The *semidirect product* $N \rtimes \Gamma$ sits by definition in a canonically split short exact sequence

$$1 \to N \to N \rtimes \Gamma \to \Gamma \to 1$$

such that for $n \in N$ and $\sigma \in \Gamma$ we have $\sigma n = \sigma(n)\sigma$, when σ is considered as an element of $N \rtimes \Gamma$ via the canonical splitting. An extension of profinite groups

$$1 \to N \to \Pi \to \Gamma \to 1 \tag{1.1}$$

is isomorphic to the semidirect product $N \rtimes \Gamma$ as an extension of Γ by N with an appropriate structure as a Γ-group, if and only if (1.1) admits a splitting $s : \Gamma \to \Pi$.

Definition 7. Let $\mathscr{S}_{\Pi \to \Gamma}$ be the set of N-conjugacy classes $[s]$ of sections s of (1.1), and let $\mathscr{S}_{N \rtimes \Gamma}$ denote the pointed set of N-conjugacy classes of sections of $N \rtimes \Gamma \twoheadrightarrow \Gamma$ with the canonical splitting as the special element.

Torsors. A Γ-*equivariant right N-torsor* is a profinite set P with a continuous Γ-action and a continuous free transitive and Γ-equivariant action by N. The Γ-action will be denoted by $(\sigma, p) \mapsto \sigma.p$. The pointed set of isomorphy classes of Γ-equivariant right N-torsor will be denote by

$$\mathrm{Tors}_\Gamma(N)$$

with $P = N$ and right translation as the special element.

1.2 Twisting, Differences and Comparison

The comparison between the three incarnations of non-abelian first cohomology starts with cocycles and torsors acting on sections.

Twisting a section by a cocycle. Let $a : \Gamma \to N$ be a continuous map. Then a is a 1-cocycle with values in N with respect to the Γ-group structure induced by conjugation through a section s of (1.1) if and only if the *twist* of the section, namely the map $^a s$ defined by

$$^a s(\sigma) = a_\sigma s(\sigma),$$

is again a section $^a s : \Gamma \to \Pi$. Two cocycles a, b are equivalent via the element $c \in N$, namely $b_\sigma = c^{-1} a_\sigma \sigma(c)$, if and only if the respective twists are conjugate via c, namely

$$c(-)c^{-1} \circ {}^b s = {}^a s.$$

Twisting a section by a pointed torsor. Let P be a Γ-equivariant right N-torsor with respect to the Γ-group structure on N induced by conjugation through a section s of (1.1). The *twist* of the section s by P and an element $q \in P$ is the section $^{P,q}s$ which is defined as follows.

We have an induced right Π-action on P via

$$p.(ns(\sigma)) = \sigma^{-1}(pn)$$

for $p \in P$, $n \in N$ and $\sigma \in \Gamma$. The stabilizer of $q \in P$ in Π is a subgroup which projects isomorphically onto Γ and thus defines the section $^{P,q}s$, which depends on the choice of $q \in P$ only up to conjugation by an element of N. If we set

$$^{P,q}s(\sigma) = a_\sigma s(\sigma)$$

with $a_\sigma \in N$, then we find

$$\sigma(q) = \sigma\big(q.(^{P,q}s(\sigma))\big) = \sigma\big(q.(a_\sigma s(\sigma))\big) = qa_\sigma,$$

and thus a formula for the cocycle $a : \Gamma \to N$ with

$$^{P,q}s = {}^a s.$$

The Difference Cocycle. Let s, t be two sections of (1.1) and regard N as a Γ-group by conjugation through s. The difference cocycle $\delta(t, s)$ of s and t is the 1-cocycle

$$\sigma \mapsto \delta(t, s)_\sigma$$

with values in N given by

$$t(\sigma) = \delta(t, s)_\sigma s(\sigma). \tag{1.2}$$

We have

$$t = {}^{\delta(t,s)}s$$

so that s is N-conjugate to t if and only if $\delta(t, s)$ is cohomologous to the trivial 1-cocycle.

The difference torsor. Let s, t be two sections of (1.1) and regard N as a Γ-group by conjugation through s. The difference torsor $\Delta(t, s)$ of s and t is the Γ-equivariant right N-torsor given as a set by the coset space

$$\Delta(t, s) = t(\Gamma)\backslash\Pi,$$

with the right N-action via right multiplication and the Γ-action

$$\sigma(t(\Gamma)p) = t(\Gamma)ps(\sigma)^{-1}.$$

The difference torsor $\Delta(t, s)$ has a canonical point $1 \in \Delta(t, s)$ given by the trivial coset $t(\Gamma)$. We have

$$t = {}^{\Delta(t,s),1} s$$

and s is N-conjugate to t if and only if $\Delta(t, s)$ is isomorphic to the trivial torsor.

Comparison. The classification of torsors and of sections are captured by the first non-abelian cohomology.

Proposition 8. *(1) The pointed sets* H$^1(\Gamma, N)$, $\mathscr{S}_{N \rtimes \Gamma}$, *and* Tors$_\Gamma(N)$ *are in natural bijection with each other.*
(2) If $s \in \mathscr{S}_{\Pi \to \Gamma}$ is a splitting of (1.1), then the maps of difference cocycle and difference torsor

$$t \mapsto \delta(t, s) \quad and \quad t \mapsto \Delta(t, s)$$

induce bijections

$$\mathscr{S}_{\Pi \to \Gamma} \xrightarrow{\sim} \mathrm{H}^1(\Gamma, N) \quad and \quad \mathscr{S}_{\Pi \to \Gamma} \xrightarrow{\sim} \mathrm{Tors}_\Gamma(N),$$

where N carries the Γ-group structure via conjugation through s. The bijections depend on s and map the class $[s]$ to the trivial cohomology class, respectively the class of the trivial torsor. The induced bijection

$$\mathrm{H}^1(\Gamma, N) = \mathrm{Tors}_\Gamma(N)$$

agrees with the natural bijection from (1) and sends the class of $\delta(t, s)$ to the class of $\Delta(t, s)$.

Proof. We first consider (2). The map $a \mapsto {}^a s$ yields an inverse to the map $t \mapsto \delta(t, s)$ as discussed in Sect. 1.2.

Secondly, the map $P \mapsto [{}^{P,q} s]$ which to a Γ-equivariant right N-torsor P assigns the class of the twist with respect to an element $q \in P$ yields a well defined map

$$\mathrm{Tors}_\Gamma(N) \to \mathscr{S}_{\Pi \to \Gamma},$$

because moving the element q conjugates ${}^{P,q} s$ by an element of N. The induced map is an inverse to the map induced by $t \mapsto \Delta(t, s)$ as discussed in Sect. 1.2.

Parts (1) and (2) are equivalent by using the section s to deduce an isomorphism $\Pi = N \rtimes \Gamma$, or conversely by using the canonical section for s.   □

Combining the above constructions we get a bijection Tors$_\Gamma(N) \to$ H$^1(\Gamma, N)$ as follows. An element $t \in P$ defines a 1-cocycle $a^{P,t} : \Gamma \to N$ by the formula

$$\sigma(t) = t a_\sigma^{P,t}.$$

Moving t replaces $a^{P,t}$ by an equivalent cocycle. Conversely, a 1-cocylce a allows to twist the trivial Γ-equivariant torsor N to a torsor ${}^a N$ which equals N as a right N torsor but has $\sigma_{.a} n = a_\sigma \sigma(n)$ as twisted Γ-action.

Generalized sections. Let $\overline{\varphi} : \Gamma \to \overline{\Gamma}$ be a continuous homomorphism and let

$$1 \to N \to E \to \overline{\Gamma} \to 1$$

be an extension of profinite groups. The set of lifts $\varphi : \Gamma \to E$ of $\overline{\varphi}$ up to conjugation by an element of N is either empty or, with the group N equipped with the conjugation action of Γ via a choice of lift φ_0, in bijection with the corresponding $H^1(\Gamma, N)$. Indeed, a lift φ corresponds uniquely to a section of the pullback exact sequence

$$1 \to N \to E \times_{\overline{\Gamma}, \overline{\varphi}} \Gamma \to \Gamma \to 1$$

and conjugation by elements of N corresponds accordingly. The claim thus follows from Proposition 8. The description of lifts via $H^1(\Gamma, N)$ is natural with respect to both Γ and N.

1.3 Long Exact Cohomology Sequence

As a well known feature of non-abelian cohomology the technical tool of long exact sequences only exists in a truncated version with range depending on the amount of commutativity one is willing to spend as an assumption.

General coefficients. To a short exact sequence of Γ-groups

$$1 \to G' \xrightarrow{i} G \xrightarrow{p} G'' \to 1$$

we can associate a (not so) long exact sequence of non-abelian cohomology

$$
\begin{array}{ccccccc}
1 & \longrightarrow & H^0(\Gamma, G') & \longrightarrow & H^0(\Gamma, G) & \longrightarrow & H^0(\Gamma, G'') \\
& & & {\scriptstyle \delta} & & & \\
& \longrightarrow & H^1(\Gamma, G') & \longrightarrow & H^1(\Gamma, G) & \longrightarrow & H^1(\Gamma, G'')
\end{array}
\qquad (1.3)
$$

with $H^0(\Gamma, M) = M^\Gamma$ being the choice of notation for the subgroup of Γ-invariant elements. The only map that needs to be explained is the map

$$\delta : H^0(\Gamma, G'') \to H^1(\Gamma, G')$$

which maps a Γ-invariant element $g'' \in G''$ to the Γ-equivariant right G''-torsor $p^{-1}(g'')$ in $\mathrm{Tors}_\Gamma(G') = H^1(\Gamma, G')$ following Proposition 8. On the level of H^0 the sequence (1.3) consists of groups and group homomorphisms, but starting with the map δ and in degree 1 the sequence (1.3) merely consists of pointed sets and is exact in that sense. The interpretation of the H^1 as isomorphism classes of Γ-equivariant torsors makes the proof of the exactness of (1.3) almost immediate.

Definition 9. Let Ext(Γ, G') be the set of isomorphy classes of *extensions* of Γ by G', i.e., the set of isomorphy classes of short exact sequences

$$1 \to G' \to E \to \Gamma \to 1$$

where isomorphisms have to induce the identity on Γ and on G'. The action of G on G' by conjugation induces by pushout an action on the set Ext(Γ, G'), the orbit set of which we denote by $[\text{Ext}(\Gamma, G')]_G$.

Using the notion of extensions as a replacement for non-abelian H^2 we can construct a map

$$\delta : \text{H}^1(\Gamma, G'') \to [\text{Ext}(\Gamma, G')]_G$$

which prolongs (1.3) in the sense that the image of

$$\text{H}^1(\Gamma, G) \to \text{H}^1(\Gamma, G'')$$

consists precisely of those cohomology classes that map to a semi-direct product extension under δ. A class in H$^1(\Gamma, G'')$ is given by a conjugacy class $[s]$ of sections s of $G'' \rtimes \Gamma \to \Gamma$ by Proposition 8. The extension $\delta(s)$ is defined as the pullback by s of the extension

$$1 \to G' \to G \rtimes \Gamma \to G'' \rtimes \Gamma \to 1$$

and changing the representative s of $[s]$ only changes $\delta(s)$ by the action of an element of G, hence the class $\delta(s) \in [\text{Ext}(\Gamma, G')]_G$ is well defined. The assertion that δ prolongs (1.3) follows obviously from the definition.

Central kernel. More can be said about (1.3) in case G' is abelian, see [Se97] I §56, or in case G' is even central in G, see [Se97] I §5.7. The notion of twisting by a 1-cocycle as developed in [Se97] I §53+4 becomes important. The best result with the most restrictive assumptions is as follows.

Proposition 10 ([Se97] Proposition 43). *Let* $1 \to Z \to G \to G'' \to 1$ *be a short exact sequence of* Γ-*groups with kernel* Z *contained in the center of* G. *Then there is a connecting map*

$$\Delta : \text{H}^1(\Gamma, G'') \to \text{H}^2(\Gamma, Z)$$

which maps the class of a cocycle $a : \sigma \mapsto a_\sigma$ *to the class of the 2-cocycle*

$$\Delta(a)_{\sigma,\tau} = a_\sigma \sigma(a_\tau)(a_{\sigma\tau})^{-1}.$$

The sequence of groups and pointed sets

$$
1 \ \longrightarrow \ H^0(\Gamma, Z) \ \xrightarrow[\delta]{} \ H^0(\Gamma, G) \ \longrightarrow \ H^0(\Gamma, G'')
$$

$$
\longrightarrow \ H^1(\Gamma, Z) \ \xrightarrow[\Delta]{} \ H^1(\Gamma, G) \ \longrightarrow \ H^1(\Gamma, G'')
$$

$$
\longrightarrow \ H^2(\Gamma, Z)
$$

is exact in the following sense.

(1) The map $\delta : H^0(\Gamma, G'') \to H^1(\Gamma, Z)$ is a group homomorphism.

(2) The part $1 \to H^0(\Gamma, Z) \to \dots \xrightarrow{\delta} H^1(\Gamma, Z)$ is an exact sequence of groups.

(3) The image of δ is the fibre above the trivial class in $H^1(\Gamma, Z) \to H^1(\Gamma, G)$.

(4) $H^1(\Gamma, Z)$ acts on $H^1(\Gamma, G)$ with orbit space given by the image in $H^1(\Gamma, G'')$.

(5) Let a_σ be a cocycle representing the class $a \in H^1(\Gamma, G'')$. Then $\Delta^{-1}(\Delta(a))$ agrees under the bijection $H^1(\Gamma, {}_aG'') \to H^1(\Gamma, G'')$ with twisting by a with the image under $H^1(\Gamma, {}_aG) \to H^1(\Gamma, {}_aG'')$. In particular $\Delta(a) = 0$ if and only if a comes from $H^1(\Gamma, G)$.

Proof. The map Δ is constructed in [Se97] I §5.6+7. Property (1) is [Se97] I §5.6 Corollary 2, and properties (2) and (3) follow from [Se97] I §5.5 Proposition 38+39. Property (4) and (5) are proved in [Se97] I §5.7 Proposition 42–44. □

1.4 Non-abelian Hochschild–Serre Spectral Sequence

We develop a certain fragment of the five-term exact sequence of low degree terms in the Hochschild–Serre spectral sequence.

Definition 11. Let $\Gamma_0 \trianglelefteq \Gamma$ be a closed normal subgroup with quotient $G = \Gamma/\Gamma_0$ and let

$$
1 \to \overline{\pi} \to \pi \xrightarrow{\mathrm{pr}} \Gamma \to 1
$$

be a short exact sequence. We abbreviate $\pi_0 = \mathrm{pr}^{-1}(\Gamma_0)$. The *centraliser in $\overline{\pi}$ of a section $s_0 : \Gamma_0 \to \pi_0$* of $\mathrm{pr} : \pi_0 \to \Gamma_0$, more precisely the centraliser in $\overline{\pi}$ of its image, is the closed subgroup

$$
Z_{\overline{\pi}}(s_0) = \{\gamma \in \overline{\pi} \ ; \ \gamma s_0(\sigma)\gamma^{-1} = s_0(\sigma) \text{ for all } \sigma \in \Gamma_0\} \subseteq \overline{\pi}.
$$

The group G acts on the set of conjugacy classes of sections of $\pi_0 \to \Gamma_0$ as follows. For $\alpha \in \pi$ and a section $s_0 : \Gamma_0 \to \pi_0$ we set

$$\alpha.(s_0) = \alpha(-)\alpha^{-1} \circ s_0 \circ \mathrm{pr}(\alpha)^{-1}(-)\mathrm{pr}(\alpha)$$

which clearly describes an action of π on the set of sections of $\pi_0 \to \Gamma_0$. For the induced action on conjugacy classes $\mathscr{S}_{\pi_0 \to \Gamma_0}$ elements of $\overline{\pi}$ act trivially by definition. In the resulting Γ-action the subgroup Γ_0 acts trivially as can be seen by acting for $\sigma \in \Gamma_0$ with the preferred preimage $s_0(\sigma)$. In the end $G = \Gamma/\Gamma_0$ acts on $\mathscr{S}_{\pi_0 \to \Gamma_0}$.

Proposition 12. *(1) With the notation as above the restriction map*

$$\mathrm{res} : \mathscr{S}_{\pi \to \Gamma} \to \mathscr{S}_{\pi_0 \to \Gamma_0} \qquad s \mapsto \mathrm{res}(s) = s|_{\Gamma_0}$$

maps conjugacy classes of sections of $\pi \to \Gamma$ to G-invariant conjugacy classes of sections of $\pi_0 \to \Gamma_0$.

(2) A G-invariant section $s_0 : \Gamma_0 \to \pi_0$ is the restriction of a section $s : \Gamma \to \pi$ if and only if the extension class $\delta(s_0) \in \mathrm{Ext}\big(G, Z_{\overline{\pi}}(s_0)\big)$ to be defined below is a semi-direct product extension.

(3) When a section $s_0 : \Gamma_0 \to \pi_0$ extends to a section $s : \Gamma \to \pi$, then the set of such extensions $\mathrm{res}^{-1}(s_0)$ with special element s is isomorphic as pointed sets with $\mathrm{H}^1\big(G, Z_{\overline{\pi}}(s_0)\big)$ where $Z_{\overline{\pi}}(s_0)$ carries the conjugation action via s.

Remark 13. (1) With the notation as above we have an 'exact sequence'

$$1 \to \mathrm{H}^1(G, Z_{\overline{\pi}}(s_0)) \to \mathscr{S}_{\pi \to \Gamma} \xrightarrow{\mathrm{res}} (\mathscr{S}_{\pi_0 \to \Gamma_0})^G \xrightarrow[s_0]{\delta} \bigsqcup \mathrm{Ext}(G, Z_{\overline{\pi}}(s_0))$$

in the sense that $\mathrm{res}^{-1}(s_0)$ has a free and transitive action by the pointed set $\mathrm{H}^1(G, Z_{\overline{\pi}}(s_0))$ if s_0 extends to a section of $\pi \to \Gamma$, and the image of res coincides with the set of those s_0 such that $\delta(s_0)$ is a semi-direct product extension.

(2) If we fix a section $s : \Gamma \to \pi$ and put $\mathrm{H}^0(G, \overline{\pi}) = Z_{\overline{\pi}}(s_0)$ as a G-group via s, then Proposition 8 extracts from Proposition 12 the non-abelian inflation–restriction exact sequence of pointed sets, see also [Se97] I §5.8,

$$1 \to \mathrm{H}^1\big(G, \mathrm{H}^0(G, \overline{\pi})\big) \xrightarrow{\mathrm{inf}} \mathrm{H}^1(\Gamma, \overline{\pi}) \xrightarrow{\mathrm{res}} \big(\mathrm{H}^1(\Gamma_0, \overline{\pi})\big)^G .$$

Proof. (1) For $\sigma \in \Gamma$ we compute

$$\sigma.\mathrm{res}(s) = s(\sigma)(-)s(\sigma)^{-1} \circ \mathrm{res}(s) \circ \sigma^{-1}(-)\sigma = \mathrm{res}(s)$$

and therefore res takes its image in $\big(\mathscr{S}_{\pi_0 \to \Gamma_0}\big)^G$.

(2) Let $s_0 : \Gamma_0 \to \pi_0$ be a section representing a G-invariant conjugacy class. The stabilizer subgroup

$$\mathscr{Z}_\pi(s_0) = \{\alpha \in \pi \; ; \; \alpha.(s_0) = s_0\}$$

of s_0 sits in a short exact sequence

$$1 \to Z_{\overline{\pi}}(s_0) \to \mathscr{Z}_\pi(s_0) \overset{\text{pr}}{\to} \Gamma \to 1 \tag{1.4}$$

with the map pr being surjective exactly because $[s_0]$ is G-invariant. The subgroup $s_0(\Gamma_0) \subset \mathscr{Z}_\pi(s_0)$ is a normal subgroup and the quotient $E(s_0)$ is canonically an extension

$$1 \to Z_{\overline{\pi}}(s_0) \to E(s_0) \to G \to 1, \tag{1.5}$$

hence induces an element $\delta(s_0) \in \text{Ext}(G, Z_{\overline{\pi}}(s_0))$. Now, if s_0 extends to a section s, then s has to take values in $\mathscr{Z}_\pi(s_0)$. But sections of (1.4) that extend s_0 are in bijection with sections of (1.5). Hence s_0 lies in the image of res if and only if $\delta(s_0)$ is a semi-direct product extension.

(3) If $s_0 = \text{res}(s)$, then, by the argument of (2) above,

$$\text{res}^{-1}(s_0) = \left\{ t : \Gamma \to \pi \quad \begin{matrix} \text{section of } \pi \to \Gamma \\ \text{with } t|_{\Gamma_0} = s_0 \end{matrix} \right\} \Big/ \text{conjugation by } Z_{\overline{\pi}}(s_0)$$

$$= \{[t] \in \mathscr{S}_{\mathscr{Z}_\pi(s_0) \to \Gamma} \; ; \; t|_{\Gamma_0} = s_0\} = \mathscr{S}_{E(s_0) \to G} = \text{H}^1(G, Z_{\overline{\pi}}(s_0)),$$

which proves the proposition.   □

Chapter 2
The Fundamental Groupoid

We recall the fundamental groupoid of a connected, quasi-compact scheme X as in [SGA1] Exposé V, with special attention towards the effect of a k-structure in case of a variety X/k. Galois invariant base points are discussed and related to the profinite Kummer map. In Sect. 2.6, we address the reformulation of the section conjecture in terms of higher étale homotopy theory.

2.1 Fibre Functors and Path Spaces

Let X be a connected, quasi-compact scheme, and let $\mathsf{Rev}(X)$ be the Galois category of finite étale covers of X, see [SGA1] V.4+5. We use the suggestive notation

$$\bar{a} : \mathsf{Rev}(X) \to \mathsf{sets} \tag{2.1}$$

for a fibre functor of $\mathsf{Rev}(X)$, which maps a finite étale cover $h : Y \to X$ to its *fibre*

$$\bar{a}(Y) = Y[\bar{a}]. \tag{2.2}$$

In case $\bar{a} : \mathrm{Spec}(\Omega) \to X$ is a geometric point we have the induced fibre functor denoted $\bar{a}$ defined by

$$Y[\bar{a}] = h^{-1}(\bar{a}) = \pi_0\big(Y \times_{X,\bar{a}} \mathrm{Spec}(\Omega)\big).$$

We use the notion of a base point of X synonymously for a fibre functor of $\mathsf{Rev}(X)$. By [SGA1] V Theorem 4.1, every fibre functor $\bar{a}$ is pro-representable by its *path space*

$$P_{\bar{a}} \in \mathsf{Pro}\text{–}\mathsf{Rev}(X) \tag{2.3}$$

and a universal point $\tilde{a} \in P[\bar{a}]$. The identification

$$\mathrm{ev}_{\tilde{a}} : \mathrm{Hom}(P_{\bar{a}}, Y) \xrightarrow{\sim} Y[\bar{a}]$$

J. Stix, *Rational Points and Arithmetic of Fundamental Groups*, Lecture Notes in Mathematics 2054, DOI 10.1007/978-3-642-30674-7_2,

comes by evaluating at the universal point $\tilde{a}$. The Yoneda lemma leads to the formula

$$\text{Hom}(P_{\bar{b}}, P_{\bar{a}}) = P_{\bar{a}}[\bar{b}] = \text{Hom}(\text{Hom}(P_{\bar{a}}, -), \bar{b}) = \text{Hom}(\bar{a}, \bar{b}) \qquad (2.4)$$

for the fibre of the path space above $\bar{b}$ with $\tilde{a} \in P[\bar{a}]$ corresponding to $\text{id}_{\bar{a}}$.

Proposition 14. *Any two fibre functors of* $\text{Rev}(X)$ *are isomorphic and any homomorphism between fibre functors is an isomorphism.*

Proof. By [SGA1] V Proposition 5.6, any two fibre functors are isomorphic. Hence it suffices to show that any endomorphism of a path space is an automorphism, see [SGA1] V.4 h).   □

Lemma 15. *Let* $f : X \to Y$ *be a map of connected, quasi-compact schemes, and let* $\bar{a}$ *be a fibre functor on* X *with path space* $P_{\bar{a}} \to X$ *and* $\bar{b}$ *a fibre functor on* Y *with path space* $P_{\bar{b}} \to Y$. *Then a morphism* $\bar{b} \xrightarrow{\sim} f_*\bar{a}$ *corresponds uniquely to a map* φ *in a commutative diagram*

$$
\begin{array}{ccc}
P_{\bar{a}} & \overset{\varphi}{\dashrightarrow} & P_{\bar{b}} \\
\downarrow & & \downarrow \\
X & \overset{f}{\longrightarrow} & Y.
\end{array}
$$

Proof. The map $\bar{b} \xrightarrow{\sim} f_*\bar{a}$ is determined by the image of the universal point in

$$P_{\bar{b}}[f_*\bar{a}] = (P_{\bar{b}} \times_Y X)[\bar{a}]$$

which itself determines uniquely a map $P_{\bar{a}} \to P_{\bar{b}} \times_Y X$ in the pro-category of $\text{Rev}(X)$. Such a map corresponds uniquely to a map φ as in the lemma.   □

2.2 The Fundamental Groupoid

For a group G we denote by G^{opp} its opposite group with the same elements but composition reversed.

Definition 16. The *fundamental groupoid* of a connected and quasi-compact X is the connected groupoid

$$\Pi_1(X)$$

of fibre functors of $\text{Rev}(X)$, see Proposition 14. The set of *étale paths* between fibre functors $\bar{a}, \bar{b}$ is

$$\pi_1(X; \bar{a}, \bar{b}) = \text{Hom}(\bar{a}, \bar{b}) = \text{Hom}(P_{\bar{b}}, P_{\bar{a}}) = P_{\bar{a}}[\bar{b}], \qquad (2.5)$$

and the *étale fundamental group* of X with base point $\bar{a}$ is

$$\pi_1(X,\bar{a}) = \operatorname{Hom}(\bar{a},\bar{a}) = \operatorname{Aut}(\bar{a}) = \operatorname{Aut}(P_{\bar{a}})^{\operatorname{opp}}. \tag{2.6}$$

The space of étale paths is naturally a profinite set, and the fundamental group is naturally a profinite group.

Proposition 17. *Any fibre functor $\bar{a}$ enhances to an equivalence of categories*

$$\bar{a} : \operatorname{Rev}(X) \xrightarrow{\sim} \pi_1(X,\bar{a}) - \text{sets}$$

of the category of finite étale covers with the category of finite discrete sets with a continuous action of the fundamental group.

Proof. [SGA1] V Theorem 4.1.   □

By composition the set of étale paths $\pi_1(X;\bar{a},\bar{b})$ becomes a right $\pi_1(X,\bar{a})$-torsor and a left $\pi_1(X,\bar{b})$-torsor, with both torsor actions commuting. Under the Yoneda identification (2.4) the following bi-torsor structures agree,

$$
\begin{array}{ccc}
\pi_1(X,\bar{b}) \times \pi_1(X;\bar{a},\bar{b}) \times \pi_1(X,\bar{a}) & \xrightarrow{\;\beta,\gamma,\alpha \,\mapsto\, \beta\circ\gamma\circ\alpha\;} & \pi_1(X;\bar{a},\bar{b}) \\[2pt]
\| & & \| \\[6pt]
\operatorname{Aut}(P_{\bar{a}}) \times \operatorname{Hom}(P_{\bar{b}},P_{\bar{a}}) \times \operatorname{Aut}(P_{\bar{b}}) & \xrightarrow{\;\varphi,f,\psi \,\mapsto\, \varphi\circ f\circ\psi\;} & \operatorname{Hom}(P_{\bar{b}},P_{\bar{a}}) \\[2pt]
\Big\| \,\operatorname{ev}_{\bar{b}} & & \Big\| \,\operatorname{ev}_{\bar{b}} \\[6pt]
\big(\operatorname{Aut}(P_{\bar{a}}) \times \pi_1(X,\bar{b})\big) \times P_{\bar{a}}[\bar{b}] & \xrightarrow{\;\varphi,\beta,b' \,\mapsto\, \varphi(\beta.b')\;} & P_{\bar{a}}[\bar{b}]
\end{array}
$$

where we have interchanged right with left actions to avoid taking opposite groups. The bi-torsor $P_{\bar{a}}[\bar{b}]$ describes via Proposition 17 the pro-object $P_{\bar{a}}$ together with the tautological action by its automorphism group $\operatorname{Aut}(P_{\bar{a}})$. Another interpretation comes via the projection maps

$$(\operatorname{pr}_{1,*}, \operatorname{pr}_{2,*}) : \pi_1(X \times X, (\bar{a},\bar{b})) \to \pi_1(X,\bar{a}) \times \pi_1(X,\bar{b})$$

and the induced left $\pi_1(X \times X, (\bar{a},\bar{b}))$-action on $\pi_1(X;\bar{a},\bar{b})$ via

$$\alpha.\gamma = \operatorname{pr}_{2,*}(\alpha) \circ \gamma \circ \operatorname{pr}_{1,*}(\alpha)^{-1}$$

for $\gamma \in \pi_1(X;\bar{a},\bar{b})$ and $\alpha \in \pi_1(X \times X, (\bar{a},\bar{b}))$. The corresponding pro-object

$$P \to X \times X$$

of $\mathsf{Rev}(X \times X)$ is independent of the base points $\bar{a}, \bar{b}$ and has fibre

$$P[(\bar{a}, \bar{b})] = \pi_1(X; \bar{a}, \bar{b}).$$

Hence P puts all étale path spaces with both base points varying into one algebraic family, while $P_{\bar{a}}$ does this for the étale path spaces with the first argument fixed to $\bar{a}$.

2.3   The Fundamental Groupoid in the Relative Case

Let k be a field. From now on we assume that X is a geometrically connected variety over k and analyse the impact of the relative map pr : $X \to \mathrm{Spec}(k)$ on the structure of the fundamental group. For this endeavor we fix once and for all a geometric point

$$* : \mathrm{Spec}(\bar{k}) \to \mathrm{Spec}(k) \tag{2.7}$$

induced by an algebraic closure $\bar{k}$ of k. The corresponding fundamental group is the absolute Galois group

$$\mathrm{Gal}_k = \mathrm{Gal}(k^{\mathrm{sep}}/k) = \pi_1(\mathrm{Spec}(k), *), \tag{2.8}$$

where $k^{\mathrm{sep}} \subseteq \bar{k}$ is the separable closure of k contained in $\bar{k}$. The base change

$$\overline{X} = X \times_k \bar{k} \tag{2.9}$$

with projection $\overline{X} \to X$ and the corresponding forgetful functor

$$\Pi_1(\overline{X}) \to \Pi_1(X)$$

of fundamental groupoids captures the content of the relative setting X/k together with the choice of base point $*$.

The fundamental exact sequence.   We will consider base points $\bar{a}$ of $\overline{X}$ and denote their projection to X by $\bar{a}$ as well. The restriction to base points of $\overline{X}$ in the study of $\Pi_1(X)$ yields that $\mathrm{pr}_* \bar{a}$ agrees canonically with $*$ as base points of $\mathrm{Spec}(k)$.

Proposition 18.   *Let $\bar{a}, \bar{b}$ be base points of $\overline{X}$. Then the natural maps*

$$1 \to \pi_1(\overline{X}, \bar{a}) \to \pi_1(X, \bar{a}) \xrightarrow{\mathrm{pr}_*} \mathrm{Gal}_k \to 1 \tag{2.10}$$

form a short exact sequence of profinite groups. Moreover, the set of étale paths $\pi_1(\overline{X}; \bar{a}, \bar{b})$ agrees with the preimage of the identity under the map

$$\mathrm{pr}_* : \pi_1(X; \bar{a}, \bar{b}) \to \pi_1(\mathrm{Spec}(k); *, *) = \mathrm{Gal}_k.$$

Proof. [SGA1] IX Theorem 6.1 proves the first assertion. The second follows from the first because all maps are compatible with composition of paths. □

Definition 19. The *fundamental exact sequence* associated to the geometrically connected X/k with base point $\bar{a} \in \overline{X}$ is the extension (2.10) and will be denoted by

$$\pi_1(X/k, \bar{a}).$$

The fundamental group as a functor on k-varieties. The fundamental group is a functor on pointed connected schemes. If the chosen base points do not match then we need to compose by an inner automorphism of the groupoid. More precisely, let $f : X \to Y$ be a map of geometrically connected k-varieties and let $\bar{a}$ (resp. $\bar{b}$) be a base point of $\overline{X}$ (resp. $\overline{Y}$). Then, using a path $\gamma \in \pi_1(\overline{Y}; f_*\bar{a}, \bar{b})$, we obtain a map

$$\pi_1(f) : \pi_1(X, \bar{a}) \xrightarrow{f_*} \pi_1(Y, f_*\bar{a}) \xrightarrow{\gamma(-)\gamma^{-1}} \pi_1(Y, \bar{b})$$

that is canonical up to conjugation by an element of $\pi_1(\overline{Y}, \bar{b})$ and compatible with the projection onto Gal_k from (2.10).

Galois action on base points and étale paths. Let $\bar{a}$ be a base point on $\overline{X}$. For $\sigma \in \mathrm{Gal}_k$ we define $\sigma(\bar{a})$ as the fibre functor

$$\sigma(\bar{a}) : \mathrm{Rev}(\overline{X}) \xrightarrow{(1 \otimes \sigma^{-1})^*} \mathrm{Rev}(\overline{X}) \xrightarrow{\bar{a}} \mathrm{sets}$$

which is nothing but $(1 \otimes \sigma^{-1})_*(\bar{a})$ induced by the automorphism $1 \otimes \sigma^{-1}$ of $\overline{X}$. Pullback via the inverse σ^{-1} instead of σ leads to a right action of Gal_k on base points. By Lemma 15 we find induced isomorphisms of path spaces that fit into the following diagram.

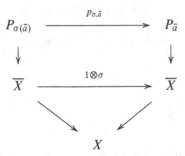

By (2.4) we can translate $p_{\sigma,\bar{a}}$ into canonical étale paths

$$p_\sigma = p_{\sigma,\bar{a}} \in \pi_1(X; \bar{a}, \sigma(\bar{a}))$$

such that

$$p_{\sigma\tau,\bar{a}} = p_{\sigma,\tau(\bar{a})} \circ p_{\tau,\bar{a}} \qquad .$$

and under the projection $\mathrm{pr}_* : \pi_1(X;\bar{a},\sigma(\bar{a})) \to \mathrm{Gal}_k$ we have $\mathrm{pr}_*(p_\sigma) = \sigma$.

The Galois action on $\overline{X}$ by structure transport leads to a Gal_k-action on étale paths that can also be described entirely in terms on the p_σ's and via path spaces as follows. We have a commutative diagram

$$
\begin{array}{ccc}
\pi_1(X;\bar{a},\bar{b}) & \xrightarrow{\;(1\otimes\sigma^{-1})_*\;} & \pi_1(X;\sigma(\bar{a}),\sigma(\bar{b})) \\[2pt]
\| & & \| \\[6pt]
P_{\bar{a}}[\bar{b}] & \xrightarrow{\;p_{\sigma,\bar{b}}(-)p_{\sigma,\bar{a}}^{-1}\;} & P_{\sigma(\bar{a})}[\sigma(\bar{b})]
\end{array}
$$

for the path space description of the map that takes $\gamma \in \pi_1(X;\bar{a},\bar{b})$ to

$$\sigma(\gamma) := (1 \otimes \sigma^{-1})_*(\gamma) = p_{\sigma,\bar{b}} \circ \gamma \circ p_{\sigma,\bar{a}}^{-1}.$$

Galois action on étale paths is compatible with composition of étale paths.

2.4   Galois Invariant Base Points

If $\bar{a}$ is a geometric point of $\overline{X}$ that sits above a k-rational point $a \in X(k)$, then the corresponding base point is Galois invariant, namely for any $\sigma \in \mathrm{Gal}_k$ we have

$$\bar{a} = \bar{a} \circ (1 \otimes \sigma^{-1})^* = \sigma(\bar{a}).$$

More generally, if a base point $\bar{a}$ is Galois invariant, then the canonical paths yield a section

$$s_a \; : \; \mathrm{Gal}_k \to \pi_1(X,\bar{a})$$

of $\mathrm{pr}_* : \pi_1(X,\bar{a}) \to \mathrm{Gal}_k$ by

$$s_a(\sigma) = p_{\sigma,\bar{a}}$$

and vice versa. It also follows that the $p_{\sigma,\bar{a}}$ describe a right action by Gal_k on the path space $P_{\bar{a}}$, the quotient of which yields a k-structure on the path space, namely a pro-object $P_a = P_{\bar{a}}/\mathrm{Gal}_k \to X$ such that

$$P_{\bar{a}} = P_a \times_k \bar{k} = P_a \times_X \overline{X}.$$

The profinite Kummer map. Let $\bar{a}$ and $\bar{b}$ be Galois invariant base points of $\overline{X}$. Then the section s_b of $\pi_1(X,\bar{b}) \to \mathrm{Gal}_k$ yields a $\pi_1(\overline{X},\bar{a})$-conjugacy class of sections $[s_b]$ of $\pi_1(X,\bar{a}) \to \mathrm{Gal}_k$ a representative of which is constructed by the choice of a $\gamma \in \pi_1(\overline{X};\bar{b},\bar{a})$ and the formula

$$s_b : \mathrm{Gal}_k \xrightarrow{s_b} \pi_1(X,\bar{b}) \xrightarrow{\gamma(-)\gamma^{-1}} \pi_1(X,\bar{a}).$$

The class of the difference cocycle $\delta(s_b, s_a)$ corresponds under the identification

$$\mathscr{S}_{\pi_1(X/k,\bar{a})} = \mathrm{H}^1(k, \pi_1(\overline{X},\bar{a})) = \mathrm{Tors}_{\mathrm{Gal}_k}(\pi_1(\overline{X},\bar{a}))$$

of Proposition 8 to the class of sections $[s_b]$ and to the isomorphism class $[\pi_1(\overline{X},\bar{a},\bar{b})]$ of the étale path space $\pi_1(\overline{X};\bar{a},\bar{b})$ as a Gal_k-equivariant right $\pi_1(X,\bar{a})$-torsor.

Definition 20. The *profinite Kummer map* is the map

$$\kappa : X(k) \to \mathrm{H}^1(k, \pi_1(\overline{X},\bar{a})) = \mathscr{S}_{\pi_1(X/k),\bar{a}} \tag{2.11}$$

defined for $b \in X(k)$ by choosing a Galois invariant base point $\bar{b}$ above b and

$$b \mapsto [\pi_1(\overline{X};\bar{a},\bar{b})].$$

The image of $b \in X(k)$ is identified with $[s_b]$, $[\delta(s_b, s_a)]$ or $[\pi_1(\overline{X},\bar{a},\bar{b})]$ accordingly.

Remark 21. In Corollary 71 we will see that our κ for abelian varieties or $\mathbb{G}_m$ is nothing but the connecting homomorphism of Galois cohomology of the respective (continuous) Kummer sequence. The analogous map in the higher non-abelian Chabauty method as introduced by Minhyong Kim bears the name of a unipotent Albanese or Kummer map. Therefore we decided to use the parallel terminology of a profinite Kummer map to describe the map from rational points to sections also in the general non-abelian setting.

Classically, Kummer theory describes cyclic extensions $k(\sqrt[n]{a})/k$ of fields containing the necessary roots of unity. More generally, the respective μ_n-torsor is described by the Kummer cocycle $\sigma \mapsto \sigma(\sqrt[n]{a}))/\sqrt[n]{a}$. This computes the connecting homomorphism $\delta(a)$ for the Kummer sequence and concludes our discussion on the etymology of the chosen terminology *profinite Kummer map*.

2.5   Abstract Sections

We pick a base point $\bar{a}$ on $\overline{X}$. To an abstract, i.e., a priori without Diophantine origin, group theoretical section $s : \mathrm{Gal}_k \to \pi_1(X, \bar{a})$ of the extension $\pi_1(X/k, \bar{a})$ of profinite groups we can associate the functor 'fibre in s'

$$s^* : \mathsf{Rev}(X) \to \mathsf{Rev}(k) = \mathsf{Rev}(\mathrm{Spec}(k))$$

which is an exact functor that fits into the following diagram.

$$
\begin{array}{ccc}
 & \mathsf{Rev}(X) & \xrightarrow{\ \bar{a}\ } & \pi_1(X, \bar{a}) - \mathrm{sets} \\[2mm]
\overset{\mathrm{pr}^*}{\nearrow} & \downarrow {\scriptstyle s^*} & & \downarrow {\scriptstyle s^*} \\[2mm]
\mathsf{Rev}(k) \ =\!\!=\ \mathsf{Rev}(k) & \xrightarrow{\ *\ } & \mathrm{Gal}_k - \mathrm{sets}.
\end{array}
$$

The functor 'fibre in s' is a retract of $\mathrm{pr}^* : \mathsf{Rev}(k) \to \mathsf{Rev}(X)$.

If $\beta : \mathsf{Rev}(X) \to \mathsf{Rev}(k)$ is a retract of pr^* and $\bar{b} = * \circ \beta$ the corresponding fibre functor, then the path space $P_{\bar{b}}$ has a canonical map $P_{\bar{b}} \to \mathrm{Spec}(\bar{k})$ coming from the identification $\mathrm{pr}_* \bar{b} = *$ of base points on $\mathrm{Spec}(k)$. Therefore the path space has a canonical structure $P_{\bar{a}} \to \overline{X}$ of a pro-object of $\mathsf{Rev}(\overline{X})$ and is the path space to a fibre functor of $\overline{X}$ that we denote by $\bar{b}$ as well.

The fact, that $\bar{b}$, as a base point of X, enriches to the functor β with values in $\mathsf{Rev}(k)$, induces a left action by Gal_k on all the fibres $Y[\bar{b}]$ for $Y \in \mathsf{Rev}(X)$ that is compatible with morphisms between covers. In particular, by using the universal points in the fibres and Lemma 22, this action translates into a right action of Gal_k on $P_{\bar{b}}$ by $\sigma \mapsto p_\sigma$ that lies above the canonical right action by $\sigma \mapsto 1 \otimes \sigma$ of Gal_k on $\overline{X}$. We conclude that for all $\sigma \in \mathrm{Gal}_k$ we can canonically identify $\sigma(\bar{b}) = \bar{b}$ and thus $\bar{b}$ is a Galois invariant base point of $\overline{X}$.

Lemma 22. *Let M be a profinite set with commuting continuous actions by a profinite group G from the left and a profinite group A from the right. If the action of A is free and transitive on M then an element $m \in M$ defines a homomorphism $\varphi : G \to A$ by the formula*

$$g.m = m.(\varphi(g))$$

for every $g \in G$.

Proof. This lemma is both elementary and standard.   □

The above discussion leads to the following chart translating between and emphasizing different points of views towards sections of the fundamental group exact sequence. Note that the left column comes without specifying a base point of $\overline{X}$ unlike the middle column that picks a base point of reference $\bar{a}$ of $\overline{X}$. Moreover, the right column even requires $\bar{a}$ to be Galois invariant and works with the Gal_k-group structure on $\pi_1(\overline{X}, \bar{a})$ induced by conjugation with the associated canonical section s_a.

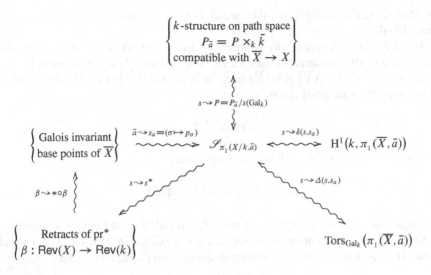

Definition 23. The *space of sections* of X/k is the set

$$\mathscr{S}_{\pi_1}(X/k) = \mathscr{S}_{\pi_1}(X/k,\bar{a})$$

that is independent of the chosen base point $\bar{a} \in \overline{X}$ in the sense that the spaces for different choices are canonically in bijection with each other.

2.6 Homotopy Fixed Points and the Section Conjecture

The idea to use higher étale homotopy theory in the context of anabelian geometry and in particular of the section conjecture occurred independently to many people, for example to Huber-Klawitter, Harpaz and Schlank [HaSch10], Pál [Pa10b], Quick [Qu10], Schmidt [Sch12], and also to the author ... just to name a few.

We review below the theory in the setup of continuous profinite homotopy theory as developed by Quick [Qu10] §3.2, which gives an interesting reformulation of the section conjecture. In particular, the conjecture is put into a more general framework available for all varieties, not only for curves. Unfortunately, to this date, this reformulation of the section conjecture has not contributed towards its proof, except for the case of $\mathbb{R}$ as the base field, see Sect. 16.1.

Homotopy fixed points. Let $\hat{\mathrm{Et}}(S)$ be the profinite étale homotopy type of a scheme S in the sense of pro-objects in simplicial profinite sets, see [Qu08]. For a connected S the étale homotopy type computes the étale fundamental group as

$$\pi_1(S) = \pi_1(\hat{\mathrm{Et}}(S)).$$

We abbreviate $\hat{\mathrm{Et}}(\mathrm{Spec}(k))$ by $\hat{\mathrm{Et}}(k)$, which is weakly equivalent to the classifying space $B\mathrm{Gal}_k$.

Let X/k be a geometrically connected variety and let $\overline{X} = X \times_k \overline{k}$ be the base change to the algebraic closure $\overline{k}$ of k. The group Gal_k acts continuously on a continuous version $\mathrm{c}\hat{\mathrm{Et}}(\overline{X})$ of $\hat{\mathrm{Et}}(\overline{X})$ from the right, see [Qu10] §3. There is a notion of a homotopy fixed point space

$$\mathrm{c}\hat{\mathrm{Et}}(\overline{X})^{h\mathrm{Gal}_k}$$

of this action, see [Qu10] §2.3, and the natural map

$$\mathrm{c}\hat{\mathrm{Et}}(\overline{X}) \times_{\mathrm{Gal}_k} \mathrm{c}\hat{\mathrm{Et}}(\overline{k}) \to \hat{\mathrm{Et}}(X) \tag{2.12}$$

is a weak equivalence, see [Qu10] Theorem 3.5. It follows that $\hat{\mathrm{Et}}(X)$ is a homotopy orbit space for the continuous Galois action of Gal_k on $\mathrm{c}\hat{\mathrm{Et}}(\overline{X})$. The natural adjunction induced by taking homotopy deduces from (2.12) a bijection

$$\mathrm{Hom}_{\hat{\mathrm{Et}}(k)}\big(\hat{\mathrm{Et}}(k), \hat{\mathrm{Et}}(X)\big) = \mathrm{Hom}_{\mathrm{Gal}_k}\big(\mathrm{c}\hat{\mathrm{Et}}(\overline{k}), \mathrm{c}\hat{\mathrm{Et}}(\overline{X})\big) = \pi_0\left(\mathrm{c}\hat{\mathrm{Et}}(\overline{X})^{h\mathrm{Gal}_k}\right)$$

where $\mathrm{Hom}_{\mathrm{Gal}_k}$ denotes the set of homotopy classes of Gal_k-equivariant maps.

Reformulation of the section conjecture. The maps induced by the functor $\hat{\mathrm{Et}}(-)$ and π_1 yield the following commutative diagram

$$
\begin{array}{ccc}
\overline{X}(\overline{k})^{\mathrm{Gal}_k} & \xleftarrow{\;\sim\;} & X(k) \\
\downarrow{\scriptstyle \mathrm{c}\hat{\mathrm{Et}}} & & \downarrow{\scriptstyle \hat{\mathrm{Et}}} \quad\searrow{\scriptstyle \kappa} \\
\pi_0\left(\mathrm{c}\hat{\mathrm{Et}}(\overline{X})^{h\mathrm{Gal}_k}\right) & \xleftarrow{\;\sim\;} & \mathrm{Hom}_{\hat{\mathrm{Et}}(k)}\big(\hat{\mathrm{Et}}(k), \hat{\mathrm{Et}}(X)\big) \xrightarrow{\;\pi_1\;} \mathscr{S}_{\pi_1(X/k)}
\end{array}
$$

which factorizes the profinite Kummer map. Now, if X is an algebraic $K(\pi,1)$-space, then $\hat{\mathrm{Et}}(X)$ is weakly equivalent to the classifying space $B\pi_1(X)$ and the map

$$\pi_1 \;:\; \mathrm{Hom}_{\hat{\mathrm{Et}}(k)}\big(\hat{\mathrm{Et}}(k), \hat{\mathrm{Et}}(X)\big) \xrightarrow{\;\sim\;} \mathscr{S}_{\pi_1(X/k)} \tag{2.13}$$

is a bijection. In particular, this applies for smooth hyperbolic curves, and the section conjecture acquires a reformulation as the following question, see [Qu10] §3.

Question 24. (1) Let X be a proper, smooth, hyperbolic curve over a number field k. Is the natural map

$$\hat{\mathrm{Et}} \;:\; X(k) \to \pi_0\left(\mathrm{c}\hat{\mathrm{Et}}(\overline{X})^{h\mathrm{Gal}_k}\right)$$

a bijection, or more verbal, does every homotopy fixed point of $\mathrm{c\hat{E}t}(\overline{X})$ for the action by Gal_k come from a true fixed point of $\overline{X}$?

(2) A natural generalization asks (1) for all proper smooth varieties, since the $K(\pi, 1)$ property has already been exploited in the bijection (2.13). But because the map in (1) is constant for $X = \mathbb{P}^1_k$, it factors for general X through the set of equivalence classes by the relation generated by images of maps $\mathbb{P}^1_k \to X$ on rational points.

Remark 25. A finite 2-group version of question (1) in the context of homotopy theory of CW-complexes is known as Sullivan's Conjecture, see [Su71] p. 179 and [Ca91] Theorem B (c) for a proof of a corrected version. Sullivan's Conjecture was proved by Miller [Mi84] for trivial group action, and independently by Dwyer, Miller and Neisendorfer [DMN89], Carlsson [Ca91], and Lannes [La92] in general. and moreover generalized to a version with finite p-groups.

The original motivation for Sullivan to state his conjecture was to study the homotopy type of the fixed point set $X(\mathbb{R}) = X(\mathbb{C})^{\mathrm{Gal}_\mathbb{R}}$ for a real algebraic variety $X/\mathbb{R}$ through the equivariant homotopy theory of its étale homotopy type of $X(\mathbb{C})$ together with its Galois action. It is possible to give a quick proof of the real section conjecture along these lines, see Sect. 16.1.

The descent spectral sequence. There is a partially defined Bousfield–Kan descent spectral sequence for homotopy fixed points, see [BoKa72] IX §4 and [Qu10] Theorem 2.16. For $\mathrm{c\hat{E}t}(\overline{X})^{h\mathrm{Gal}_k}$ it reads

$$E^i_{2;j} = \mathrm{H}^i\left(k, \pi_j(\overline{X})\right) \implies \pi_{j-i}\left(\mathrm{c\hat{E}t}(\overline{X})^{h\mathrm{Gal}_k}\right)$$

with zero entries outside $j \geq i \geq 0$ by definition and with a differential

$$d^i_{2;j} : E^i_{2;j} \to E^{i+2}_{2;j+1}.$$

For a $K(\pi, 1)$-space $\overline{X}$ the spectral sequence degenerates at the E_2-level and yields

$$\pi_0\left(\mathrm{c\hat{E}t}(\overline{X})^{h\mathrm{Gal}_k}\right) = \mathscr{S}_{\pi_1(X/k)},$$
$$\pi_1\left(\mathrm{c\hat{E}t}(\overline{X})^{h\mathrm{Gal}_k}\right) = \mathrm{H}^0\left(k, \pi_1(\overline{X})\right),$$

which is the centraliser of the section used to define the Galois action on $\pi_1(\overline{X})$, so vanishes in arithmetically interesting cases, see Proposition 104, and

$$\pi_i\left(\mathrm{c\hat{E}t}(\overline{X})^{h\mathrm{Gal}_k}\right) = 0 \qquad \text{for } i \geq 2.$$

Consequently, the space $\mathrm{c\hat{E}t}(\overline{X})^{h\mathrm{Gal}_k}$ is weakly equivalent to a collection of contractible components which are in natural bijection with $\mathscr{S}_{\pi_1(X/k)}$. But please beware that the descent spectral sequence has subtle convergence issues, especially along the line $i = j$, which we have ignored.

Chapter 3
Basic Geometric Operations in Terms of Sections

Geometric structures that apply to rational points can sometimes be formulated in terms of sections: functorial properties of the space of sections, abelianization and base change. In favourable circumstances we establish Galois descent for sections, see Proposition 28. We furthermore study the behaviour of sections under fibrations and finite étale covers between varieties. The notion of the anabelian fibre above a section is introduced.

The results of [Sx10a] on the behaviour of the space of sections under Weil restriction of scalars are summarized in Sect. 3.5.

3.1 Functoriality in the Space and Abelianization

Let $f : X \to Y$ be a morphism of geometrically connected k-varieties. We fix a base point $\bar{a}$ of $\overline{X}$ and a base point $\bar{b}$ of $\overline{Y}$. The induced map

$$f_* = \pi_1(f) : \pi_1(X, \bar{a}) \to \pi_1(Y, \bar{b})$$

is canonical up to conjugation by elements of $\pi_1(\overline{Y}, \bar{b})$ and thus defines functorially a map

$$f_* : \mathscr{S}_{\pi_1(X/k)} \to \mathscr{S}_{\pi_1(Y/k)},$$

$$s \mapsto f_*(s) = f_* \circ s.$$

Abelianization. For a profinite group Γ we call the quotient

$$\Gamma^{\mathrm{ab}} = \Gamma/[\Gamma, \Gamma] \tag{3.1}$$

by the closure of the commutator subgroup $[\Gamma, \Gamma]$, or more precisely the quotient map

$$\Gamma \twoheadrightarrow \Gamma^{\mathrm{ab}}$$

the *abelianization* of Γ.

J. Stix, *Rational Points and Arithmetic of Fundamental Groups*, Lecture Notes in Mathematics 2054, DOI 10.1007/978-3-642-30674-7_3,

Definition 26. (1) The *maximal (geometrically) abelian quotient extension* is the pushout $\pi_1^{ab}(X/k)$ of $\pi_1(X/k)$ by the abelization map

$$\pi_1(\overline{X}) \twoheadrightarrow \pi_1^{ab}(\overline{X}) := \left(\pi_1(\overline{X})\right)^{ab}.$$

(2) The *abelianization* s^{ab} of a section s is the image under the natural map

$$\mathcal{S}_{\pi_1(X/k)} \to \mathcal{S}_{\pi_1^{ab}(X/k)}.$$

Albanese and abelianization. Let X be a smooth, projective curve with Albanese variety

$$\text{Alb}_X = \text{Pic}_X^0 \tag{3.2}$$

and universal Albanese torsor map, see [Wi08],

$$\alpha_X : X \to \text{Alb}_X^1 = \text{Pic}_X^1. \tag{3.3}$$

It is a well known formulation of geometric class field theory that the natural map

$$\pi_1(\alpha_X) : \pi_1(\overline{X}, \bar{a}) \to \pi_1(\overline{\text{Pic}_X^1}, \alpha_X(\bar{a}))$$

agrees with the maximal abelian quotient $\pi_1^{ab}(\overline{X}, \bar{a})$, see Proposition 69. In the case of smooth, projective curves, the abelianization s^{ab} of a section s agrees with the image under the map on spaces of sections

$$\mathcal{S}_{\pi_1(X/k)} \to \mathcal{S}_{\pi_1^{ab}(X/k)} = \mathcal{S}_{\pi_1(\text{Pic}_X^1/k)}$$

$$s \mapsto \alpha_{X,*}(s) = s^{ab}$$

functorially induced by the Albanese torsor map. For more on the abelianization of the fundamental group see Sect. 7.1.

3.2 Base Change

Let X/k be a geometrically connected variety over a field k with algebraic closure $\bar{k}$. We assume that X/k is proper or k has characteristic 0. For a field extension K/k with algebraic closure $\bar{k} \subseteq \bar{K}$ the projection $X_K = X \times_k K \to X$ induces an isomorphism, see [Sz09] Remark 5.7.8,

$$\pi_1(X_K \times_K \bar{K}, \bar{a}) \xrightarrow{\sim} \pi_1(X \times_k \bar{k}, \bar{a})$$

where $\bar{a}$ is a base point of $\overline{X_K} = X_K \times_K \bar{K}$ or its image in $\overline{X} = X \times_k \bar{k}$. Thus the extension $\pi_1(X_K/K, \bar{a})$ is the pullback of $\pi_1(X/k, \bar{a})$ via the restriction map

$$\text{res}_{K/k} : \text{Gal}_K \to \text{Gal}_k \tag{3.4}$$

and the natural map provides an isomorphism

$$\pi_1(X_K, \bar{a}) \xrightarrow{\sim} \pi_1(X, \bar{a}) \times_{\mathrm{Gal}_k} \mathrm{Gal}_K.$$

Definition 27. The *base change* $s_K = s \otimes K$ of a section s with respect to the field extension K/k is defined by the defining property of a fibre product as

$$s_K = (s \circ \mathrm{res}_{K/k}, \mathrm{id}) : \mathrm{Gal}_K \to \pi_1(X, \bar{a}) \times_{\mathrm{Gal}_k} \mathrm{Gal}_K = \pi_1(X_K, \bar{a}).$$

Base change of sections yields a *base change map* for spaces of sections

$$\mathscr{S}_{\pi_1}(X/k) \to \mathscr{S}_{\pi_1}(X_K/K),$$
$$s \mapsto s_K = s \otimes K$$

where we have neglected the base point from our notation. We set

$$\mathscr{S}_{\pi_1}(X/k)(K) = \mathscr{S}_{\pi_1}(X_K/K)$$

which describes a covariant functor on field extensions K/k with values in sets.

In particular, the space of sections forms a presheaf of sets on $\mathrm{Spec}(k)_{\mathrm{\acute{e}t}}$ by

$$k'/k \mapsto \mathscr{S}_{\pi_1}(X/k)(k') = \mathscr{S}_{\pi_1}(X_{k'}/k')$$

where k'/k is a finite separable field extension. The presheaf $\mathscr{S}_{\pi_1}(X/k)$ on $\mathrm{Spec}(k)_{\mathrm{\acute{e}t}}$ also exists for X in characteristic $p > 0$ that is not necessarily proper, because, for k'/k finite algebraic, the base change $\overline{X_{k'}}$ agrees with $\overline{X}$.

3.3 Centralisers and Galois Descent for Sections

Let s be a section of a short exact sequence of profinite groups

$$1 \to \overline{\pi} \to \pi \to \Gamma \to 1.$$

We recall from Sect. 1.4 the notion of the *centraliser of the section s* as the subgroup $Z_{\overline{\pi}}(s)$ of elements $\gamma \in \overline{\pi}$ which centralise the image $s(\Gamma)$ in π. If t is another section that is conjugate to s via $\alpha \in \overline{\pi}$, namely $t = \alpha(-)\alpha^{-1} \circ s$, then the centralisers are conjugates of each other as well, namely

$$Z_{\overline{\pi}}(t) = \alpha(Z_{\overline{\pi}}(s))\alpha^{-1}.$$

In particular, the property of having trivial centraliser is a property of a conjugacy class of sections.

Proposition 28. *Let X/k be a geometrically connected variety. Let us assume that for any finite separable extension k' every section $s \in \mathscr{S}_{\pi_1(X/k)}(k')$ has trivial centraliser.*

Then the space of sections defines a sheaf of sets on $\mathrm{Spec}(k)_{\acute{e}t}$. *In other words, conjugacy classes of sections satisfy Galois descent, i.e., for a Galois extension E/F of finite extensions of k the natural map*

$$\mathscr{S}_{\pi_1(X/k)}(F) \to \mathscr{S}_{\pi_1(X/k)}(E)$$

is injective and has the $\mathrm{Gal}(E/F)$-*invariant conjugacy classes as its image.*

Proof. This is an immediate consequence of Proposition 12. $\qquad\qquad\qquad\square$

Remark 29. (1) Actually, Proposition 12 is more precise. The presheaf $\mathscr{S}_{\pi_1(X/k)}$ is separated if and only if for all finite separable extension E/F finite over k with E/F Galois and a section $s \in \mathscr{S}_{\pi_1(X/k)}(F)$ the non-abelian

$$\mathrm{H}^1\big(E/F, Z_{\pi_1(\overline{X})}(s|_{\mathrm{Gal}_E})\big)$$

vanishes, where the coefficients inherit its $\mathrm{Gal}(E/F)$-action via conjugation by lifts via s. And the natural map of presheaves into the sheafification

$$\mathscr{S}_{\pi_1(X/k)}(-) \to \big(\mathscr{S}_{\pi_1(X/k)}\big)^{\#}(-)$$

is surjective if and only if certain elements in non-abelian $\mathrm{H}^2(E/F, -)$ vanish.

(2) We will see later in Sect. 9.4 that when X/k is a smooth geometrically connected curve and k is either an algebraic number field or a finite extension of $\mathbb{Q}_p$ then the assumption on sections of $\pi_1(X/k)$ having trivial centraliser holds.

3.4 Fibres Above Sections

We study fibres above sections in two setups: for a fibration and for finite étale covers.

The fibre in a fibration. Let $f : X \to S$ be a universally submersive morphism, see [SGA1] IX Definition 2.1, of geometrically connected k-varieties with geometrically connected fibres. We fix compatible base points $\overline{a} \in \overline{X}$ and $\overline{b} \in \overline{S}$ and consider the short exact sequence

$$1 \to \overline{\pi} \to \pi_1(X, \overline{a}) \xrightarrow{\pi_1(f)} \pi_1(S, \overline{b}) \to 1$$

with the implicitly defined kernel $\overline{\pi}$. The map $\pi_1(f)$ is surjective, because otherwise there would be a nontrivial connected finite étale cover $S' \to S$ such that f lifts to a map $f' : X \to S'$ which contradicts the assumption on the fibres if $\deg(S'/S) > 1$, see [SGA1] IX Corollary 5.6.

Definition 30. The *anabelian fibre* in the section $s : \mathrm{Gal}_k \to \pi_1(S, \bar{b})$ along f as above is the extension

$$1 \to \overline{\pi} \to \pi_s \to \mathrm{Gal}_k \to 1$$

in which the group π_s is simply the fibre product by s

$$\pi_s = \pi_1(X, \bar{a}) \times_{\pi_1(S, \bar{b})} \mathrm{Gal}_k$$

and the maps are the obvious ones.

The notion of an anabelian fibre becomes reasonable at least when $f : X \to S$ is a smooth, proper map and $\pi_2(S) = 0$ in the sense of Artin–Mazur, e.g, if S is a good Artin neighbourhood in a smooth k-variety. Then the homotopy sequence of the fibration reads

$$1 \to \pi_1(X_{\bar{b}}, \bar{a}) \to \pi_1(X, \bar{a}) \to \pi_1(S, \bar{b}) \to 1$$

and we can identify $\overline{\pi}$ with the fundamental group of the geometric fibre $X_{\bar{b}}$ above $\bar{b}$ endowed with the base point $\bar{a}$ induced by the fibre product property. If the section $s : \mathrm{Gal}_k \to \pi_1(S, \bar{b})$ comes from a rational point, which for simplicity we assume to be $b \in S(k)$ underlying the base point $\bar{b}$, then π_s and $\pi_1(X_b/k)$ are naturally isomorphic as extensions. Here we have used the notation $X_b = X \times_{S,b} \mathrm{Spec}(k)$ for the fibre in the rational point in the sense of schemes.

The anabelian fibre allows to discuss the effect of f on sections as follows.

Proposition 31. *With the notation as above we have a long exact sequence of non-abelian cohomology groups*

$$1 \to \mathrm{H}^0(k, \overline{\pi}) \to \mathrm{H}^0(k, \pi_1(\overline{X})) \to \mathrm{H}^0(k, \pi_1(\overline{S}))$$

$$\to \mathscr{S}_{\pi_s \to \mathrm{Gal}_k} \to \mathscr{S}_{\pi_1(X/k)} \xrightarrow{f_*} \mathscr{S}_{\pi_1(S/k)} \xrightarrow{\delta} [\mathrm{Ext}(\mathrm{Gal}_k, \overline{\pi})]_{\pi_1(\overline{X})}$$

which has to be read and is made more precise as follows.

(1) A section $s \in \mathscr{S}_{\pi_1(S/k)}$ maps to the extension

$$\delta(s) = [1 \to \overline{\pi} \to \pi_s \to \mathrm{Gal}_k \to 1]$$

and lifts to $t \in \mathscr{S}_{\pi_1(X/k)}$ if and only if $\delta(s)$ is a semi-direct product extension.

(2) From now on we assume the existence of a section $t \in \mathscr{S}_{\pi_1(X/k)}$. Let $s = f_*(t)$ be its image in $\mathscr{S}_{\pi_1(S/k)}$. The pointed set $f_*^{-1}(s)$ is the image of the natural map

$$\mathscr{S}_{\pi_s \to \mathrm{Gal}_k} \to \mathscr{S}_{\pi_1(X/k)}.$$

(3) The section t defines a section of $\pi_s \to \mathrm{Gal}_k$. The group $\mathrm{H}^0(k, \pi_1(\overline{S}))$ is by definition the centraliser of the image of s and acts on sections of $\pi_s \to \mathrm{Gal}_k$ through conjugation by lifts to $\pi_1(\overline{X})$.

(4) The groups $H^0\left(k, \pi_1(\overline{X})\right)$ *and* $H^0(k, \overline{\pi})$ *are the centralisers of the image of* t *as a section of* $\pi_1(X/k)$ *or of* $\pi_s \to \mathrm{Gal}_k$ *respectively.*

Proof. This is merely an explicit application of Sect. 1.3.   □

The fibre of a finite étale map. Let $h : Y \to X$ be a finite étale cover of connected, quasi-compact schemes. Let $\bar{b}$ be a base point of $\overline{Y}$ and $\bar{a} = h(\bar{b})$ its image in $\overline{X}$. By Proposition 17, the cover $Y \to X$ together with its base point $\bar{b}$ is described by the $\pi_1(X, \bar{a})$-set

$$M = \pi_1(X, \bar{a})/\pi_1(Y, \bar{b}).$$

Definition 32. The *(anabelian) fibre* $h^{-1}(s)$ of the finite étale cover $h : Y \to X$ described by the $\pi_1(X, \bar{a})$-set M above a morphism $s : \mathrm{Gal}_k \to \pi_1(X, \bar{a})$ is the Gal_k-set

$$h^{-1}(s) = s^*M$$

which is the set M endowed with the Gal_k-action via s. The associated finite locally constant sheaf of sets on $\mathrm{Spec}(k)_{\text{ét}}$ has values

$$\underline{s^*M}(k') = M^{\mathrm{Gal}_{k'}} = \{m \in M \; ; \; \sigma.m = m \text{ for all } \sigma \in \mathrm{Gal}_{k'}\}$$

for a finite separable subextension k'/k of $\bar{k}/k$.

To a section $m \in \underline{s^*M}(k')$ we associate the lift

$$s_m = m^{-1}(-)m \circ s|_{\mathrm{Gal}_{k'}} \; : \; \mathrm{Gal}_{k'} \to \pi_1(Y, \bar{b})$$

of the morphism s up to conjugation by an element of $\pi_1(X, \bar{a})$. Indeed, the map

$$\underline{s^*M}(k') \twoheadrightarrow \left\{ \begin{matrix} \text{lifts of } s|_{\mathrm{Gal}_{k'}} \text{ to } \pi_1(Y, \bar{b}) \\ \text{up to conjugation} \end{matrix} \right\} \tag{3.5}$$
$$m \mapsto s_m$$

is well defined and surjective, because any lift has necessarily the form s_m for some $m \in M$ and

$$m^{-1}s(\mathrm{Gal}_{k'})m \subset \pi_1(Y, \bar{b})$$

is equivalent to m being fixed by $s(\mathrm{Gal}_{k'})$.

By Proposition 8, the Gal_k-set s^*M corresponds to a finite étale algebra A_s/k such that $\mathrm{Spec}(A_s)$ represents $\underline{s^*M}$. If $s = \pi_1(a)$ comes by applying the functor π_1 to an actual map $a : \mathrm{Spec}(k) \to X$, then $\mathrm{Spec}(A_s)$ is isomorphic to the schematic fibre $Y \times_{X,a} \mathrm{Spec}(k)$ and so the anabelian fibre describes the actual scheme theoretic fibre.

The fibre of a geometrically connected étale cover. We discuss the previous paragraph in the special case of varieties. Let

$$h : Y \to X$$

be a finite étale cover of geometrically connected k-varieties, and let $\bar{b}$ be a base point of $\overline{Y}$ and $\bar{a} = h(\bar{b})$ its image in $\overline{X}$. By Proposition 17 and the assumption on being geometrically connected, the cover $Y \to X$ together with its base point $\bar{b}$ is described by the $\pi_1(X, \bar{a})$-set

$$M = \pi_1(X, \bar{a})/\pi_1(Y, \bar{b}) = \pi_1(\overline{X}, \bar{a})/\pi_1(\overline{Y}, \bar{b}). \tag{3.6}$$

Let s be a section of $\pi_1(X/k, \bar{a})$. If we replace s by a conjugate section

$$t = \gamma(-)\gamma^{-1} \circ s$$

for some $\gamma \in \pi_1(\overline{X}, \bar{a})$, then as Gal_k-sets the translation by γ yields an isomorphism

$$s^* M \cong t^* M,$$

so that the isomorphy class of the anabelian fibre $h^{-1}(s)$ as a Gal_k-set only depends on the class $[s]$.

We consider the preimage $h_*^{-1}(s)$ under the map of presheaves on $\mathrm{Spec}(k)_{\mathrm{ét}}$

$$h_* \; : \; \mathscr{S}_{\pi_1(Y/k, \bar{b})} \to \mathscr{S}_{\pi_1(X/k, \bar{a})}$$

of the global section $[s]$ considered as the image of the constant one point sheaf. As in (3.5) there is a well defined map of presheaves

$$h^{-1}(s) \to h_*^{-1}(s)$$

defined by

$$m \mapsto s_m = \overline{m}^{-1}(-)\overline{m} \circ s|_{\mathrm{Gal}_{k'}}$$

where we choose a representative $\overline{m}$ of m from $\pi_1(\overline{X}, \bar{a})$ by (3.6). Under an anabelian hypothesis we can say more.

Proposition 33. *Let $h : Y \to X$ be a finite étale cover of geometrically connected k-varieties.*

(1) For a section s of $\pi_1(X/k)$ the map $m \mapsto [s_m]$ is a surjective map of presheaves

$$h^{-1}(s) \twoheadrightarrow h_*^{-1}(s).$$

(2) If in addition for any finite separable extension k'/k the restriction $s|_{\mathrm{Gal}_{k'}}$ has trivial centraliser, then

$$h^{-1}(s) \xrightarrow{\sim} h_*^{-1}(s)$$

is an isomorphism of finite locally constant sheaves on $\mathrm{Spec}(k)_{\mathrm{ét}}$.

Proof. (1) Let k'/k be a finite separable subextension of $\bar{k}/k$ and consider a section $t : \mathrm{Gal}_{k'} \to \pi_1(Y_{k'}, \bar{b})$ with $h_*([t]) = [s|_{\mathrm{Gal}_{k'}}]$. Thus we can pick $\gamma \in \pi_1(\overline{X}, \bar{a})$ such that

$$t = \gamma^{-1}(-)\gamma \circ s|_{\mathrm{Gal}_k'}$$

and $t = s_y$. Since t has values in $\pi_1(Y, \bar{b})$, the image $s(\mathrm{Gal}_{k'})$ lies in $\gamma \pi_1(Y, \bar{b}) \gamma^{-1}$ and therefore stabilizes the class of γ in

$$M = \pi_1(\overline{X}, \bar{a})/\pi_1(\overline{Y}, \bar{b}) = \pi_1(X, \bar{a})/\pi_1(Y, \bar{b}).$$

Hence $\gamma \in h^{-1}(s)(k')$ is a section which maps to t and $h^{-1}(s) \to h_*^{-1}(s)$ is surjective.

(2) Let $\alpha, \beta \in \pi_1(\overline{X}, \bar{a})$ be representatives of classes in

$$h^{-1}(s)(k') \subseteq \pi_1(\overline{X}, \bar{a})/\pi_1(\overline{Y}, \bar{b})$$

such that

$$[s_\alpha] = [s_\beta] \in \mathscr{S}_{\pi_1(Y/k)}(k').$$

Hence, there is a $\gamma \in \pi_1(\overline{Y}, \bar{b})$ such that

$$\alpha^{-1}(-)\alpha \circ s|_{\mathrm{Gal}_k'} = \gamma(-)\gamma^{-1} \circ \beta^{-1}(-)\beta \circ s|_{\mathrm{Gal}_k'}$$

as actual maps. This means that $\alpha\gamma\beta^{-1}$ centralises $s(\mathrm{Gal}_{k'})$ and therefore $\alpha\gamma = \beta$ by the assumption on the centralisers. Consequently, the classes of α and β agree. $\square$

Corollary 34. *Let $h : Y \to X$ be a finite étale cover of geometrically connected k-varieties.*

(1) For a section $s = s_x$ of $\pi_1(X/k, \bar{a})$ associated to a rational point $x \in X(k)$ the fibre

$$h^{-1}(s_x)$$

of s_x is represented by the fibre $Y \times_{X,x} \mathrm{Spec}(k)$ in the sense of schemes.

(2) The section s_x associated to a rational point $x \in X(k)$ lifts to a section of $\pi_1(Y/k, \bar{b})$ if and only if the rational point lifts to a rational point $y \in Y(k)$.

Proof. (1) This is obvious by the naturality of the construction and Proposition 8.

(2) If $x \in X(k)$ lifts to $y \in Y(k)$, then obviously s_y lifts s_x. Conversely, if s is a lift of s_x, then, because the map

$$h^{-1}(s) \twoheadrightarrow h_*^{-1}(s)$$

is surjective as a map of presheaves, we find a global section

$$y \in h^{-1}(s_x)(k)$$

such that $[s_y] = [s]$. By part (1), the element y corresponds to a k-rational point of the schematic fibre $Y \times_{X,x} \mathrm{Spec}(k)$ which is nothing but a lift of x to a $y \in Y(k)$.

$\square$

Centralisers, fibres and fixed points. We extend the notion of isogenous abelian varieties to arbitrary connected varieties in the following way.

Definition 35. An *anabelian isogeny* of two geometrically connected varieties X, Y is a finite étale cover $h : Y \to X$. Two varieties are *anabelian–isogenous* if they can be connected by a chain of anabelian–isogenies, or equivalently, the varieties X and Y are anabelian–isogenous if they have a common geometrically connected finite étale cover $Z \to X$ and $Z \to Y$.

Note that the anabelian isogeny class of an abelian variety contains its isogeny class but in general can be bigger. In particular, for surfaces, the anabelian isogeny class of an abelian surface may contain also hyperelliptic surfaces.

Proposition 36. *Let X/k be a geometrically connected variety. Then the following assertions on the family of geometrically connected varieties X'/k' over finite extensions k'/k that are anabelian–isogenous to $X_{k'} = X \times_k k'$ are equivalent.*

(a) *For every section $s' \in \mathscr{S}_{\pi_1(X'/k')}$, the centraliser $Z_{\pi_1(\overline{X}')}(s')$ is trivial.*

(b) *For every geometrically connected finite étale cover $h : X'' \to X'$ and for every section $s' \in \mathscr{S}_{\pi_1(X'/k')}$, the natural map*

$$\underline{h^{-1}(s')} \to h_*^{-1}(s')$$

from the anabelian fibre to the fibre of h_* is bijective.

(c) *For every faithful k'-linear group action $G \times X' \to X'$ by a finite group G on the k'-variety X' the induced action of G on $\mathscr{S}_{\pi_1(X'/k')}$ is faithful.*

Proof. That (a) implies (b) is the content of Proposition 33. Let us assume (b) and show (c). Without loss of generality we may assume that $k' = k$ and X is the quotient $G \backslash X'$. By faithfulness, the quotient map is a geometrically connected finite étale cover $h : X' \to X$. For a section s' of $\pi_1(X'/k)$ with image $h_*(s') = s \in \mathscr{S}_{\pi_1(X/k)}$, we have the natural map

$$G = \pi_1(X)/\pi_1(X) = \underline{h^{-1}(s)} \to h_*^{-1}(s)$$

which maps $g \in G$ to $g_*(s')$. Hence (b) implies (c).

Let us now assume (c) and show (a). Without loss of generality we may consider a section $s \in \mathscr{S}_{\pi_1(X/k)}$ and argue by contradiction. If $\gamma \in Z_{\pi_1(\overline{X})}(s)$ is a nontrivial element, then we can find a finite extension k'/k and an open normal subgroup

$$\gamma \notin H \trianglelefteq \pi_1(X_{k'})$$

and $s(\mathrm{Gal}_{k'}) \subset H$. The corresponding geometrically connected finite étale Galois cover $X' \to X_{k'}$ has Galois group $\pi_1(X_{k'})/H$, so that γ induces a nontrivial covering automorphism $g \in \mathrm{Aut}(X'/X_{k'})$. By (c) we have

$$s' = s|_{\mathrm{Gal}_{k'}} \neq g_*(s') = \gamma(s')\gamma^{-1}$$

which agrees with s' because γ centralises s and hence s'. This yields the desired contradiction and thus (c) implies (a). $\square$

3.5 Weil Restriction of Scalars and Sections

As a first step we need a non-abelian version of induction for extensions of groups.

Definition 37. The category $\mathrm{Ext}[G]$ of extensions of a pro-finite group G has as objects continuous extensions of G with arbitrary kernel

$$1 \to N \to E \to G \to 1$$

and morphisms are equivalence classes of maps of extensions that are the identity on G. Two such maps of extensions are said to be equivalent if they differ by an inner automorphisms from an element of the kernel.

Non-abelian induction. Let G be a profinite group. Let H be a open subgroup of G. The restriction functor from G to H

$$\mathrm{res}_H^G : \mathrm{Ext}[G] \to \mathrm{Ext}[H],$$

$$E \mapsto \mathrm{res}_H^G(E) = E|_H$$

admits a right adjoint Ind_H^G which by analogy is called *(non-abelian) induction* from H to G. The construction of Ind_H^G mimics an explicit version of Shapiro's Lemma and can be found in [Sx10a] 2.4.

The fundamental group of a Weil restriction. Let L/K be a finite separable field extension and let X/L be a geometrically connected quasi-projective variety. The Weil restriction of scalars

$$\mathrm{R}_{L|K} X \tag{3.7}$$

of X/L with respect to L/K represents the functor on K-varieties

$$Y \mapsto \mathrm{Hom}_L(Y \times_K L, X)$$

For the construction and properties of the Weil restriction functor we refer the reader to [BLR90] 7.6.

The non-abelian induction functor explains the structure of the fundamental group of the Weil restriction of a variety as follows.

Theorem 38 ([Sx10a] Theorem 17). *Let* K *be of characteristic* 0 *or assume that* X/L *is a projective variety. Then we have*

$$\pi_1(R_{L|K} X / K) = \operatorname{Ind}_{\operatorname{Gal}_L}^{\operatorname{Gal}_K} \left(\pi_1(X/L) \right)$$

as extensions of Gal_K.   □

The non-abelian Shapiro lemma. By Proposition 8, conjugacy classes of sections are in bijection with non-abelian H^1. We may therefore address the following theorem as the non-abelian Shapiro Lemma in degrees 1 and partially in degree 2.

Theorem 39 ([Sx10a] Corollary 15). *Let* H *be an open subgroup of* G *and let*

$$E = [1 \to N \to E \to H \to 1]$$

be an extension in $\operatorname{Ext}[H]$ *with induction in* $\operatorname{Ext}[G]$

$$\operatorname{Ind}_H^G(E) = [1 \to M \to \operatorname{Ind}_H^G(E) \to G \to 1].$$

(1) The extension E *splits if and only if* $\operatorname{Ind}_H^G(E)$ *splits.*
(2) Adjointness induces a bijection between N*-conjugacy classes of sections of* $E \twoheadrightarrow H$ *and* M*-conjugacy classes of sections of* $\operatorname{Ind}_H^G(E) \twoheadrightarrow G$.   □

The space of sections and Weil restriction. The behaviour of the space of sections under Weil restriction of scalars is now described by the following theorem.

Theorem 40. *Let* L/K *be a finite separable field extension and let* X/L *be a quasi-projective, geometrically connected variety. Let* K *have characteristic* 0 *or let* X/L *be projective. Then there is a natural bijection of spaces of sections:*

$$\mathscr{S}_{\pi_1}(X/L) = \mathscr{S}_{\pi_1}(R_{L|K} X/K).$$

Proof. This follows from Theorem 39 together with Theorem 38.   □

The above translates in terms of the profinite Kummer map and the section conjecture to the following.

Corollary 41 ([Sx10a] Theorem 4). *Let* $s \in \mathscr{S}_{\pi_1}(X/L)$ *and* $t \in \mathscr{S}_{\pi_1}(R_{L|K} X/K)$ *be sections that correspond to each other under the bijection of Theorem 40.*

(1) Then $s = s_a$ *for a rational point* $a \in X(L)$ *if and only if* $t = s_b$ *is the section associated to a rational point* $b \in R_{L|K} X(K)$. *Moreover, in this case we may choose* a *and* b *so that they correspond to each other via the identification* $R_{L|K} X(K) = X(L)$.

(2) Applying the functor π_1 yields a bijective profinite Kummer map

$$\kappa_X : X(L) \to \mathscr{S}_{\pi_1(X/L)}$$

if and only if π_1 yields a bijective profinite Kummer map

$$\kappa_{R_{L|K}X} : R_{L|K}X(K) \to \mathscr{S}_{\pi_1(R_{L|K}X/K)}.$$

Proof. Theorem 40 shows that there is a natural commutative diagram

$$
\begin{array}{ccc}
R_{L|K}X(K) & \xrightarrow{\;\cong\;} & X(L) \\
\downarrow & & \downarrow \\
\mathscr{S}_{\pi_1(R_{L|K}X/K)} & \xrightarrow{\;\cong\;} & \mathscr{S}_{\pi_1(X/L)}
\end{array}
$$

whose horizontal maps are bijections. Properties (1) and (2) follow at once.   □

Corollary 42. *The section conjecture holds for smooth projective curves of genus at least 2 over number fields if it holds for smooth, projective algebraic $K(\pi,1)$ spaces over $\mathbb{Q}$ (see [Sx02] Appendix A), which embed into their Albanese variety and have non-vanishing Euler–Poincaré characteristic.*   □

This corollary explains the title of [Sx10a]. We have lowered the degree of the number field in the section conjecture to 1 at the expense of working with varieties of dimension exceeding 1. Of course, these higher dimensional varieties of interest are simply $\mathbb{Q}$-forms of products of smooth, projective hyperbolic curves and so the trade might be marginal. But on the one hand, we are not required to limit the section conjecture to curves or products of curves, and secondly, the arithmetic of $\mathbb{Q}$ and so presumably also sections over $\mathrm{Gal}_{\mathbb{Q}}$ are arithmetically much simpler than for more general algebraic number fields. So the base has become simpler.

Chapter 4
The Space of Sections as a Topological Space

In this chapter we shall shift our focus from an individual section to the space of all sections at once. The space of sections naturally forms a pro-discrete topological space, see Lemma 44, which allows important limit arguments in arithmetically relevant cases, see Lemma 48. The fundamental notion of a neighbourhood of a section is introduced and used to describe the decomposition tower of a section.

4.1 Sections and Closed Subgroups

Let π be a profinite group. The set $\mathrm{Sub}(\pi)$ of closed subgroups of π allows a description as a profinite set

$$\mathrm{Sub}(\pi) \xrightarrow{\sim} \varprojlim_{\pi \twoheadrightarrow G} \mathrm{Sub}(G)$$

via the projection to the finite quotients G of π. We consider $\mathrm{Sub}(\pi)$ as a topological space with respect to this profinite topology.

For a short exact sequence $1 \to \overline{\pi} \to \pi \to \Gamma \to 1$ of profinite groups the set

$$\mathrm{Sub}(\pi)/\overline{\pi}$$

of $\overline{\pi}$-conjugacy classes of closed subgroups of π allows similarly a description as a profinite space

$$\mathrm{Sub}(\pi)/\overline{\pi} \xrightarrow{\sim} \varprojlim_{\varphi:\pi \twoheadrightarrow G} \mathrm{Sub}(G)/\varphi(\overline{\pi}).$$

The space of sections. The set of conjugacy classes of sections $\mathscr{S}_{\pi \to \Gamma}$ is naturally a subspace

$$\mathscr{S}_{\pi \to \Gamma} \subseteq \mathrm{Sub}(\pi)/\overline{\pi}$$

J. Stix, *Rational Points and Arithmetic of Fundamental Groups*, Lecture Notes in Mathematics 2054, DOI 10.1007/978-3-642-30674-7_4,
© Springer-Verlag Berlin Heidelberg 2013

by assigning to a class of sections $[s]$ the conjugacy class of images $s(\Gamma) \subseteq \pi$. We endow $\mathscr{S}_{\pi \to \Gamma}$ with the induced topology. Unfortunately, the subspace $\mathscr{S}_{\pi \to \Gamma}$ is not necessarily closed, in other words compact, as the following lemma illustrates.

Lemma 43. *If $\overline{\pi}$ is finite, then $\mathscr{S}_{\pi \to \Gamma}$ is a discrete topological space.*

Proof. Let s be a section of $\pi \to \Gamma$. The kernel Γ_0 of the Γ-action on $\overline{\pi}$ by conjugation via s yields a normal subgroup $s(\Gamma_0)$ of π. The corresponding quotient $G = \pi/s(\Gamma_0)$ sits in a short exact sequence of finite groups

$$1 \to \overline{\pi} \to G \to \Gamma/\Gamma_0 \to 1.$$

The only class $[t] \in \mathscr{S}_{\pi \to \Gamma}$ with a representative t such that $t(\Gamma)$ maps to the subgroup

$$s(\Gamma)/s(\Gamma_0) \subseteq G$$

is the class of s.   □

Characteristic quotients. A topologically finitely generated profinite group $\overline{\pi}$ is in several natural ways a projective limit of characteristic finite quotients along the index system $(\mathbb{N}, <)$. For example, we may set

$$Q_n(\overline{\pi}) = \overline{\pi}/\bigcap_{\varphi} \ker(\varphi) \tag{4.1}$$

where φ ranges over all continuous homomorphisms $\overline{\pi} \to G$ with $\#G \leq n$. Another option is to ask $\#G \mid n$, but by passing in the former to the subsystem of indices $n!$, we see that both options are equivalent in the pro-sense. Then $Q_n(\overline{\pi})$ is finite, the natural map

$$\overline{\pi} \to \varprojlim_n Q_n(\overline{\pi})$$

is an isomorphism and

$$\ker\left(\overline{\pi} \twoheadrightarrow Q_n(\overline{\pi})\right)$$

is preserved under any continuous automorphism of $\overline{\pi}$, i.e., is characteristic in $\overline{\pi}$.

The necessary assumption on $\overline{\pi}$ for this construction is being of *finite corank*. A group of finite corank has for every $n \in \mathbb{N}$ only finitely many continuous quotients of order $\leq n$. If in the short exact sequence

$$1 \to \overline{\pi} \to \pi \to \Gamma \to 1$$

of profinite groups the kernel $\overline{\pi}$ is of finite corank, then we can push the extension by the characteristic quotient map $\overline{\pi} \twoheadrightarrow Q_n(\overline{\pi})$ to obtain an extension

$$1 \to Q_n(\overline{\pi}) \to E_n(\pi) \to \Gamma \to 1$$

with

$$E_n(\pi) = \pi / \ker(\overline{\pi} \twoheadrightarrow Q_n(\overline{\pi})).$$

Lemma 44. *Let $\overline{\pi}$ be of finite corank. The natural map*

$$\mathscr{S}_{\pi \to \Gamma} \to \varprojlim_n \mathscr{S}_{E_n(\pi) \to \Gamma}$$

is a homeomorphism. The space of sections $\mathscr{S}_{\pi \to \Gamma}$ is in a natural way a pro-discrete topological space.

Proof. We first show that the map is bijective. Let s, t be representatives of classes in $\mathscr{S}_{\pi \to \Gamma}$ that agree at every level $E_n(\pi) \to \Gamma$. Let $M_n \subseteq \overline{\pi}$ be the nonempty set of elements which conjugate s into t at level $E_n(\pi) \to \Gamma$. Then $\varprojlim_n M_n$ is a projective limit of non-empty compact sets and is therefore non-empty, see [Bo98] I §9.6 Proposition 8. Any element in the limit conjugates s into t. The surjectivity is clear.

The assertion of being a homeomorphism follows immediately from the definition. The structure as a pro-discrete space follows from Lemma 43. □

Corollary 45. *If $\overline{\pi}$ is of finite corank and Γ is topologically finitely generated, then $\mathscr{S}_{\pi \to \Gamma}$ is profinite and thus compact.*

Proof. It suffices to show that $E_n(\pi) \to \Gamma$ admits only finitely many classes of sections. But as a section s is determined by its values on a set $\gamma_1, \ldots, \gamma_r$ of topological generators and two different values $s(\gamma_i)$ differ by an element in the finite group $Q_n(\overline{\pi})$, there are only finitely many sections and moreover classes of sections. □

4.2 Topology on the Space of Sections

We specialise to fundamental group extensions. Let X/k be a geometrically connected variety over the field k with algebraic closure $\overline{k}$. The group $\pi_1(\overline{X})$ is topologically finitely generated if k is of characteristic 0 or X/k is proper, so that the discussion of Sect. 4.1 applies. The extension

$$1 \to Q_n(\pi_1(\overline{X})) \to E_n(\pi_1(X)) \to \mathrm{Gal}_k \to 1 \qquad (4.2)$$

will be denoted by $Q_n(\pi_1(X/k))$.

Definition 46. A *neighbourhood* of a section $s \in \mathscr{S}_{\pi_1(X/k)}$ is an open subgroup H of $\pi_1(X)$ together with a representative of s whose image is contained in H, considered up to conjugation by $H \cap \overline{\pi}_1(X)$. Equivalently, a neighbourhood is a finite étale map $h : X' \to X$ with X' geometrically connected over k and a section

$s' \in \mathscr{S}_{\pi_1(X'/k)}$ with $h_* s' = s$. We sometimes abbreviate this by saying that the pair (X', s') is a neighbourhood of s.

When $h : X' \to X$ runs through the geometrically connected finite étale covers of X the image sets

$$U_h = \mathrm{im}\big(h_* : \mathscr{S}_{\pi_1(X'/k)} \to \mathscr{S}_{\pi_1(X/k)}\big)$$

form the open sets of a basis of a topology on $\mathscr{S}_{\pi_1(X/k)}$. We will consider $\mathscr{S}_{\pi_1(X/k)}$ as a topological space endowed with this topology.

Lemma 47. *Both topologies on $\mathscr{S}_{\pi_1(X/k)} = \mathscr{S}_{\pi_1(X) \to \mathrm{Gal}_k}$ agree. In particular, the space of sections carries a natural pro-discrete topology.*

Proof. As $\pi_1(X') \subseteq \pi_1(X)$ is an open subgroup, the space

$$\mathrm{Sub}(\pi_1(X')) \subseteq \mathrm{Sub}(\pi_1(X))$$

is an open subset. It follows that the map

$$h_* : \mathscr{S}_{\pi(X') \to \mathrm{Gal}_k} \to \mathscr{S}_{\pi_1(X) \to \mathrm{Gal}_k}$$

is open. In particular, U_h is open and the identity map

$$\mathscr{S}_{\pi_1(X) \to \mathrm{Gal}_k} \to \mathscr{S}_{\pi_1(X/k)} \tag{4.3}$$

is continuous.

It follows from Lemma 44 that $\mathscr{S}_{\pi(X) \to \mathrm{Gal}_k}$ has a basis of open sets given by the set of sections which agree with a given section s at the $Q_n(\pi_1(\overline{X}))$-level, i.e., the set of all section t such that the images of s and t agree for the quotient extension $Q_n(\pi_1(X/k))$. Let

$$h : X_n(s) \to X$$

be the finite étale cover which corresponds to the open subgroup

$$\ker\big(\overline{\pi} \twoheadrightarrow Q_n(\overline{\pi})\big) \cdot s(\mathrm{Gal}_k) \subseteq \pi_1(X).$$

Then the basic open set just described is nothing but U_h in the notation above. It follows that (4.3) is also open. □

4.3 Limits of Sections

As we have equipped the space of sections $\mathscr{S}_{\pi_1(X/k)}$ with a topology it makes sense to speak of convergent series (s_i) of (conjugacy classes of) sections and its limit section. Because $\mathscr{S}_{\pi_1(X/k)}$ is Hausdorff as a pro-discrete topological space, the limit if it exists is unique.

Lemma 48. *The following are equivalent.*

(a)   The sequence (s_i) converges to the section s in the sense of the topology on $\mathscr{S}_{\pi_1(X/k)}$.

(b)   For every neighbourhood $X' \to X$ of s the sections s_i lift to sections of $\pi_1(X'/k)$ for $i \gg 0$.

(c)   For every $n \in \mathbb{N}$ the sections s_i and s for $i \gg 0$ agree modulo Q_n, i.e., induce the same section in $\mathscr{S}_{Q_n(\pi_1(X/k))}$.

Proof. Lemma 47 shows that (a) implies (b). Property (c) is a special case of (b), while Lemma 44 shows that (c) implies (a).   □

Example 49. Not every sequence of sections must have a convergent subsequence. For example, the cyclotomic semi direct product

$$1 \to \mu_n \to \mu_n \rtimes \mathrm{Gal}_k \to \mathrm{Gal}_k \to 1$$

has a discrete space of sections $\mathrm{H}^1(k, \mu_n) = k^*/(k^*)^n$ which may be infinite. However, we will see later in Sect. 9.1 that in the arithmetic context for a proper, smooth variety X/k the space of sections $\mathscr{S}_{\pi_1(X/k)}$ is even compact, so that every sequence of sections has a convergent subsequence.

Lemma 50. *The topological space $\mathscr{S}_{\pi_1(X/k)}$ has the following properties.*

(1)   Every $s \in \mathscr{S}_{\pi_1(X/k)}$ has a countable basis of neighbourhoods.

(2)   The space of sections $\mathscr{S}_{\pi_1(X/k)}$ is compact if and only if every sequence of sections has a convergent subsequence.

(3)   If $\mathscr{S}_{\pi_1(X/k)}$ is compact, then it is profinite and has a countable basis of open subsets.

Proof. Assertion (1) follows immediately from the description as an $\mathbb{N}$-indexed pro-discrete space of Lemma 44. For (2) it suffices to show that sequentially compactness implies compactness. If for some n the image $\mathscr{S}_n$ of

$$\mathscr{S}_{\pi_1(X/k)} \to \mathscr{S}_{Q_n(\pi_1(X/k))}$$

is infinite, then a sequence s_i that maps injectively to $\mathscr{S}_n$ has no convergent subsequence. But otherwise, if all the $\mathscr{S}_n$ are finite, then

$$\mathscr{S}_{\pi_1(X/k)} = \varprojlim_n \mathscr{S}_n$$

is a profinite space. Assertion (3) follows from the description of $\mathscr{S}_{\pi_1(X/k)}$ as an $\mathbb{N}$-indexed profinite space in case it is compact.   □

4.4   The Decomposition Tower of a Section

Now we pass to the limit with respect to all neighbourhoods of a section.

Definition 51. The *decomposition tower* of a section $s : \mathrm{Gal}_k \to \pi_1(X)$ is the pro-étale cover

$$X_s = (X')$$

of all neighbourhoods $h : X' \to X$ with $s' \in h^{-1}(s)$ and all transfer maps respect these sections s'. Alternatively, the pro-cover X_s is nothing but the k-structure associated to s in Sect. 2.5 on the universal pro-étale cover $\tilde{X} \to X$ that serves as the path space for the fibre functor with respect to which we describe the section and the extension $\pi_1(X/k)$.

A closed subgroup in a profinite group is the intersection of the open subgroups it is contained in. Hence the image $s(\mathrm{Gal}_k)$ of the section s coincides with

$$\pi_1(X_s) = \bigcap_{X'} \pi_1(X')$$

in $\pi_1(X)$ where the X' ranges over all the neighbourhoods X' in the pro-system X_s.

Lemma 52. *A section s of $\pi_1(X/k)$ equals the section s_a associated to $a \in X(k)$ if and only if a belongs to the image of the natural map $X_s(k) \to X(k)$.*

Proof. If $s = s_a$, then a lifts canonically to a compatible system of rational points of all neighbourhoods of s, hence to a point in $X_s(k)$, see Corollary 34. Conversely, if a lifts to a point in $X_s(k)$, then the map $s_a = \pi_1(a)$ factors over $\pi_1(X_s) = s(\mathrm{Gal}_k)$ and s equals s_a. □

Lemma 53. *Let s be a section of $\pi_1(X/k)$. The following are equivalent.*

(a)   *The section s belongs to the closure $\overline{\kappa(X(k))}$ of the image of the Kummer map, i.e., s is the limit of s_{a_i} for a sequence of rational points $a_i \in X(k)$.*
(b)   *Every neighbourhood X' of s has $X'(k) \neq \emptyset$.*

Proof. Let $U_h \subseteq \mathscr{S}_{\pi_1(X/k)}$ be the open set associated to a neighbourhood

$$h : X' \to X$$

of s. Then $s_a \in U_h$ if and only if a lifts to a rational point $a' \in X'(k)$. The lemma follows at once. □

The core of the following well-known limit argument goes back at least to Neukirch, and was introduced to anabelian geometry by Nakamura, while Tamagawa [Ta97] Proposition 0.7 emphasized its significance to the section conjecture, see also [Fa98] Theorem 3, [Ko05] Lemma 1.7 and [Sx10b] Appendix C.

Proposition 54 (weak vs. strong). *Let X/k be a geometrically connected variety.*

(1)   *The Kummer map* $\kappa : X(k) \to \mathscr{S}_{\pi_1(X/k)}$ *is injective if and only if for every rational point* $a \in X(k)$ *we have*

$$\mathrm{im}\big(X_{s_a}(k) \to X(k)\big) = \{a\}.$$

(2)   *The Kummer map* $\kappa : X'(k) \to \mathscr{S}_{\pi_1(X'/k)}$ *is injective for all geometrically connected finite étale covers of* X *if and only if for every* $a \in X(k)$ *we have*

$$\#X_{s_a}(k) = 1.$$

(3)   *The Kummer map* $\kappa : X(k) \to \mathscr{S}_{\pi_1(X/k)}$ *is surjective if and only if for every section* s *we have* $X_s(k) \neq \emptyset$.

(4)   *The Kummer map* $\kappa : X(k) \to \mathscr{S}_{\pi_1(X/k)}$ *is surjective if and only if* $\kappa(X(k))$ *is closed in* $\mathscr{S}_{\pi_1(X/k)}$ *and for every section* s *and every neighbourhood* X' *of* s *we have* $X'(k) \neq \emptyset$.

(5)   *The weak section property, i.e., for every neighbourhood* $X' \to X$ *of a section of* $\pi_1(X/k)$ *we have* $X'(k) \neq \emptyset$, *is equivalent to the map*

$$\kappa : X(k) \to \mathscr{S}_{\pi_1(X/k)}$$

having a dense image.

Proof. The assertions (1) and (3) follow at once from Lemma 52, and (2) follows from (1) by

$$X_s(k) = \varprojlim X'(k)$$

where the limit ranges over all neighbourhoods of s. And the assertions (4) and (5) follow from Lemma 53.   □

Remark 55. (1) Grothendieck in his letter to Faltings, [Gr83] page 8, describes the content of Proposition 54 in terms of a fixed point property of Galois action as follows.

Let $\tilde{X} \to \overline{X}$ be the universal pro-étale cover of X which is a path space for the fibre functor $\bar{a}$ of $\overline{X}$ used in $\pi_1(X, \bar{a})$. As such $\tilde{X}$ carries a continuous $\pi_1(X)$-action from the right. We let Gal_k act on $\tilde{X}$ from the right via the section $s : \mathrm{Gal}_k \to \pi_1(X, \bar{a})$. We obtain a fibre product diagram of quotient maps

$$
\begin{array}{ccccc}
\tilde{X} & \longrightarrow & \overline{X} & \longrightarrow & \mathrm{Spec}(\bar{k}) \\
\Big\downarrow {\scriptstyle /s(\mathrm{Gal}_k)} & & \Big\downarrow & & \Big\downarrow \\
X_s & \longrightarrow & X & \longrightarrow & \mathrm{Spec}(k).
\end{array}
$$

Now the Galois action extends to a Galois action on $\tilde{X}(\bar{k}) = X_s(\bar{k})$ with set of fixed points equal to $X_s(k)$. Proposition 54 therefore translates to the following. The section s belongs to a rational point $a \in X(k)$ if and only if the Galois action via s on $\tilde{X}(\bar{k})$ has a fixed point. Moreover, the rational point a with $s = s_a$ is unique for s and all restrictions of s to neighbourhoods of s if and only if there is a unique fixed point for the Galois action on $\tilde{X}(\bar{k})$.

(2) A more precise discussion at this point also takes care of rational points in the boundary of a good and in particular minimal smooth compactification, see [Gr83] page 8. These yield potentially fixed points beyond rational points of X and thus explain certain sections of $\pi_1(X/k)$ which we will call cuspidal and which are very well understood in the case of hyperbolic curves over arithmetically relevant fields, see Chap. 18.

(3) The weak section conjecture hinted at in Proposition 54 (5) will be stated as Conjecture 100 later and discussed further in the arithmetic case in Sect. 9.2

Chapter 5
Evaluation of Units

In the spirit of the section conjecture, a section should behave like a rational point. In particular, we should be able to evaluate a function in a section. At least for invertible functions this can be achieved via Kummer theory, see Definition 57, if we accept that the values will be taken in a certain completion of the multiplicative group of the ground field.

5.1 Kummer Theory with Finite Coefficients

Let U/k be a geometrically connected variety. For $n \in \mathbb{N}$ not divisible by the characteristic of k, the Kummer sequence determines a homomorphism

$$\kappa : \mathcal{O}^*(U) \to \mathrm{H}^1(U, \mu_n) = \mathrm{H}^1(\pi_1 U, \mu_n), \tag{5.1}$$

$$f \mapsto \kappa_f.$$

The element

$$\kappa_f \in \mathrm{H}^1(U, \mu_n)$$

is addressed as the *Kummer torsor (modulo n)* associated to f, which is the cohomological description of the finite étale cover of nth roots of f above U.

Definition 56. The *evaluation map* (more precisely the evaluation map modulo n) for a conjugacy class of sections $[s] \in \mathscr{S}_{\pi_1(U/k)}$ is the composite

$$\mathrm{ev}_s = \mathrm{ev}_{[s]} : \mathcal{O}^*(U) \to k^*/(k^*)^n = \mathrm{H}^1(k, \mu_n) = \mathrm{H}^1(\mathrm{Gal}_k, \mu_n)$$

defined by

$$\mathrm{ev}_s(f) = f(s) = s^*(\kappa_f).$$

J. Stix, *Rational Points and Arithmetic of Fundamental Groups*, Lecture Notes in Mathematics 2054, DOI 10.1007/978-3-642-30674-7_5,
© Springer-Verlag Berlin Heidelberg 2013

The map ev_s is well defined, namely independent of the representative s in $[s]$, because $\pi_1(\overline{U})$ acts trivially on μ_n and then inner automorphisms act trivial on cohomology, see [Se79] VII.5 Proposition 3.

The functoriality of the Kummer sequence shows that for a map $h : U \to V$ of geometrically connected k-varieties, a unit $f \in \mathcal{O}^*(V)$ and a section s of $\pi_1(U/k)$, the evaluation map is compatible with pullback:

$$(h^*f)(s) = s^*(\kappa_{h^*f}) = s^*(h^*\kappa_f) = s^*\pi_1(h)^*(\kappa_f) = (h_*s)^*(\kappa_f) = f(h_*s).$$

In particular, for a section s_a associated to a k-rational point $a \in U(k)$ we have

$$f(s_a) = f(a_*\mathrm{id}_{\mathrm{Gal}_k}) = (a^*f)(\mathrm{id}_{\mathrm{Gal}_k}) = f(a)(\mathrm{id}_{\mathrm{Gal}_k}) = f(a) \mod (k^*)^n$$

and so the name *evaluation* is justified.

Another consequence of the functoriality of the Kummer sequence is that evaluation in the section s only depends on its abelianization s^{ab}. For $f \in \mathcal{O}^*(U)$ we consider the corresponding map

$$f : U \to \mathbb{G}_m$$

so that $f = f^*(T)$ for the standard coordinate T on $\mathbb{G}_m$. Then

$$f(s) = (f^*T)(s) = T(f_*s)$$

which only depends on the image section f_*s of $\pi_1(\mathbb{G}_m/k)$ and moreover the geometrically prime to p part of $\pi_1(\mathbb{G}_m/k)$. As the geometrically prime to p part of $\pi_1(\overline{\mathbb{G}_m})$ is abelian, the value $f(s)$ only depends on

$$f_*(s^{\mathrm{ab}}) \in \mathscr{S}_{\pi_1^{\mathrm{ab}}(\mathbb{G}_m/k)}.$$

Galois equivariant evaluation. Let k'/k be a finite Galois extension with Galois group $G = \mathrm{Gal}(k'/k)$. For a section s of $\pi_1(U/k)$ the restriction $s' = s|_{\mathrm{Gal}_{k'}}$, which in characteristic 0 agrees with the base change $s \otimes k'$, induces a G-equivariant map

$$s'^* : \mathrm{H}^1(\pi_1 U_{k'}, \mu_n) \to \mathrm{H}^1(k', \mu_n)$$

because for $\sigma \in G$ we have

$$s'^* \circ \mathrm{H}^1(\pi_1(\mathrm{id} \times \sigma)) = s'^* \circ \big(s(\sigma)^{-1}(-)s(\sigma)\big)^* = \big(s(\sigma)^{-1}s'(-)s(\sigma)\big)^*$$

$$= \big(s' \circ (\sigma^{-1}(-)\sigma)\big)^* = \big(\sigma^{-1}(-)\sigma\big)^* \circ s'^* = \mathrm{H}^1(\pi_1(\sigma)) \circ s'^*.$$

Evaluation and norms. More generally, for a finite field extension k'/k the projection $h : U_{k'} \to U$ is finite flat and thus allows a norm map

$$N : h_*\mathbb{G}_m \to \mathbb{G}_m.$$

The induced map

$$h_* \mu_n \to \mu_n$$

yields the corestriction map on cohomology. Thus, with $s' = s|_{\mathrm{Gal}_{k'}}$, the following diagram commutes

$$
\begin{array}{ccccccc}
\mathcal{O}^*(U_{k'}) & \xrightarrow{\ \kappa\ } & \mathrm{H}^1(\pi_1 U_{k'}, \mu_n) & \xrightarrow{\ s'^*\ } & \mathrm{H}^1(k', \mu_n) & = & k'^*/(k'^*)^n \\
\downarrow{\scriptstyle N} & & \downarrow{\scriptstyle \mathrm{cor}} & & \downarrow{\scriptstyle \mathrm{cor}} & & \downarrow{\scriptstyle N_{k'/k}} \\
\mathcal{O}^*(U) & \xrightarrow{\ \kappa\ } & \mathrm{H}^1(\pi_1 U, \mu_n) & \xrightarrow{\ s^*\ } & \mathrm{H}^1(k, \mu_n) & = & k^*/(k^*)^n
\end{array}
$$

so that for $f \in \mathcal{O}^*(U_{k'})$

$$N(f)(s) = N_{k'/k}(f(s')).$$

5.2   Kummer Theory with Profinite Coefficients

Let $p \geq 0$ be the characteristic of k.

The continuous Kummer torsor and evaluation. We work with continuous cohomology in the sense of Jannsen, see [Ja88]. Our index system will be

$$\mathbb{N}' = \{n \in \mathbb{N}\,;\ p \nmid n\} \tag{5.2}$$

and partial ordering with respect to divisibility. The system $\mathbb{N}'$ has a cofinal subsystem isomorphic to $(\mathbb{N}, <)$ so that the homological algebra of pro-systems as in [Ja88] applies. The sheaf $\hat{\mathbb{Z}}'(1)$ is by definition the pro-system $(\mu_n)_{n \in \mathbb{N}'}$. The Kummer sequence of continuous cohomology yields a map

$$\kappa \ :\ \mathcal{O}^*(U) \to \mathrm{H}^1(U, \hat{\mathbb{Z}}'(1)) = \mathrm{H}^1(\pi_1(U), \hat{\mathbb{Z}}'(1)),$$

$$f \mapsto \kappa_f.$$

Definition 57. The profinite *evaluation map* for $[s] \in \mathscr{S}_{\pi_1(U/k)}$ is the composite

$$\mathrm{ev}_s = \mathrm{ev}_{[s]} \ :\ \mathcal{O}^*(U) \to \varprojlim k^*/(k^*)^n = \mathrm{H}^1(k, \hat{\mathbb{Z}}'(1)) = \mathrm{H}^1(\mathrm{Gal}_k, \hat{\mathbb{Z}}'(1))$$

defined by

$$\mathrm{ev}_s(f) = f(s) = s^*(\kappa_f).$$

The map is again well defined, namely independent of the representative s in $[s]$, by the same reason as before. For ease of notation we abbreviate the value group by

$$\widehat{k^*} = \varprojlim_{n \in \mathbb{N}'} k^*/(k^*)^n. \tag{5.3}$$

An alternative formula for the value at a section is obtained by functoriality from the universal case as follows.

Lemma 58. *Let U/k be a geometrically connected variety. Let $f \in \mathcal{O}^*(U)$ be a unit and $f : U \to \mathbb{G}_m$ the corresponding map. For a section s of $\pi_1(U/k)$ we have*

$$f(s) = \delta(f_* s, s_1) \in \mathrm{H}^1(k, \widehat{\mathbb{Z}}'(1)) = \widehat{k^*},$$

where $\delta(t, s)$ is the difference cocyle for two sections and s_1 is the standard section of $\pi_1(\mathbb{G}_m/k)$ associated to $1 \in \mathbb{G}_m(k)$.

Proof. By naturality it suffices to treat the universal case $U = \mathbb{G}_m$ and $f = T$ is the standard coordinate. Let $\mathrm{pr} : \mathbb{G}_m \to \mathrm{Spec}(k)$ be the projection. We compute

$$\kappa_T = \left(\gamma \mapsto \left(\gamma(\sqrt[n]{T}/\sqrt[n]{T} \right)_{n \in \mathbb{N}'} \right) = \delta(\mathrm{id}, s_1 \circ \mathrm{pr}_*) \in \mathrm{H}^1(\pi_1(\mathbb{G}_m), \widehat{\mathbb{Z}}'(1))$$

where we have exploited the difference cocycle for generalized sections of $\pi_1(\mathbb{G}_m/k)$ in the sense of Sect. 1.2 for the map $\pi_1(\mathbb{G}_m) \to \mathrm{Gal}_k$. If we pullback with the section $s : \mathrm{Gal}_k \to \pi_1(\mathbb{G}_m)$ we obtain

$$T(s) = s^*(\kappa_T) = s^*(\delta(\mathrm{id}, s_1 \circ \mathrm{pr}_*)) = \delta(s, s_1 \circ \mathrm{pr}_* \circ s) = \delta(s, s_1) = \delta(T_* s, s_1)$$

as claimed by the lemma.   □

Values for the sheafified evaluation. For fixed $n \in \mathbb{N}'$ the presheaf

$$\mathcal{H}^1(\mu_n) = \left(k' \mapsto k'^*/(k'^*)^n \right)$$

on $\mathrm{Spec}(k)_{\text{ét}}$ has trivial sheafification. Instead, we consider the pro-presheaf

$$\widehat{\mathbb{G}}_{\mathrm{m},k} = \left(\mathcal{H}^1(\mu_n) \right)_{n \in \mathbb{N}'} \tag{5.4}$$

and consider it as a presheaf with values in the category $\mathrm{Ab}_{\mathrm{ML}}^{\mathbb{N}'}$ of $\mathbb{N}'$-systems of abelian groups localised at Mittag–Leffler-zero objects, see [Ja88] §1. Performing the profinite limit we obtain the presheaf $\varprojlim \widehat{\mathbb{G}}_{\mathrm{m},k}$ of abelian groups

$$\left(\varprojlim \widehat{\mathbb{G}}_{\mathrm{m},k} \right)(k') = \widehat{k'^*}.$$

Proposition 59. *Let k be a field of characteristic $p \geq 0$ such that for all $\ell \neq p$ the ℓ-adic cyclotomic character has infinite image.*

(1) Then the pro-system $\widehat{\mathbb{G}}_{m,k}$ yields a sheaf on $\mathrm{Spec}(k)_{\text{ét}}$ with values in the Mittag–Leffler localised category $\mathrm{Ab}_{\mathrm{ML}}^{\mathbb{N}'}$ of pro-$\mathbb{N}'$-systems of abelian groups.

(2) In particular, the presheaf

$$\varprojlim \widehat{\mathbb{G}}_{m,k}$$

is actually a sheaf.

Proof. The condition on the ℓ-adic cyclotomic character is equivalent to the finiteness of the group $\mu_{\ell^\infty}(k')$ of ℓ-power roots of unity in any fixed finite separable extension k'/k.

(1) We have to show that for a Galois extension E/F of finite separable extensions of k the natural map

$$(F^*/(F^*)^n)_{n \in \mathbb{N}'} \to (E^*/(E^*)^n)_{n \in \mathbb{N}'}^{\mathrm{Gal}(E/F)}$$

is a Mittag–Leffler isomorphism. The Hochschild–Serre spectral sequence yields a description of kernel and cokernel levelwise via the exact sequence

$$0 \to \mathrm{H}^1(E/F, \mu_n(E)) \to \mathrm{H}^1(F, \mu_n) \to \mathrm{H}^1(E, \mu_n)^{\mathrm{Gal}(E/F)} \xrightarrow{d_n} \mathrm{H}^2(E/F, \mu_n(E)).$$

By assumption the sequence $\mu_{n^r}(E)$ stabilizes for $r \to \infty$, and therefore the map

$$\mu_{n^{2r}}(E) \to \mu_{n^r}(E)$$

which is raising to the n^rth power vanishes for $r \gg 0$. This is the transfer map in the pro-systems to be considered. Therefore $(\mathrm{H}^i(E/F, \mu_n(E))_{n \in \mathbb{N}'}$ is a Mittag–Leffler zero system for all $i \geq 0$ and thus also $(\mathrm{im}(d_n))_{n \in \mathbb{N}'}$. This proves (1), and (2) is an immediate consequence. $\square$

Corollary 60. *Let U/k be a geometrically connected variety over a field k of characteristic $p \geq 0$ such that for all $\ell \neq p$ the ℓ-adic cyclotomic character has infinite image. Then*

$$k' \mapsto \left(\mathrm{H}^1(U_{k'}, \mu_n)\right)_{n \in \mathbb{N}'}$$

is a sheaf on $\mathrm{Spec}(k)_{\text{ét}}$ with values in $\mathrm{Ab}_{\mathrm{ML}}^{\mathbb{N}'}$.

Proof. As $\mathrm{H}^0(U_E, \mu_n) = \mu_n(E)$ the proof of Proposition 59 applies. $\square$

Evaluation as a map of sheaves. Let $\mathrm{pr} : U \to \mathrm{Spec}(k)$ denote the structure map. Under the assumptions of Proposition 59 we can now describe the evaluation map for a section s of $\pi_1(U/k)$ as a map

$$\mathrm{ev}_s \;:\; \mathrm{pr}_* \mathbb{G}_m \xrightarrow{\kappa} \left(k' \mapsto \left(\mathrm{H}^1(U_{k'}, \mu_n)\right)_{n \in \mathbb{N}'}\right) \xrightarrow{s^*} \widehat{\mathbb{G}}_{m,k}$$

$$f \mapsto \kappa_f \mapsto ((s|_{\mathrm{Gal}_{k'}})^*(\kappa_f))_{n \in \mathbb{N}'}$$

or in the limit as a map

$$\text{ev}_s \; : \; \text{pr}_* \mathbb{G}_m \to \varprojlim \widehat{\mathbb{G}}_{m,k}$$

of sheaves with values in the appropriate categories.

Evaluation as a retraction. The evaluation map is the composite of the Kummer map with a retraction s^* of the natural map

$$\text{pr}^* \; : \; \widehat{\mathbb{G}}_{m,k} = \big(k' \mapsto \text{H}^1(k', \hat{\mathbb{Z}}'(1)) \big) \to \big(k' \mapsto \text{H}^1(U_{k'}, \hat{\mathbb{Z}}'(1)) \big).$$

For an abelian group M we define the *prime to p Tate module* as

$$\text{T}'(M) = \prod_{\ell \neq p} \text{Hom}(\mathbb{Q}_\ell / \mathbb{Z}_\ell, M). \tag{5.5}$$

Lemma 61. *The group* $\text{H}^1(U_{k'}, \hat{\mathbb{Z}}'(1))$ *has the following description.*

(1) The group $\text{H}^1(U_{k'}, \hat{\mathbb{Z}}'(1))$ *sits in a short exact sequence*

$$1 \to \widehat{\mathcal{O}^*}(U_{k'}) \xrightarrow{\hat{\kappa}} \text{H}^1(U_{k'}, \hat{\mathbb{Z}}'(1)) \to \text{T}'(\text{Pic}(U_{k'})) \to 0$$

where the map $\hat{\kappa}$ is the pro-$\mathbb{N}'$ completion of the map κ and

$$\widehat{\mathcal{O}^*}(U_{k'}) = \varprojlim_{n \in \mathbb{N}'} \mathcal{O}^*(U_{k'})/(\mathcal{O}^*(U_{k'}))^n.$$

(2) In particular, if the torsion subgroup $\text{Pic}(U_{k'})_{\text{tors}}$ *has a trivial group of divisible elements, then $\hat{\kappa}$ is an isomorphism*

$$\hat{\kappa} \; : \; \widehat{\mathcal{O}^*}(U_{k'}) \xrightarrow{\sim} \text{H}^1(U_{k'}, \hat{\mathbb{Z}}'(1)).$$

Proof. (1) The Kummer sequence yields a short exact sequence

$$1 \to \mathcal{O}^*(U_{k'})/(\mathcal{O}^*(U_{k'}))^n \to \text{H}^1(U_{k'}, \mu_n) \to \text{Pic}(U_{k'})[n] \to 0$$

where $\text{Pic}(U_{k'})[n]$ is the n-torsion in $\text{Pic}(U_{k'})$. In the projective limit over $\mathbb{N}'$ we obtain the claim of the lemma because

$$\text{H}^1(U_{k'}, \hat{\mathbb{Z}}'(1)) = \varprojlim_{n \in \mathbb{N}'} \text{H}^1(U_{k'}, \mu_n)$$

by [Ja88] Proposition 1.6 and the Mittag–Leffler condition for the system

$$(\mu_n(U_{k'}))_{n \in \mathbb{N}'}.$$

(2) The group

$$T'(\mathrm{Pic}(U_{k'}))$$

depends only on the subgroup of divisible elements in $\mathrm{Pic}(U_{k'})_{\mathrm{tors}}$ and thus vanishes in (2). The assertion then follows from (1).   □

The computation of Lemma 61 shows that when $\mathrm{Pic}(U_{k'})_{\mathrm{tors}}$ has no nontrivial divisible elements, then the pullback map s^* agrees with the pro-$\mathbb{N}'$ completion of the evaluation map ev_s and therefore does not contain more information.

Let us stress that in general the evaluation map does not factorize through

$$\mathrm{R}^1\mathrm{pr}_*\hat{\mathbb{Z}}'(1) = \left(\mathrm{R}^1\mathrm{pr}_*\mu_n\right)_{n\in\mathbb{N}'}$$

which agrees with the Gal_k-module $\mathrm{H}^1(\overline{U},\hat{\mathbb{Z}}'(1))$. Indeed, for every k' we have the natural exact sequence from the Leray spectral sequence for $\mathrm{pr} : U_{k'} \to \mathrm{Spec}(k')$

$$1 \to \widehat{k'^*} \xrightarrow{\mathrm{pr}^*} \mathrm{H}^1(U_{k'},\hat{\mathbb{Z}}'(1)) \to \mathrm{H}^1(\overline{U},\hat{\mathbb{Z}}'(1))^{\mathrm{Gal}_{k'}}.$$

The map ev_s, being a retraction of pr^*, cannot factorize through $\mathrm{H}^1(\overline{U},\hat{\mathbb{Z}}'(1))$ in general.

Chapter 6
Cycle Classes in Anabelian Geometry

Cycle classes appear in the context of anabelian geometry in different incarnations, see for example [Pa90, Mo99, Mo03, Mo07, EsWi09]. After recalling and comparing several known constructions we describe yet another construction of the cycle class of a section.

Our construction relies on the principle that for a hyperbolic curve, or more generally for an algebraic $K(\pi, 1)$-space, see Sect. 6.2 below, cohomological arguments could take place on the finite étale site. This version of the cycle class originates again from the class of the diagonal, which we consider as the universal cycle class for points on the curve. The cycle class of a section arises as the evaluation of this universal family of cycle classes in the section just as if the section were a closed point. The construction by evaluation extends to subvarieties and also to the relative case of cycles in families.

As an application we partially present Parshin's proof of the geometric Mordell Theorem using fundamental groups and hyperbolic geometry [Pa90] from an algebraic viewpoint, at least concerning the use of fundamental groups and cycle classes.

6.1 Various Definitions of Cycle Classes of a Section

We recall various constructions of the cycle class of a section.

The class of graphs [after Esnault and Wittenberg]. Let X/k be a smooth algebraic $K(\pi, 1)$-space of dimension d, and let s be a section of $\pi_1(X/k)$. For n invertible in k we define the *cycle class modulo n* as the unique class

$$\mathrm{cl}_s^{\mathrm{graph}} = \mathrm{cl}_{s,n}^{\mathrm{graph}} \in \mathrm{H}_c^{2d}\left(X, \mathbb{Z}/n\mathbb{Z}(d)\right) \tag{6.1}$$

which, for fine enough neighbourhoods $h : Y \to X$ of s, maps under pullback by
the projection $\mathrm{pr}_2 : Y \times_k X \to X$ to the class

$$\mathrm{cl}_{\Gamma(h)} \in \mathrm{H}_c^{2d}\left(Y \times_k X, \mathbb{Z}/n\mathbb{Z}(d)\right)$$

of the graph of h

$$\Gamma(h) \subset Y \times_k X.$$

We refer to [EsWi09] Theorem 2.6 for details.

Norm compatible classes. Let X/k be a smooth algebraic $\mathrm{K}(\pi, 1)$-space of
dimension d, and let s be a section of $\pi_1(X/k)$. For n invertible in k and $i > 0$
the $\mathrm{K}(\pi, 1)$-property immediately implies that

$$\varinjlim_{h^*} \mathrm{H}^i\left(\overline{Y}, \mathbb{Z}/n\mathbb{Z}\right) = 0$$

in the limit over all neighbourhoods $Y \to X$ of s with pullback as transfer maps.
By duality we deduce that for $i < 2d$ also

$$\varprojlim_{h_*} \mathrm{H}_c^i\left(\overline{Y}, \mathbb{Z}/n\mathbb{Z}(d)\right) = 0 \qquad (6.2)$$

with push forward h_* as transfer maps. The Leray spectral sequences

$$\mathrm{E}_2^{i,j} = \mathrm{H}^i\left(k, \mathrm{H}_c^j\left(\overline{Y}, \mathbb{Z}/n\mathbb{Z}(d)\right)\right) \Longrightarrow \mathrm{H}_c^{i+j}\left(Y, \mathbb{Z}/n\mathbb{Z}(d)\right)$$

for the various neighbourhoods Y of s are compatible with h_*. For the class

$$1 \in \mathrm{E}_2^{0,2d} = \mathbb{Z}/n\mathbb{Z}$$

we find that

$$d_r^{0,2d}(1)$$

is a norm compatible system, hence vanishes for all $r \geq 2$ by (6.2). More
precisely, (6.2) says that the rows with $j < 2d$ form Mittag–Leffler zero systems.
Consequently, the edge map

$$\mathrm{deg} : \varprojlim_{h_*} \mathrm{H}_c^{2d}\left(Y, \mathbb{Z}/n\mathbb{Z}(d)\right) \to \varprojlim_{h_*} \mathrm{H}^0\left(k, \mathrm{H}_c^{2d}\left(\overline{Y}, \mathbb{Z}/n\mathbb{Z}(d)\right)\right) = \mathbb{Z}/n\mathbb{Z}$$

is an isomorphism. The cycle class

$$\mathrm{cl}_s^{\mathrm{norm}} \in \mathrm{H}_c^{2d}\left(X, \mathbb{Z}/n\mathbb{Z}(d)\right) \qquad (6.3)$$

is by definition the image under the natural projection of $\mathrm{deg}^{-1}(1)$, the only norm
compatible class of degree 1 in the tower of all neighbourhoods of the section s.

Connection to the logarithm sheaf. Let ℓ be invertible in k. If we let $\pi_1(X)$ act on $\pi_1(\overline{X})$ via left translation and the identification

$$\pi_1(\overline{X}) = \pi_1(X)/s(\mathrm{Gal}_k),$$

we obtain a pro-locally constant sheaf on $X_{\text{ét}}$

$$\mathbb{Z}_\ell[[\pi_1(\overline{X})]] = \varprojlim h_*\mathbb{Z}_\ell,$$

which is called the *logarithm sheaf* $\mathcal{L}og_{X,\mathbb{Z}_\ell}$ in [Ki09] §1.1. By (6.2), the map

$$\mathrm{H}_c^{2d}\left(X, \mathbb{Z}_\ell[[\pi_1(\overline{X})]](d)\right) \to \varprojlim_{h_*} \mathrm{H}_c^{2d}\left(Y, \mathbb{Z}_\ell(d)\right) \to \varprojlim_{h_*} \mathrm{H}^0\left(k, \mathrm{H}_c^{2d}\left(\overline{Y}, \mathbb{Z}_\ell(d)\right)\right) = \mathbb{Z}_\ell$$

is an isomorphism, and this connects the definition via norm compatible classes to the theory of the logarithm sheaf, see [Ki09] Theorem 1.1.4.

An explicit extension [after Mochizuki]. Let ℓ be a prime number that is invertible in the field k. Let X/k be smooth hyperbolic curve and assume that X is proper or k has characteristic 0. For a section s of $\pi_1(X/k)$, we can define the cycle class of s as the class

$$\mathrm{cl}_s^{\mathrm{group}} \in \mathrm{H}^2\left(\pi_1(X), \mathbb{Z}_\ell(1)\right) = \mathrm{H}^2\left(X, \mathbb{Z}_\ell(1)\right) \tag{6.4}$$

of an explicit extension of profinite groups, see [Mo03] and [Mo07].

Let $\mathbb{L}^*$ be the $\mathbb{G}_m$-torsor associated to the line bundle $\mathcal{O}_{X \times_k X}(\Delta)$ where Δ is the diagonal. By pushing the homotopy sequence for

$$\mathbb{L}^* \to X \times_k X$$

with the homomorphism $\pi_1(\overline{\mathbb{G}_m}) \twoheadrightarrow \mathbb{Z}_\ell(1)$ we obtain an exact sequence

$$1 \to \mathbb{Z}_\ell(1) \to \pi_1(\mathbb{L}^*)/\ker\left(\pi_1(\overline{\mathbb{G}_m}) \twoheadrightarrow \mathbb{Z}_\ell(1)\right) \to \pi_1(X \times_k X) \to 1. \tag{6.5}$$

The class

$$\mathrm{cl}_s^{\mathrm{group}} \in \mathrm{H}^2\left(\pi_1(X), \mathbb{Z}_\ell(1)\right) = \mathrm{H}^2\left(X, \mathbb{Z}_\ell(1)\right)$$

is represented by the pullback of (6.5) along

$$s \times \mathrm{id} : \pi_1(X) \to \pi_1(X) \times_{\mathrm{Gal}_k} \pi_1(X) = \pi_1(X \times_k X).$$

By projecting with $\mathbb{Z}_\ell(1) \to \mathbb{Z}/\ell^n\mathbb{Z}(1)$ we obtain also here cycle classes with finite coefficients.

Duality. In case the base field k is a finite extension of $\mathbb{Q}_p$, a definition via duality becomes available. Let X/k be a smooth $K(\pi, 1)$-space of dimension d. Then

$$H_c^{2d+2}\left(X, \mathbb{Z}/n\mathbb{Z}(d+1)\right) = H^2\left(k, H_c^{2d}\left(\overline{X}, \mathbb{Z}/n\mathbb{Z}(d)\right)(1)\right) = H^2\left(k, \mathbb{Z}/n\mathbb{Z}(1)\right) = \mathbb{Z}/n\mathbb{Z}$$

and the cup-product pairing

$$H_c^{2d}\left(X, \mathbb{Z}/n\mathbb{Z}(d)\right) \times H^2\left(X, \mathbb{Z}/n\mathbb{Z}(1)\right) \to \mathbb{Z}/n\mathbb{Z}$$

is a perfect pairing of finite groups. Hence, we can define the cycle class associated to a section s of $\pi_1(X/k)$ as the class

$$\mathrm{cl}_s^{\mathrm{dual}} \in H_c^{2d}\left(X, \mathbb{Z}/n\mathbb{Z}(d)\right) \tag{6.6}$$

that induces by the cup-product pairing the pullback map via the section s:

$$s^* : H^2\left(X, \mathbb{Z}/n\mathbb{Z}(1)\right) = H^2\left(\pi_1(X), \mathbb{Z}/n\mathbb{Z}(1)\right) \to H^2\left(k, \mathbb{Z}/n\mathbb{Z}(1)\right) = \mathbb{Z}/n\mathbb{Z}.$$

In fact, the $K(\pi, 1)$-property is only necessary in cohomological degree 2.

Comparison of definitions. By [EsWi09] Theorem 2.6, the class $\mathrm{cl}_s^{\mathrm{graph}}$ is norm compatible with respect to neighbourhoods of the section s and

$$\deg(\mathrm{cl}_s^{\mathrm{graph}}) = 1,$$

so that

$$\mathrm{cl}_s^{\mathrm{graph}} = \mathrm{cl}_s^{\mathrm{norm}}.$$

When it is defined, the class $\mathrm{cl}_s^{\mathrm{dual}}$ is compatible with norms due to adjointness of h^* and h_*. Namely, if $h : X' \to X$ is a neighbourhood of the section s with a lift s', then for $\alpha \in H^2\left(X, \mathbb{Z}/n\mathbb{Z}(1)\right)$ we have

$$\langle \mathrm{cl}_s^{\mathrm{dual}}, \alpha \rangle = s^*(\alpha) = s'^*(h^*(\alpha)) = \langle \mathrm{cl}_{s'}^{\mathrm{dual}}, h^*(\alpha) \rangle = \langle h_*(\mathrm{cl}_{s'}^{\mathrm{dual}}), \alpha \rangle.$$

The degree map
$$\deg : H_c^{2d}\left(X, \mathbb{Z}/n\mathbb{Z}(d)\right) \to \mathbb{Z}/n\mathbb{Z}$$

is adjoint to pullback by the projection $\mathrm{pr} : X \to \mathrm{Spec}(k)$. Therefore, for $\alpha \in H^2(k, \mathbb{Z}/n\mathbb{Z}(1))$ we compute

$$\langle \mathrm{pr}_*(\mathrm{cl}_s^{\mathrm{dual}}), \alpha \rangle = s^* \circ \mathrm{pr}^*(\alpha) = \alpha,$$

so that

$$\deg(\mathrm{cl}_s^{\mathrm{dual}}) = 1.$$

Consequently, we have $\mathrm{cl}_s^{\mathrm{dual}} = \mathrm{cl}_s^{\mathrm{norm}}$ whenever both classes are defined.

By [Mo07] Proposition 1.5, the class of the extension (6.5) is nothing but the arithmetic first Chern class of the diagonal, the graph of the identity

$$\Delta = \Gamma(\mathrm{id}) \subset X \times X.$$

Let n be invertible in k and let $h : X' \to X$ be a neighbourhood of s that is fine enough to define $\mathrm{cl}_s^{\mathrm{graph}}$ as a class modulo n. Let s' be the lift of s to X'. It follows (for the version of $\mathrm{cl}_s^{\mathrm{group}}$ with finite coefficients) that

$$\mathrm{cl}_s^{\mathrm{group}} = (s \times \mathrm{id})^*(\mathrm{cl}_\Delta) = (s' \times \mathrm{id})^*(h \times \mathrm{id})^*(\mathrm{cl}_\Delta) = (s' \times \mathrm{id})^*(\mathrm{cl}_{\Gamma(h)})$$

$$= (s' \times \mathrm{id})^* \mathrm{pr}_2^*(\mathrm{cl}_s^{\mathrm{graph}}) = \mathrm{cl}_s^{\mathrm{graph}}$$

so that both classes agree whenever they are defined.

Definition 62. The *cycle class* cl_s of a section s is any of the above defined cycle classes whenever the specific definition is applicable.

6.2 Algebraic K(π, 1) and Continuous Group Cohomology

In this section we prepare the new construction of the cycle class of a section given in Sect. 6.3 below.

The comparison map. The *finite étale site* $X_{\mathrm{f\acute{e}t}}$ of a scheme X consists of maps $Y \to X$ as objects which are finite étale, morphisms are maps relative X, and open coverings are given by surjective families of such maps. For a connected scheme X, a geometric point $\bar{x} \in X$ defines an equivalence

$$X_{\mathrm{f\acute{e}t}} \cong \mathrm{B}\pi_1(X, \bar{x})$$

of the finite étale site with the classifying site $\mathrm{B}\pi_1(X, \bar{x})$ of $\pi_1(X, \bar{x})$. The inclusion defines a canonical continuous map of sites

$$X_{\mathrm{\acute{e}t}} \to X_{\mathrm{f\acute{e}t}}$$

from the small étale site to the finite étale site. The composite will be denoted by

$$\gamma = \gamma_X = \gamma_{X,\bar{x}} : X_{\mathrm{\acute{e}t}} \to \mathrm{B}\pi_1(X, \bar{x}).$$

Definition 63. An *algebraic K(π, 1)-space* is a connected scheme X such that the adjoint map

$$\eta = \eta_A : A \to \mathrm{R}\,\gamma_* \gamma^* A$$

is an isomorphism for sheaves A of finite abelian groups on $\mathrm{B}\pi_1(X, \bar{x})$.

Change of group. Let $\varphi : H \to G$ be a continuous group homomorphism of profinite groups. There is an associated continuous map

$$\varphi : \mathrm{B}H \to \mathrm{B}G$$

on classifying sites. The push forward $\varphi_* A$ of a continuous H-module A is the induced module

$$\varphi_* A = \mathscr{H}om_H(G, A).$$

Here $\mathscr{H}om$ is the inner Hom, namely the set of maps which are invariant under some open subgroup of G under the right translation action on the first argument. The pullback $\varphi^* B$ of a continuous G-module B is given by H acting via φ on B. As usual, push forward and pullback are adjoint functors:

$$\mathrm{Hom}_H(\varphi^* B, A) = \mathrm{Hom}_G(B, \varphi_* A).$$

If φ happens to be a surjection $\varphi : \Pi \twoheadrightarrow \Gamma$ with kernel $\overline{\Pi}$, then $\varphi_* A$ consists of the invariants of $A^{\overline{\Pi}}$ and the higher direct images $\mathrm{R}^q \varphi_* A$ are continuous group cohomology $\mathrm{H}^q(\overline{\Pi}, A)$ together with its natural Γ action induced via conjugation of lifts and action on A.

The evaluation map Let $s : \Gamma \to \Pi$ be a section of $\varphi : \Pi \twoheadrightarrow \Gamma$, so s is a continuous homomorphism such that $\varphi \circ s = \mathrm{id}$. We define the *evaluation map*

$$\mathrm{ev}_s : \mathrm{R}\varphi_*(\varphi^* A) \to A$$

along s as follows.

$$\mathrm{ev}_s = \mathrm{R}\varphi_*(\eta_{\varphi^* A}) : \mathrm{R}\varphi_*(\varphi^* A) \to \mathrm{R}\varphi_*(\mathrm{R}s_* s^* \varphi^* A) = \mathrm{R}\ \mathrm{id}_* A = A,$$

Here $\eta_{\varphi^* A}$ is the corresponding adjoint map for the pair s_*, s^*.

6.3   The Cycle Class of a Section Via Evaluation

Let Λ be a finite ring of order invertible in the field k.

Cycle class with support. Let U/k be a smooth variety of dimension d and let $i : Z \to U$ be a regular closed immersion of pure codimension r and let $j : U \subset X$ be an open immersion into a smooth variety X/k such that $j \circ i$ is still closed with complement

$$i_Y : Y = X - U \hookrightarrow X.$$

The adjoint map

$$(j \circ i)_* \mathrm{R}(j \circ i)^! \Lambda(r) \to \Lambda(r)$$

factors over $j_!\Lambda(r) \subset \Lambda(r)$ as the cokernel $i_{Y,*}\Lambda(r)$ has support disjoint of the support of $(j \circ i)_* R(j \circ i)^! \Lambda(r)$ and hence there are no nontrivial Hom and Ext. The cycle class

$$\mathrm{cl}_Z \in \mathrm{H}_Z^{2r}(X, \Lambda(r))$$

thus maps via $\mathrm{H}^{2r}(X, -)$ applied to the above factorisation to a cycle class

$$\mathrm{cl}^U(Z) \in \mathrm{H}^{2r}(X, j_!\Lambda(r)).$$

In particular, for the diagonal $\Delta : U \to U \times_k U$ with image $\Delta = \Delta(U)$, we note that

$$(j \times \mathrm{id}) \circ \Delta : U \to X \times_k U$$

is still a regular closed immersion. Hence we have a cycle class with support for the diagonal

$$\mathrm{cl}_\Delta \in \mathrm{H}_\Delta^{2d}(X \times_k U, \Lambda(d)).$$

The cycle class of a section.   In addition to the above, we assume now that U is a geometrically connected algebraic $\mathrm{K}(\pi, 1)$-space and that X/k is proper. We will abusively also denote by j the inclusion

$$U \times_k U \subset X \times_k U,$$

and the ith projection will be denoted by pr_i. The map on fundamental groups induced by the projection $f : U \to \mathrm{Spec}(k)$ will be denoted by $\varphi : \pi_1(U) \to \mathrm{Gal}_k$. By smooth base change and the algebraic $\mathrm{K}(\pi, 1)$ property we have canonical isomorphisms

$$\mathrm{H}^{2d}(X \times_k U, j_!\Lambda(d)) = \mathrm{H}_c^{2d}(U, \mathrm{Rpr}_{1,*}\Lambda(d)) = \mathrm{H}_c^{2d}(U, f^*\mathrm{R}f_*\Lambda(d))$$

$$= \mathrm{H}_c^{2d}(U, f^*\mathrm{R}\varphi_*\mathrm{R}\gamma_*\gamma^*\varphi^*\Lambda(d)) = \mathrm{H}_c^{2d}(U, f^*\mathrm{R}\varphi_*\varphi^*\Lambda(d)).$$

Here we have silently identified

$$\mathrm{Spec}(k)_{\text{ét}} = \mathrm{Spec}(k)_{\text{fét}} = \mathrm{BGal}_k.$$

The evaluation map associated to a section s of φ, i.e., of $\pi_1(U/k)$, yields a composite map s^* in the following diagram.

$$
\begin{array}{ccc}
\mathrm{H}_\Delta^{2d}(X \times_k U, \Lambda(d)) & \longrightarrow & \mathrm{H}^{2d}(X \times_k U, j_!\Lambda(d)) \\
\Big\downarrow {\scriptstyle s^*} & & \Big\| \\
\mathrm{H}_c^{2d}(U, \Lambda(d)) & \xleftarrow{\ \mathrm{ev}_s\ } & \mathrm{H}_c^{2d}(U, f^*\mathrm{R}\varphi_*\varphi^*\Lambda(d)).
\end{array}
$$

We define the *cycle class of the section s* as

$$\mathrm{cl}^U(s) = s^*(\mathrm{cl}_\Delta).$$

If $s = s_a$ comes from a rational point $a \in U(k)$, then the pullback functoriality of cycle classes shows that

$$\mathrm{cl}^U(s_a) = \mathrm{cl}^U(a).$$

ℓ-adic cycle classes à la Jannsen. We use the theory of continuous or ℓ-adic cohomology as in [Ja88], where ℓ is assumed to be invertible in the field k. As an abbreviation, $\mathbb{Z}_\ell$ stands for the pro-system of $(\mathbb{Z}/\ell^n\mathbb{Z})_{n \in \mathbb{N}}$. The diagonal has an ℓ-adic cycle class, see [Ja88] Theorem 3.23,

$$\mathrm{cl}_\Delta \in \mathrm{H}^{2d}_\Delta(X \times_k U, \mathbb{Z}_\ell(d)).$$

The evaluation maps

$$\mathrm{ev}_s : \mathrm{R}\varphi_* \varphi^* \mathbb{Z}/\ell^n\mathbb{Z}(d) \to \mathbb{Z}/\ell^n\mathbb{Z}(d)$$

for various n are compatible and yield an ℓ-adic evaluation map

$$\mathrm{ev}_s : \mathrm{R}\varphi_* \varphi^* \mathbb{Z}_\ell(d) = \left(\mathrm{R}\varphi_* \varphi^* \mathbb{Z}/\ell^n\mathbb{Z}(d)\right)_{n \in \mathbb{N}} \to \mathbb{Z}_\ell(d)$$

by means of which we can define as above a map

$$s^* : \mathrm{H}^{2d}_\Delta(X \times_k U, \mathbb{Z}_\ell(d)) \to \mathrm{H}^{2d}_c(U, \mathbb{Z}_\ell(d)),$$

and an *ℓ-adic cycle class* of a section s by

$$\mathrm{cl}^U(s) = s^*(\mathrm{cl}_\Delta).$$

6.4 Anabelian Cycle Class for Subvarieties

We push the idea of a cycle class in anabelian geometry from Sect. 6.3 further to a cycle class of a subvariety.

The cycle class of a subvariety. Let $j : U \subset X$ be as in Sect. 6.3 and consider the closed immersion $i : Z \hookrightarrow U$ of pure codimension r of a connected, smooth and proper variety $g : Z \to \mathrm{Spec}(k)$. We will abusively also denote the inclusion

$$U \times_k Z \to X \times_k Z$$

by j. In the following diagram

$$\begin{array}{ccccc}
H^{2d}(X \times_k U, j_!\mathbb{Z}_\ell(d)) & \xrightarrow{(1\times h)^*} & H^{2d}(X \times_k Z, j_!\mathbb{Z}_\ell(d)) & \xrightarrow{\mathrm{pr}_{1,*}} & H^{2r}_c(U, \mathbb{Z}_\ell(r)) \\
\| & & \| & & \| \\
H^{2d}_c(U, f^*Rf_*\mathbb{Z}_\ell(d)) & \xrightarrow{\ i^*\ } & H^{2d}_c(U, f^*Rg_*\mathbb{Z}_\ell(d)) & \longrightarrow & H^{2r}_c(U, \mathbb{Z}_\ell(r)),
\end{array}$$

$$(6.7)$$

the vertical canonical isomorphisms are due to smooth base change, and $\mathrm{pr}_{1,*}$ comes from the adjoint map

$$\mathrm{tr}_g \ : \ Rg_*\mathbb{Z}_\ell(d) = Rg_!\big(g^!\mathbb{Z}_\ell(r)[2r-2d]\big) \to \mathbb{Z}_\ell(r)[2r-2d],$$

or, as one wishes, is the natural covariant map of H^*_c for proper maps.

More important is the fact, that the map i^* can be expressed as follows. We denote

$$\zeta = \pi_1(i) : \pi_1(Z) \to \pi_1(U)$$

and the map induced by g is denoted by

$$\psi : \pi_1(Z) \to \mathrm{Gal}_k.$$

In the diagram

$$\begin{array}{ccccccc}
Rf_*\mathbb{Z}_\ell(d) & \xrightarrow{\ \eta\ } & Rf_*Ri_*i^*\mathbb{Z}_\ell(d) & = & Rg_*\mathbb{Z}_\ell(d) & = & R\psi_*R\gamma_{Z,*}\gamma_Z^*\mathbb{Z}_\ell(d) \\
\| & & \uparrow & & & & \uparrow {\scriptstyle R\psi_*(\eta)} \\
R\varphi_*\mathbb{Z}_\ell(d) & \xrightarrow{\ \eta\ } & R\varphi_*R\zeta_*\zeta^*\mathbb{Z}_\ell(d) & = & = & & R\psi_*\mathbb{Z}_\ell(d).
\end{array}$$

the composite in the lower row is

$$\zeta^* : R\varphi_*\mathbb{Z}_\ell(d) \to R\psi_*\mathbb{Z}_\ell(d)$$

and thus

$$i^* = H^{2d}_c(U, f^*R\psi_*(\eta) \circ f^*\zeta^*).$$

The cycle class $\mathrm{cl}^U(Z)$ of Z in U is given via the image in $H^{2r}_c(U, \mathbb{Z}_\ell(r))$ along the top row in (6.7) of the class cl_Δ of the diagonal. The intermediate image is the class of the graph of i. Thus we find

$$\mathrm{cl}^U(Z) = \big(H^{2d}_c\big(U, f^*(\mathrm{tr}_g \circ R\psi_*(\eta) \circ \zeta^*)\big)\big)(\mathrm{cl}_\Delta).$$

The anabelian cycle class. The discussion in the preceding paragraph shows that $\mathrm{cl}^U(Z)$ only depends on the map ζ and the variety Z itself. The dependence on the way Z is embedded into U is exclusively through the map ζ. More precisely, the variety Z sponsors a map

$$\mathrm{tr}_\psi = \mathrm{tr}_g \circ \mathrm{R}\psi_*(\eta) \; : \; \mathrm{R}\psi_* \mathbb{Z}_\ell(d) \to \mathbb{Z}_\ell(r)[2r - 2d]$$

which together with ζ determines the cycle class.

Consequently, we define the *anabelian cycle class* of a pair consisting of a continuous homomorphism

$$\zeta : \pi_1(Z) \to \pi_1(U)$$

and a trace map tr_ψ as above as the class

$$\mathrm{cl}^U(\zeta, \mathrm{tr}_\psi) = \left(\mathrm{H}_c^{2d}\left(U, f^*(\mathrm{tr}_\psi \circ \zeta^*) \right) \right)(\mathrm{cl}_\Delta).$$

Remark 64. (1) If the pair $(\zeta, \mathrm{tr}_\psi)$ comes form geometry as above, then the consistency of the anabelian cycle class with the traditional geometric cycle class, i.e.,

$$\mathrm{cl}^U(Z) = \mathrm{cl}^U(\zeta, \mathrm{tr}_\psi),$$

follows from our presentation of the cycle class of the cycle $Z \hookrightarrow U$.

(2) If Z is also an algebraic $\mathrm{K}(\pi, 1)$, then the map tr_ψ is intrinsic to $\pi_1(Z)$ via the duality pairing. In this case we take this canonical choice and obtain a cycle class $\mathrm{cl}^U(\zeta)$ for any continuous homomorphism $\zeta : \pi_1(Z) \to \pi_1(U)$.

(3) In particular, we have a cycle class for a proper smooth curve $C \hookrightarrow U$ of genus ≥ 1 which depends only on the map $\pi_1(C) \to \pi_1(U)$. More precisely, by (2), even the datum of a continuous homomorphism

$$\zeta : \pi_1(C) \to \pi_1(U),$$

which a priori need not come from geometry, suffices to construct the cycle class. This observation might help in higher dimensional anabelian geometry.

6.5 Parshin's π_1-Approach to the Geometric Mordell Theorem

Parshin [Pa90] gave a proof of the geometric Mordell Theorem using fundamental groups and hyperbolic geometry. We present Parshin's method from an algebraic point of view, at least concerning the use of fundamental groups and cycle classes. For another recent survey of Parshin's proof we refer to [Sz12] §2.4.

Relative cycle class. Let X/k be a proper variety and let $f : X \to B$ be a proper flat map of k-varieties of pure relative dimension d. Let $U \subset B$ be a dense open subvariety such that

$$f|_V : V = f^{-1}(U) \to U$$

is smooth. For a section t of $f|_V$, as in the diagram

$$
\begin{array}{ccc}
V & \subset & X \\
t \left(\Big\downarrow f|_V \right. & & \Big\downarrow f \\
U & \subset & B,
\end{array}
$$

the image of s is a closed subscheme $W \hookrightarrow U$. By abuse of notation we denote all open immersions by j, for example $V \times_U V \subseteq X \times_B V$. The relative diagonal

$$\Delta : V \to V \times_U V$$

is a regular immersion and has image $\Delta = \Delta(V)$ which is still closed in $X \times_B V$. We obtain a cycle class with support

$$\mathrm{cl}_\Delta \in \mathrm{H}^{2d}_\Delta (X \times_B V, \mathbb{Z}_\ell(d))$$

which maps along the composition in the following diagram

$$
\begin{array}{ccc}
\mathrm{H}^{2d}_\Delta (X \times_B V, \mathbb{Z}_\ell(d)) & \longrightarrow & \mathrm{H}^{2d}(X \times_B V, j_!\mathbb{Z}_\ell(d)) & \!\!=\!\! & \mathrm{H}^{2d}(X, j_!\mathrm{Rpr}_{1,*}\mathbb{Z}_\ell(d)) \\
\Big\downarrow & & & & \Big\| \ \text{smooth bc} \\
\mathrm{H}^{2d}_c (V, \mathbb{Z}_\ell(d)) & \!\!=\!\! & \mathrm{H}^{2d}(X, j_!(f|_V)^*\mathbb{Z}_\ell(d)) & \xleftarrow{\ cv_t\ } & \mathrm{H}^{2d}(X, j_!(f|_V)^*\mathrm{R}(f|_V)_*\mathbb{Z}_\ell(d))
\end{array}
$$

to the cycle class

$$\mathrm{cl}^V(W) \in \mathrm{H}^{2d}_c (V, \mathbb{Z}_\ell(d)).$$

Let us assume in addition that X/k is smooth and that t extends to a section of f with image C. Then we also have a cycle class

$$\mathrm{cl}^X(C) \in \mathrm{H}^{2d}(X, \mathbb{Z}_\ell(d)).$$

The construction being contravariant functorial in the pair (X, V) we find that $\mathrm{cl}^V(W)$ and $\mathrm{cl}^X(C)$ map to the same class in $\mathrm{H}^{2d}(V, \mathbb{Z}_\ell(d))$ under the maps

$$\mathrm{H}^{2d}_c (V, \mathbb{Z}_\ell(d)) \to \mathrm{H}^{2d}(X, \mathbb{Z}_\ell(d)) \to \mathrm{H}^{2d}(V, \mathbb{Z}_\ell(d)).$$

Relative anabelian cycles. With $\varphi = \pi_1(f|_V) : \pi_1(V) \to \pi_1(U)$ the diagram of sites

$$
\begin{array}{ccc}
V_{\text{ét}} & \xrightarrow{\ \gamma\ } & B\pi_1(V) \\
\downarrow{\scriptstyle f|_V} & & \downarrow{\scriptstyle \varphi} \\
U_{\text{ét}} & \xrightarrow{\ \gamma\ } & B\pi_1(U)
\end{array}
$$

induces a base change map

$$
\mathrm{bc} : \gamma^* R\varphi_* \mathscr{F} \to R(f|_V)_*(\gamma^* \mathscr{F}),
$$

which we assume to be an isomorphism for $\mathscr{F} = \mathbb{Z}_\ell(d)$. This scenario occurs when the fibres of $V \to U$ are algebraic $K(\pi, 1)$-spaces. In this case the evaluation map only depends on $s = \pi_1(t)$ as a section of φ because the following diagram commutes.

$$
\begin{array}{ccc}
\gamma^* R\varphi_* \mathbb{Z}_\ell(d) & \xrightarrow{\ \mathrm{bc}\ } & R(f|_V)_*(\gamma^* \mathbb{Z}_\ell(d)) \\
\downarrow{\scriptstyle \gamma^* \mathrm{ev}_s} & & \downarrow{\scriptstyle \mathrm{ev}_t} \\
\gamma^* \mathbb{Z}_\ell(d) & =\!=\!=\!=\!= & \mathbb{Z}_\ell(d).
\end{array}
$$

Consequently the cycle class $\mathrm{cl}^V(W)$ does only depend on the section $s = \pi_1(t)$ on the level of fundamental groups. Moreover, for any such section s of φ we can define a cycle class

$$
\mathrm{cl}^V(s) \in \mathrm{H}^{2d}_c(V, \mathbb{Z}_\ell(d)). \tag{6.8}
$$

Sketch of proof of the geometric Mordell theorem. Thanks to Parshin [Pa90] we have a π_1-proof of the following important and well known theorem.

Theorem 65 (Geometric Mordell; Manin, Grauert, Parshin, see [Co90]). *Let K be the function field of a smooth, projective curve $B/\mathbb{C}$ and X/K a smooth, projective curve of genus ≥ 2. Then exactly one of the following holds.*

(a) X/K is isotrivial.
(b) For all finite extensions L/K the set $X(L)$ of L-rational points is finite.

Proof. If X/K satisfies (a), then for suitable L/K the curve $X \times_K L$ comes by scalar extension from a curve $X_0/\mathbb{C}$ and thus $X(L) \supseteq X_0(\mathbb{C})$ is infinite, whence (b) fails. The theorem now follows from Theorem 66 by potentially a finite branched base change, the semistable reduction theorem and by the valuative criterion of properness.   □

Theorem 66. *Let B be a smooth, projective curve over $\mathbb{C}$ and let $f : X \to B$ be a projective, semistable, generically smooth, relative curve with $f_* \mathcal{O}_X = \mathcal{O}_B$ and generic fibres of genus ≥ 2, such that $X/\mathbb{C}$ is smooth. If X/B is not isotrivial then the map $X \to B$ has only finitely many sections.*

We will sketch Parshin's strategy for proving Theorem 66. Let $U \subset B$ be a dense open hyperbolic curve such that $f : V = f^{-1}(U) \to U$ is smooth and projective. In particular, U, V and the fibres of $V \to U$ are algebraic $K(\pi, 1)$-spaces. We set $S = B - U$ and let $\sum_\alpha F_\alpha$ be the reduced divisor with support $X - V$ and contemplate about the following diagram.

$$X(B) = V(U) \xrightarrow{\ \pi_1\ } \mathscr{S}_{\pi_1(V) \twoheadrightarrow \pi_1(U)} \qquad (6.9)$$

$$\Big\downarrow \mathrm{cl}^X(-) \qquad\qquad\qquad\quad \Big\downarrow \mathrm{cl}^V(-)$$

$$\bigoplus_\alpha \mathrm{H}^2_{F_\alpha}(X, \mathbb{Z}_\ell(1)) \longrightarrow \mathrm{H}^2(X, \mathbb{Z}_\ell(1)) \longrightarrow \mathrm{H}^2(V, \mathbb{Z}_\ell(1)),$$

where $\mathrm{cl}^X(-)$ is the cycle class of the image of a section of $X \to B$, and where $\mathrm{cl}^V(-)$ for sections of $\varphi : \pi_1(V) \twoheadrightarrow \pi_1(U)$ was constructed in (6.8).

Lemma 67. *The set of sections $t \in X(B)$ with fixed induced section $s = \pi_1(t|_U)$ of φ is finite.*

Proof. Let us assume that t_1 and t_2 with images C_1 and C_2 induce the same section

$$s = \pi_1(t_1|_U) = \pi_1(t_2|_U)$$

of the map $\pi_1(V) \twoheadrightarrow \pi_1(U)$ on fundamental groups. Then, by (6.9), the respective classes differ by a linear combination

$$\mathrm{cl}^X(C_1) - \mathrm{cl}^X(C_2) = \sum_\alpha m_\alpha \mathrm{cl}^X(F_\alpha)$$

of classes with support in $X - V$. There are only finitely many choices for the intersection numbers $(C_i \bullet F_\alpha) \in \{0, 1\}$. We may thus assume for all F_α that

$$(C_1 \bullet F_\alpha) = (C_2 \bullet F_\alpha).$$

Because the intersection pairing depends only on the cycle classes, we find

$$0 = (\sum_\alpha m_\alpha F_\alpha \bullet C_1 - C_2) = (C_1 - C_2 \bullet C_1 - C_2) = (C_1 \bullet C_1) - 2(C_1 \bullet C_2) + (C_2 \bullet C_2).$$

By Arakelov's Theorem [Ar71], see [Sz79] Corollary to Theorem 1 p. 177, the nonisotriviality of X/B implies that $(C_i \bullet C_i) < 0$ so that $(C_1 \bullet C_2) < 0$, which is only possible if C_1 agrees with C_2.   □

Lemma 67 reduces the proof of Theorem 66 to the problem of proving the finiteness of the set of sections of $\varphi : \pi_1(V) \twoheadrightarrow \pi_1(U)$ of geometric origin. Note that whenever $U \neq B$ is not proper, then $\pi_1(U)$ is free profinite and hence there is an abundance of group theoretic sections. The task is to single out the geometric ones and to show that there can be only finitely many of those. Parshin accomplishes this task exploiting complex hyperbolic geometry, more explicitly the Kobayashi metric and the finiteness of closed geodesics with bounded Kobayashi-length.

Part II
Basic Arithmetic of Sections

Chapter 7
Injectivity in the Section Conjecture

We recall the well known injectivity of the profinite Kummer map in the arithmetically relevant cases. There are at least two approaches towards injectivity. The abelian approach relies on the determination of the Kummer map for abelian varieties and their arithmetic, see Corollary 71, and also on the computation of the maximal abelian quotient extension $\pi_1^{ab}(X/k)$, see Proposition 69, which for later use in Sect. 13.5 we carefully revise also for smooth projective varieties of arbitrary dimension. The second approach is intrinsically anabelian and due to Mochizuki, see Theorem 76.

In general, for a geometrically connected variety, the injectivity of the Kummer map (after arbitrary finite scalar extension) implies that the fundamental group must be large in the sense of Kollár, see Proposition 77, which imposes strong geometric constraints on possible higher dimensional anabelian varieties.

7.1 Injectivity via Arithmetic of Abelian Varieties

This approach exploits essentially the cohomology of the Kummer sequence of an abelian variety. The crucial arithmetic input in the case of a number field as base field comes form the Mordell–Weil Theorem. Injectivity of the profinite Kummer map for smooth projective curves of genus ≥ 2 based on the Mordell–Weil Theorem was known to Grothendieck, see [Gr83] p. 4.

The abelian weight -1 quotient. Let U/k be a smooth geometrically connected variety with a smooth proper completion X/k. Such a completion always exists in characteristic 0 by Nagata's embedding theorem and Hironaka's resolution of singularities, or at least if $\dim U \leq 2$. The birational invariance of the fundamental group [SGA1] X Corollary 3.4 shows that the quotient

$$\pi_1(U) \twoheadrightarrow \pi_1(X)$$

is independent of the choice of X.

J. Stix, *Rational Points and Arithmetic of Fundamental Groups*, Lecture Notes in Mathematics 2054, DOI 10.1007/978-3-642-30674-7_7,
© Springer-Verlag Berlin Heidelberg 2013

Definition 68. We define the *weight* -1 *quotient* of $\pi_1^{\mathrm{ab}}(\overline{U})$ as the quotient

$$\mathrm{W}_{-1}\pi_1^{\mathrm{ab}}(\overline{U}) = \pi_1^{\mathrm{ab}}(\overline{X})$$

and the *weight* -1 *quotient extension* of $\pi_1(X/k)$ as

$$\mathrm{W}_{-1}\pi_1^{\mathrm{ab}}(U/k) = \pi_1^{\mathrm{ab}}(X/k).$$

The abelianized fundamental group. Let Alb_U be the Albanese variety of U. The Albanese torsor map, see Wittenberg [Wi08],

$$\alpha = \alpha_U : U \to \mathrm{Alb}_U^1 \tag{7.1}$$

factors over the inclusion $U \subset X$ into a proper smooth completion, and the resulting map $\alpha_X : X \to \mathrm{Alb}_U^1$ agrees with the Albanese torsor map of X, see [Mi86] Theorem 3.1. We recall the connection of the Albanese torsor map with $\mathrm{W}_{-1}\pi_1^{\mathrm{ab}}(\overline{U})$. We define the *Néron–Severi group scheme* NS_X over $\mathrm{Spec}(k)$ of X/k by the exact sequence

$$0 \to \left(\mathrm{Pic}_X^0\right)_{\mathrm{red}} \to \mathrm{Pic}_X \to \mathrm{NS}_X \to 0. \tag{7.2}$$

It is known, that $\mathrm{NS}_X(\bar{k})$ is a finitely generated abelian group, hence the torsion subgroup scheme $\mathrm{NS}_{X,\mathrm{tors}}$, namely the preimage of the finite étale torsion subgroup-scheme of $\pi_0(\mathrm{NS}_X)$ under the projection

$$\mathrm{NS}_X \to \pi_0(\mathrm{NS}_X),$$

is a finite flat commutative group scheme over $\mathrm{Spec}(k)$.

We denote by G^D the Cartier dual of a finite flat group scheme G, and by $G^{\mathrm{\acute{e}t}}$ its maximal étale quotient group scheme.

Proposition 69. *Let X/k be a smooth projective and geometrically connected variety. Then we have an exact sequence*

$$0 \to \left(\mathrm{NS}_{X,\mathrm{tors}}\right)^D(\bar{k}) \to \pi_1^{\mathrm{ab}}(\overline{X}) \to \pi_1(\overline{\mathrm{Alb}_X}) \to 0 \tag{7.3}$$

of Gal_k-modules. In particular, the torsion subgroup of $\pi_1^{\mathrm{ab}}(\overline{X})$ is the finite group

$$\left(\mathrm{NS}_{X,\mathrm{tors}}\right)^D(\bar{k}).$$

Proof. This can essentially be found in [Mi80] III §4 Corollary 4.19, or [Mi82] §1. Here are the necessary corrections and modifications that prove the more precise result.

Since $(\mathrm{Pic}_X^0)_{\mathrm{red}}$ is an abelian variety and thus divisible, the kernel of multiplication by n yields from (7.2) the short exact sequence of finite flat group schemes

$$0 \to \left(\mathrm{Pic}_X^0\right)_{\mathrm{red}}[n] \to \mathrm{Pic}_X[n] \to \mathrm{NS}_X[n] \to 0.$$

If n is divisible by the order of $\mathrm{NS}_{X,\mathrm{tors}}$, then by taking the Cartier dual we obtain

$$0 \to (\mathrm{NS}_{X,\mathrm{tors}})^D \to \left(\mathrm{Pic}_X[n]\right)^D \to \mathrm{Alb}_X[n] \to 0 \qquad (7.4)$$

because by the Poincaré-pairing $\left(\mathrm{Pic}_X^0\right)_{\mathrm{red}}[n]$ and $\mathrm{Alb}_X[n]$ are mutually Cartier dual. With the projection map $\mathrm{pr} : X \to \mathrm{Spec}(k)$ we obtain from [Mi80] III §4 Prop 4.16 that as sheaves on $\mathrm{Spec}(k)_{\mathrm{fppf}}$

$$R^1\, \mathrm{pr}_{\mathrm{fppf},*}\,\mathbb{Z}/n\mathbb{Z} = \underline{\mathrm{Hom}}_{X_{\mathrm{fppf}}}(\mu_n, \mathrm{Pic}_X) = \underline{\mathrm{Hom}}_{X_{\mathrm{fppf}}}(\mu_n, \mathrm{Pic}_X[n])$$

which furthermore by Cartier duality agrees with

$$= \underline{\mathrm{Hom}}_{X_{\mathrm{fppf}}}(\mathrm{Pic}_X[n]^D, \mathbb{Z}/n\mathbb{Z}) = \underline{\mathrm{Hom}}_{X_{\mathrm{fppf}}}(\mathrm{Pic}_X[n]^{D,\mathrm{\acute{e}t}}, \mathbb{Z}/n\mathbb{Z}).$$

Let the comparison map of topologies be denoted by

$$\beta : \mathrm{Spec}(k)_{\mathrm{fppf}} \to \mathrm{Spec}(k)_{\mathrm{\acute{e}t}}.$$

Because étale and flat cohomology agree for smooth coefficients we deduce

$$R^1\, \mathrm{pr}_{\mathrm{\acute{e}t},*}\,\mathbb{Z}/n\mathbb{Z} = \beta_* R^1\, \mathrm{pr}_{\mathrm{fppf},*}\,\mathbb{Z}/n\mathbb{Z}$$

$$= \beta_*\left(\underline{\mathrm{Hom}}_{X_{\mathrm{fppf}}}(\mathrm{Pic}_X[n]^{D,\mathrm{\acute{e}t}}, \mathbb{Z}/n\mathbb{Z})\right) = \underline{\mathrm{Hom}}_{X_{\mathrm{\acute{e}t}}}(\mathrm{Pic}_X[n]^{D,\mathrm{\acute{e}t}}, \mathbb{Z}/n\mathbb{Z}).$$

It follows by Pontrjagin duality that as finite étale group schemes we have

$$\pi_1^{\mathrm{ab}}(\overline{X})/n\pi_1^{\mathrm{ab}}(\overline{X}) = \mathrm{Pic}_X[n]^{D,\mathrm{\acute{e}t}}$$

so that the maximal étale quotient of (7.4) translates into the short exact sequence

$$0 \to (\mathrm{NS}_{X,\mathrm{tors}})^D(\bar{k}) \to \pi_1^{\mathrm{ab}}(\overline{X})/n\pi_1^{\mathrm{ab}}(\overline{X}) \to \mathrm{Alb}_X[n]^{\mathrm{\acute{e}t}} \to 0. \qquad (7.5)$$

The sequence (7.3) agrees with the projective limit over n of (7.5) which completes the proof of the proposition. □

The Kummer map for abelian varieties. Let A/k be an abelian variety and let $\varphi : B \to A$ an étale isogeny. We consider the isogeny sequence

$$0 \to \ker(\varphi) \to B \xrightarrow{\varphi} A \to 0. \qquad (7.6)$$

For $\varphi = [n]$ the multiplication by n map, this sequence is also known as the Kummer sequence. The isogeny φ corresponds to a homomorphism of Gal_k-modules, which by abuse of notation we also denote by

$$\varphi : \pi_1(\overline{A}) \to \ker(\varphi).$$

We can push $\pi_1(A/k)$ by φ to obtain the *modulo φ extension* $\varphi_*(\pi_1(A/k))$

$$1 \to \ker(\varphi) \to \pi_1^{(\varphi)}(A) \to \mathrm{Gal}_k \to 1.$$

As usual, a rational point $a \in A(k)$ yields a section s_a, or better a conjugacy class of sections, which we may push to a section $\varphi_*(s_a)$ of $\varphi_*(\pi_1(A/k))$.

Proposition 70. *The boundary of the isogeny sequence (7.6)*

$$\delta : A(k)/\varphi(B(k)) \hookrightarrow \mathrm{H}^1(k, \ker(\varphi))$$

maps $a \in A(k)$ to the class of the difference cocycle

$$\delta(\varphi_*(s_a), \varphi_*(s_0)) : \sigma \mapsto s_a(\sigma)s_0(\sigma)^{-1}$$

of the section s_a and the section s_0 associated to the origin.

Proof. The cocycle $\delta(a)$ describes the Gal_k-equivariant $\ker(\varphi)$-torsor $\varphi^{-1}(a)$. Let $b \in B(\bar{k})$ with $\varphi(b) = a$ be a preimage. Then, following Proposition 8, a cocycle description is given by

$$\delta(a) : \sigma \mapsto \delta(a)_\sigma = \sigma(b) - b.$$

Following Sect. 1.2 we define a right $\pi_1(A)$-action on $\varphi^{-1}(a)(\bar{k})$ and let s_a be the representative of its conjugacy class, that acts on $\varphi^{-1}(a)(\bar{k})$ by fixing b. Then

$$b = b.(s_a(\sigma)) = b.\big(\delta(s_a, s_0)_\sigma s_0(\sigma)\big) = \sigma^{-1}\big(b + \delta(s_a, s_0)_\sigma\big)$$

so that by acting with σ we obtain $\delta(s_a, s_0)_\sigma = \sigma(b) - b = \delta(a)_\sigma$. $\qquad\square$

Corollary 71. *Let A/k be an abelian variety. Then the following diagram commutes*

$$
\begin{array}{ccc}
 & A(k) & \\
{\scriptstyle \kappa}\swarrow & & \searrow{\scriptstyle \delta_{\mathrm{kum}}} \\
\mathscr{S}_{\pi_1(A/k)} \xrightarrow[\;s \mapsto \delta(s,s_0)\;]{\sim} & & \mathrm{H}^1(k, \pi_1(\overline{A}))
\end{array}
$$

where κ is the profinite Kummer map for A and δ_{kum} is the boundary map of the continuous Kummer sequence for A/k.

Proof. We simply have to perform the projective limit over all étale isogenies $\varphi : B \to A$ and apply Proposition 70 to every φ to realize $\delta(s_a, s_0) = \delta_{\mathrm{kum}}(a)$. $\qquad\square$

Abelian separable varieties. For an abelian separable variety, the profinite Kummer map κ is injective, as follows immediately from the following definition.

Definition 72. An *abelian separable* variety over k is a smooth geometrically connected variety U/k with a smooth projective completion X/k such that for all finite extensions k'/k the composition

$$U(k') \xrightarrow{\kappa} \mathscr{S}_{\pi_1(U/k)}(k') \to \mathscr{S}_{\pi_1^{\text{ab}}(X/k)}(k')$$

is injective.

Proposition 73. *Let U be a nonempty open in a smooth projective geometrically connected variety X/k. We consider the following properties.*

(a) U is abelian separable.
(b) The Albanese torsor map $\alpha : U \to A = \text{Alb}_U^1$ is injective on $\bar{k}$-points.
($*$) For every finite extension k'/k the group $A(k')$ has a trivial group of divisible elements

$$\text{div}(A(k')) = \bigcap_{n \geq 1} n A(k').$$

Then (a) implies (b) and the converse holds if ($*$) is true, for example, if k is an algebraic number field or a finite extension of $\mathbb{Q}_p$.

Proof. As $\text{NS}_{X,\text{tors}}^{D,\text{ét}}$ is finite, we have

$$\varinjlim_{k'/k} \text{H}^1\left(k', \text{NS}_{X,\text{tors}}^{D,\text{ét}}\right) = 0$$

where the limit ranges over all finite extensions k'/k. The cohomology sequence applied to (7.3) thus reads in the limit over all k'/k as

$$0 = \varinjlim_{k'/k} \text{H}^1\left(k', \text{NS}_{X,\text{tors}}^{D,\text{ét}}\right) \to \varinjlim_{k'/k} \text{H}^1\left(k', \pi_1^{\text{ab}}(\overline{X})\right) \to \varinjlim_{k'/k} \text{H}^1\left(k', \pi_1(\overline{A})\right).$$

We conclude that in the commutative diagram, see Corollary 71,

$$
\begin{array}{ccc}
U(\bar{k}) & \xrightarrow{\quad \alpha \quad} & A(\bar{k}) \\
\downarrow{\scriptstyle \kappa_U} & & \downarrow{\scriptstyle \kappa_A} \qquad \searrow{\scriptstyle \delta_{\text{kum}}} \\
\varinjlim_{k'/k} \mathscr{S}_{W_{-1}\pi_1^{\text{ab}}(U/k)}(k') \;\hookrightarrow\; & \varinjlim_{k'/k} \mathscr{S}_{\pi_1(A/k)}(k') \;\xrightarrow{\sim}\; & \varinjlim_{k'/k} \text{H}^1(k', \pi_1(\overline{A}))
\end{array}
$$

the first arrow in the bottom row is injective. Thus (a) clearly implies (b). Conversely, if ($*$) holds, then

$$\delta_{\text{kum}} : A(k') \to \text{H}^1(k', \pi_1(\overline{A}))$$

is injective for every finite k'/k and (b) implies (a).

Algebraic number fields k satisfy property $(*)$ by the Mordell–Weil Theorem. For p-adic local fields k we deduce $(*)$ from the Theorem of Mattuck–Tate [Ma55] that $A(k)$ is a topologically finitely generated abelian profinite group. $\qquad\square$

The following is an immediate corollary.

Corollary 74. *Let U/k be a smooth geometrically connected variety with a projective completion X/k over a field k that satisfies $(*)$ for all abelian varieties, and moreover assume that the Albanese torsor map $U \to \mathrm{Alb}^1_U$ is injective. Then the profinite Kummer map $\kappa : U(k) \to \mathscr{S}_{\pi_1(U/k)}$ is injective. This applies to smooth, projective curves of genus ≥ 1 over number fields or p-adic local fields.* $\qquad\square$

With the last proposition of this section we include affine hyperbolic curves with completion of genus 0.

Proposition 75. *Let U/k be a smooth geometrically connected variety such that a geometrically connected finite étale cover $V \to U$ is abelian separable. Moreover, we assume that $(*)$ holds for all abelian varieties over k. Then the profinite Kummer map*

$$\kappa : U(k) \to \mathscr{S}_{\pi_1(U/k)}$$

is injective. In particular, this applies to hyperbolic curves over number fields or p-adic local fields.

Proof. Let $a, b \in U(k)$ be rational points with $[s_a] = [s_b]$. We choose representatives such that $s_a = s_b = s$. After a finite extension k'/k we may assume that $V \to U$ is a neighbourhood of s. Without loss of generality we may assume that $k' = k$ since with V also $V_{k'}$ is abelian separable. Let $t : \mathrm{Gal}_k \to \pi_1(V)$ be a lift of s, and let $a', b' \in V(k)$ be lifts of a, b such that $s_{a'} = t = s_{b'}$, see Proposition 33 (1). Since V/k is abelian injective, we deduce that $a' = b'$ and consequently $a = b$. $\qquad\square$

7.2 Injectivity via Anabelian Assumptions

The following reformulates a result of Mochizuki, [Mo99] Theorem 19.1, because hyperbolic curves over sub-p-adic fields satisfy the anabelian assumption.

Let p be a prime number. For a geometrically connected variety X/k, let

$$\pi_1^{\mathrm{pro}\text{-}p}(X/k)$$

denote the characteristic quotient extension of $\pi_1(X/k)$

$$1 \to \pi_1^{\mathrm{pro}\text{-}p}(\overline{X}) \to \pi_1^{(\mathrm{pro}\text{-}p)}(X) \to \mathrm{Gal}_k \to 1 \qquad (7.7)$$

obtained by pushing with the maximal pro-p quotient

$$\pi_1(\overline{X}) \to \pi_1^{\mathrm{pro}\text{-}p}(\overline{X}),$$

of the geometric fundamental group $\pi_1(\overline{X})$. The extension $\pi_1^{\text{pro-}p}(X/k)$ canonically defines a pro-p outer Galois representation

$$\rho_{X/k} : \text{Gal}_k \to \text{Out}\left(\pi_1^{\text{pro-}p}(\overline{X})\right) \tag{7.8}$$

by conjugation with preimages, see for example [Sx05] §2.

Let us temporarily say that a collection of hyperbolic curves $\{U\}$ over the field k is pro-p-anabelian if the natural map

$$\text{Isom}_k(U_1, U_2) \to \text{Isom}_{\text{Gal}_k}^{\text{out}}\left(\pi_1^{\text{pro-}p}(\overline{U}_1), \pi_1^{\text{pro-}p}(\overline{U}_2)\right)$$

for U_1, U_2 from the collection, is a bijection.

Theorem 76 (after Mochizuki). *Let k be of characteristic 0 and let p be a prime number. Let X/k be a smooth hyperbolic curve, such that the collection of complements $U_a = X \setminus \{a\}$ for the k-rational points $a \in X(k)$ is pro-p-anabelian. Then the pro-p Kummer map*

$$\kappa_p : X(k) \to \mathscr{S}_{\pi_1^{\text{pro-}p}(X/k)}$$

and hence also the profinite Kummer map $\kappa : X(k) \to \mathscr{S}_{\pi_1(X/k)}$ is injective.

Proof. For the convenience of the reader we recall the proof of [Mo99] Theorem 19.1. Let

$$U - X \times_k X \setminus \Delta(X)$$

be the complement of the diagonal. We consider the family $\text{pr}_2 : U \to X$ of hyperbolic curves which has U_a/k as its fibre above $a \in X(k)$. Let $\bar{a}$ be the canonical geometric point above $a \in X(k)$ and let $\bar{u}$ be a lift to a base point on $\overline{U}_a$. Because we are in characteristic 0, the sequence

$$1 \to \pi_1(\overline{U}_a, \bar{u}) \to \pi_1(U, \bar{u}) \to \pi_1(X, \bar{a}) \to 1$$

is exact, see [Sx05] Proposition 2.3. The resulting outer pro-p monodromy representation factors as

$$\rho_{U/X,\bar{a}} : \pi_1^{(\text{pro-}p)}(X, \bar{a}) \to \text{Out}\left(\pi_1^{\text{pro-}p}(\overline{U}_a, \bar{u})\right). \tag{7.9}$$

Indeed, the kernel of

$$\text{Out}\left(\pi_1^{\text{pro-}p}(\overline{U}_a, \bar{u})\right) \to \text{GL}\left(\pi_1^{\text{ab}}(\overline{U}_a) \otimes \mathbb{Z}_p\right)$$

is a pro-p group by a theorem of Hall, and the geometric monodromy action on $\pi_1^{\text{ab}}(\overline{U}_a)$ is unipotent with respect to the filtration defined by the exact sequence

$$\mathbb{Z}_p(1) \to \pi_1^{\text{ab}}(\overline{U}_a) \otimes \mathbb{Z}_p \to \pi_1^{\text{ab}}(\overline{X}) \otimes \mathbb{Z}_p \to 0,$$

hence factors through a geometrically pro-p group.

The naturality of the outer monodromy representations, see [Sx05] Proposition 2.2 (2), shows that we have

$$\rho_{U_a/k} = \rho_{U/X,\bar{a}} \circ s_a,$$

where s_a is the canonical section $s_a : \mathrm{Gal}_k \to \pi_1(X, \bar{a})$. By the factorisation (7.9), the representation $\rho_{U_a/k}$ only depends on the pro-p section

$$s_a^p = \kappa_p(a).$$

Let $a, b \in X(k)$ be rational points such that $\kappa_p(a) = \kappa_p(b)$. We choose an étale path $\gamma : \bar{a} \rightsquigarrow \bar{b}$ such that under the identification by the isomorphism

$$\gamma(-)\gamma^{-1} : \pi_1(X, \bar{a}) \xrightarrow{\sim} \pi_1(X, \bar{b})$$

the canonical sections s_a and s_b coincide. In addition to $\bar{u}$ over $\bar{a}$ we fix a geometric point $\bar{v}$ on $\overline{U}_b$ above $\bar{b}$. Let

$$\delta : \bar{u} \rightsquigarrow \bar{v}$$

be an étale path on $\overline{U}$ that lifts γ such that we have a commutative diagram

$$
\begin{array}{ccccccccc}
1 & \longrightarrow & \pi_1(\overline{U}_a, \bar{u}) & \longrightarrow & \pi_1(U, \bar{u}) & \longrightarrow & \pi_1(X, \bar{a}) & \longrightarrow & 0 \\
 & & \downarrow{\scriptstyle \varphi} & & \downarrow{\scriptstyle \delta(-)\delta^{-1}} & & \downarrow{\scriptstyle \gamma(-)\gamma^{-1}} & & \\
1 & \longrightarrow & \pi_1(\overline{U}_b, \bar{v}) & \longrightarrow & \pi_1(U, \bar{v}) & \longrightarrow & \pi_1(X, \bar{b}) & \longrightarrow & 0
\end{array}
$$

which implicitly defines the isomorphism φ that has the property that

$$
\begin{array}{ccc}
\pi_1(\overline{U}_a, \bar{u}) & \xrightarrow{\ \varphi\ } & \pi_1(\overline{U}_b, \bar{v}) \\
\downarrow & & \downarrow \\
\pi_1(\overline{X}, \bar{u}) & \xrightarrow{\ \psi\ } & \pi_1(\overline{X}, \bar{v})
\end{array}
\qquad (7.10)
$$

commutes with ψ being a representative of $\pi_1(\mathrm{id}_X)$. If we set Out^* for the subgroup of outer automorphisms which respect the inertia subgroups at a resp. b, then we find the following commutative diagram.

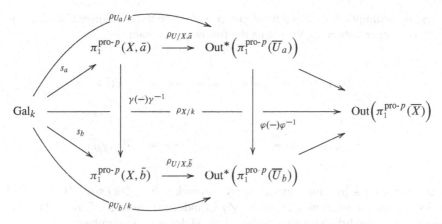

Hence the map φ is an isomorphism of outer Gal_k-representations. By the hypothesis of pro-p-anabelianity of U_a and U_b we extract a k-isomorphism $f : U_a \to U_b$ such that $\pi_1(f) = \varphi$. The commutativity of (7.10) shows that the rational map $f : X \to X$ maps $f(a) = b$ and thus is actually an isomorphism of X. Moreover, $\pi_1(f)$ acts as the identity on $\pi_1(\overline{X})$ which by well known arguments, due to X being hyperbolic, implies that $f = \mathrm{id}_X$. As a result we have $b = f(a) = a$ and the proof of the theorem is complete. $\qquad\square$

7.3 Large Fundamental Group in the Sense of Kollár

Following Kollár [Ko95] IV, a variety X/k has *large algebraic fundamental group* if for every normal irreducible variety $\overline{Z}/\overline{k}$ and every non-constant map $f : \overline{Z} \to \overline{X}$ of $\overline{k}$-varieties the map

$$f_* : \pi_1(\overline{Z}) \to \pi_1(\overline{X})$$

has infinite image.

Proposition 77. *Let X/k be a geometrically connected variety such that for every finite extension k'/k the profinite Kummer map*

$$\kappa : X(k') \to \mathscr{S}_{\pi_1(X/k)}(k')$$

is injective. Then X has large algebraic fundamental group.

Proof. We argue by contradiction. Replacing k with a finite extension, we may assume that a witness of the failure of having a large fundamental group is given by the map $f : Z \to X$ of k-varieties. We denote the image of $\pi_1(Z/k)$ in $\pi_1(X/k)$ by

$$1 \to \overline{\pi}_{Z|X} \to \pi_{Z|X} \to \mathrm{Gal}_k \to 1 \qquad (7.11)$$

where by assumption $\overline{\pi}_{Z|X}$ is a finite group. Therefore the centraliser C of $\overline{\pi}_{Z|X}$ in $\pi_{Z|X}$ is an open subgroup. We obtain the following diagram.

$$
\begin{array}{ccc}
Z(k) & \xrightarrow{\;\;\;f\;\;\;} & X(k) \\
\Big\downarrow{\scriptstyle \kappa_Z} & & \Big\downarrow{\scriptstyle \kappa_X} \\
\mathscr{S}_{\pi_1(Z/k)} & \longrightarrow\;\; \mathscr{S}_{\pi_{Z|X}\to\mathrm{Gal}_k} \;\;\longrightarrow & \mathscr{S}_{\pi_1(X/k)}
\end{array}
$$

Again replacing k by a finite extension we can pick $a, b \in Z(k)$ with $f(a) \neq f(b)$ and such that the sections s_a, s_b of $\pi_1(Z/k)$ induce sections $\tilde{s}_a, \tilde{s}_b$ of (7.11) with image in C. The difference cocycle $\delta_{\tilde{s}_a, \tilde{s}_b}$ thus yields a homomorphism

$$
\delta : \mathrm{Gal}_k \to \overline{\pi}_{Z|X}
$$

with image in the center of $\overline{\pi}_{Z|X}$

$$
Z(\overline{\pi}_{Z|X}) = C \cap \overline{\pi}_{Z|X}
$$

and which therefore is independent of the chosen representatives of sections in their conjugacy classes. By possibly extending k a third time we may assume that $\delta = 0$ and a forteriori

$$
s_{f(a)} = s_{f(b)}
$$

which contradicts the injectivity of κ_X in light of $f(a) \neq f(b)$.   □

Corollary 78. *Let X/k be a geometrically connected variety such that for every finite extension k'/k the profinite Kummer map*

$$
\kappa : X(k') \to \mathscr{S}_{\pi_1(X/k)}(k')
$$

is injective. Then X does not contain rational curves, i.e., every map $\mathbb{P}^1_k \to \overline{X}$ is constant. More generally, every map $\overline{Z} \to \overline{X}$ from a simply connected variety $\overline{Z}$ must be constant.

Proof. $\mathbb{P}^1$ is simply connected, and a variety with a large algebraic fundamental group does only allow constant maps from simply connected varieties.   □

Corollary 79. *Let X/k be a smooth projective geometrically connected variety such that for every finite extension k'/k the profinite Kummer map*

$$
\kappa : X(k') \to \mathscr{S}_{\pi_1(X/k)}(k')
$$

is injective. Then X does not contain movable pointed curves, i.e., for every proper smooth curve $C/\bar{k}$ with a closed point $c \in C$ and every map $f : C \times T \to \overline{X}$ with constant restriction $f|_{c \times T}$, the family f is constant, i.e., factors over the projection $\mathrm{pr}_1 : C \times T \to C$.

In particular, X is minimal in the sense of the minimal model program of birational geometry.

Proof. Otherwise the family would break by Mori's bend and break lemma, see [De01] Proposition 3.1, and produce a rational curve in X. The asserted minimality is equivalent to the canonical divisor ω_X of X being numerically effective, which follows from Miyaoka–Mori's Theorem [De01] Theorem 3.6, because otherwise again X would accommodate a rational curve. $\square$

Chapter 8
Reduction of Sections

Let $S = \mathrm{Spec}(R)$ be the spectrum of an excellent henselian discrete valuation ring with generic point $\mathrm{Spec}(k)$ and closed point $\mathrm{Spec}(\mathbb{F})$ with $\mathbb{F}$ a perfect field of characteristic $p \geq 0$. For a geometrically connected, proper variety X/k, which is the generic fibre $j : X \subseteq \mathscr{X}$ of a proper flat model $f : \mathscr{X} \to S$ with $f_* \mathcal{O}_{\mathscr{X}} = \mathcal{O}_S$, we have a specialisation map

$$X(k) = \mathscr{X}(R) \to Y(\mathbb{F}),$$

where $Y = \mathscr{X}_{\mathbb{F},\mathrm{red}}$ is the underlying reduced subscheme of the special fibre $\mathscr{X}_{\mathbb{F}}$.

The corresponding structure for sections of $\pi_1(X/k)$ distinguishes between sections that specialise like rational points, namely the sections with vanishing ramification, see Definition 82, and the ramified sections. In the case of curves, the ramication must be purely wild, see Proposition 91, but the case of wild ramification has not been excluded yet.

8.1 Specialisation

The specialisation homomorphism for fundamental groups was introduced and studied in [SGA1]. Here we study this map in the context of sections.

The specialisation map. It follows from [SGA1] X Theorem 2.1 and Artin approximation that the inclusion $i : Y \hookrightarrow \mathscr{X}$ induces an isomorphism

$$\pi_1(i) : \pi_1(Y) \xrightarrow{\sim} \pi_1(\mathscr{X}).$$

The specialisation map of fundamental groups, see [SGA1] X Corollary 2.4, is the map

$$\mathrm{sp} = \pi_1(i)^{-1} \circ \pi_1(j) \tag{8.1}$$

J. Stix, *Rational Points and Arithmetic of Fundamental Groups*, Lecture Notes in Mathematics 2054, DOI 10.1007/978-3-642-30674-7_8,
© Springer-Verlag Berlin Heidelberg 2013

which is part of the following map of extensions

$$
\begin{array}{ccccccccc}
1 & \longrightarrow & \pi_1(\overline{X}) & \longrightarrow & \pi_1(X) & \longrightarrow & \mathrm{Gal}_k & \longrightarrow & 1 \qquad (8.2)\\
& & \downarrow{\scriptstyle \mathrm{sp}} & & \downarrow{\scriptstyle \mathrm{sp}} & & \downarrow & & \\
1 & \longrightarrow & \pi_1(\overline{Y}) & \longrightarrow & \pi_1(Y) & \longrightarrow & \mathrm{Gal}_{\mathbb{F}} & \longrightarrow & 1.
\end{array}
$$

The *inertia group* I_k is the kernel of the natural map $\mathrm{Gal}_k \twoheadrightarrow \mathrm{Gal}_{\mathbb{F}}$, which is the absolute Galois group of the field of fractions k^{nr} of the strict henselisation R^{sh}.

Lemma 80. *If $\mathscr{X}$ is normal, then $\mathrm{sp}(\pi_1(\overline{X}))$ is a normal subgroup in $\pi_1(Y)$ of finite index in $\pi_1(\overline{Y})$.*

Proof. This is contained in [Ra70] Proposition 6.3.5(i). As an alternative, we argue by inspection of (8.2). Since $\mathscr{X}$ is normal, the map $\pi_1(j) : \pi_1(X) \twoheadrightarrow \pi_1(\mathscr{X})$ and thus the specialisation map

$$
\mathrm{sp} : \pi_1(X) \twoheadrightarrow \pi_1(Y)
$$

are surjective. The assertion of the image being a normal subgroup follows and finiteness of the index will be deduced in Section 8.3 from Lemma 86 below. $\square$

The cokernel of the geometric specialisation homomorphism. From now on we assume that $\mathscr{X}$ is normal. We set

$$
R_{\mathscr{X}/S} = \pi_1(\overline{Y})/\mathrm{sp}(\pi_1(\overline{X})) \qquad (8.3)
$$

and consider the pushout of $\pi_1(Y/\mathbb{F})$ via $\pi_1(\overline{Y}) \twoheadrightarrow R_{\mathscr{X}/S}$, namely the extension

$$
1 \to R_{\mathscr{X}/S} \to G_{\mathscr{X}/S} \to \mathrm{Gal}_{\mathbb{F}} \to 1.
$$

The group $R_{\mathscr{X}/S}$ is the main object of study in Wolfrath's thesis [Wo09] and is called there the "Scheingeometrische Fundamentalgruppe". It was previously studied by Raynaud in [Ra70].

Lemma 81. *Diagram (8.2) induces canonically a diagram*

$$
\begin{array}{ccccccccc}
1 & \longrightarrow & I_k & \longrightarrow & \mathrm{Gal}_k & \longrightarrow & \mathrm{Gal}_{\mathbb{F}} & \longrightarrow & 1 \qquad (8.4)\\
& & \downarrow{\scriptstyle \rho} & & \downarrow & & \| & & \\
1 & \longrightarrow & R_{\mathscr{X}/S} & \longrightarrow & G_{\mathscr{X}/S} & \longrightarrow & \mathrm{Gal}_{\mathbb{F}} & \longrightarrow & 1
\end{array}
$$

in which the map $\rho = \rho_{\mathscr{X}/S}$ is surjective. The group $R_{\mathscr{X}/S}$ is an extension of a cyclic prime to p group by a p group.

Proof. We quotient out $\pi_1(\overline{X})$ and its image from (8.2) and perform a diagram chase. Since the group I_k is the extension of $\hat{\mathbb{Z}}'(1)$ by a pro-p group, the claim on the structure of $R_{\mathscr{X}/S}$ follows.   □

8.2   The Ramification of a Section and Unramified Sections

Proper behaviour of a section under specialisation is controlled by the restriction to the inertia subgroup.

Definition 82. The *ramification* of a section $s : \mathrm{Gal}_k \to \pi_1(X)$ is the map

$$r = \mathrm{ram}(s) : I_k \to \pi_1(\overline{Y})$$

defined as the restriction of $\mathrm{sp} \circ s$ to I_k.

The ramification of a section vanishes exactly for those sections which specialise to a section $s_0 : \mathrm{Gal}_{\mathbb{F}} \to \pi_1(Y)$, i.e., allow a commutative diagram

$$
\begin{array}{ccccccccc}
1 & \longrightarrow & \pi_1(\overline{X}) & \longrightarrow & \pi_1(X) & \overset{\overset{s}{\longleftarrow}}{\longrightarrow} & \mathrm{Gal}_k & \longrightarrow & 1 \\
 & & \downarrow{\scriptstyle \mathrm{sp}} & & \downarrow{\scriptstyle \mathrm{sp}} & {\scriptstyle s_0} & \downarrow & & \\
1 & \longrightarrow & \pi_1(\overline{Y}) & \longrightarrow & \pi_1(Y) & \overset{\longleftarrow}{\longrightarrow} & \mathrm{Gal}_{\mathbb{F}} & \longrightarrow & 1.
\end{array}
$$

The ramification of a section s sits in a commutative diagram

$$
\begin{array}{ccccccccc}
1 & \longrightarrow & I_k & \longrightarrow & \mathrm{Gal}_k & \longrightarrow & \mathrm{Gal}_{\mathbb{F}} & \longrightarrow & 1 \\
 & & \downarrow{\scriptstyle \mathrm{ram}_s} & & \downarrow{\scriptstyle \mathrm{sp}\circ s} & & \| & & \\
1 & \longrightarrow & \pi_1(\overline{Y}) & \longrightarrow & \pi_1(Y) & \longrightarrow & \mathrm{Gal}_{\mathbb{F}} & \longrightarrow & 1.
\end{array}
$$

Consequently, the ramification ram_s is a $\mathrm{Gal}_{\mathbb{F}}$-equivariant homomorphism.

Let $h : X' \to X$ be a neighbourhood of the section s with a section s' as a lift. Let $\mathscr{X}' \to \mathscr{X}$ be the normalization of $\mathscr{X}$ in X', and let Y' be the reduced special fibre of $\mathscr{X}'$. The basic naturality of the ramification homomorphism is given by the following commutative diagram.

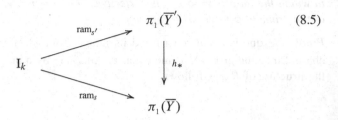

$$\text{(8.5)}$$

Definition 83. An *unramified section* is a section $s : \text{Gal}_k \to \pi_1(X)$ such that its ramification is the zero map, or equivalently, a section which specialises. The space of all $\pi_1(\overline{X})$-conjugacy classes of unramified sections of $\pi_1(X/k)$ will be denoted by $\mathscr{S}^{\text{nr}}_{\pi_1(X/k)}$.

The subset $\mathscr{S}^{\text{nr}}_{\pi_1(X/k)}$ is closed in $\mathscr{S}_{\pi_1(X/k)}$, because it is the fibre above the trivial group of the continuous map

$$\mathscr{S}_{\pi_1(X/k)} \to \text{Sub}(\pi_1(\overline{Y}))/\pi_1(\overline{X}) - \text{conjugacy}$$

which maps a section s to $\text{ram}_s(I_k)$.

Lemma 84. *The section s_{a_k} associated to a rational point $a_k \in X(k)$ is unramified:*

$$X(k) \subseteq \mathscr{S}^{\text{nr}}_{\pi_1(X/k)} \subseteq \mathscr{S}_{\pi_1(X/k)}.$$

Proof. The rational point a_k extends by the valuative criterion of properness to an S-valued point $a \in \mathscr{X}(S)$. Therefore s_{a_k} specialises to $s_{a_{\mathbb{F}}}$, for the specialisation $a_{\mathbb{F}} \in Y(\mathbb{F})$ of a_k, by functoriality of π_1.   □

Lemma 85. *Let $\mathscr{X} \to S$ be a relative curve. A section $s : \text{Gal}_k \to \pi_1(X)$ is unramified if its restriction $s' = s \otimes k' = s|_{\text{Gal}_{k'}}$ for a finite extension k'/k is unramified.*

Proof. Let Y' be the reduced special fibre of the normalization $\mathscr{X}'$ of $\mathscr{X}$ in $X \times_k k'$. Then we have a commutative diagram

$$
\begin{array}{ccc}
I_{k'} & \xrightarrow{\text{ram}_{s'}} & \pi_1(\overline{Y'}) \\
\downarrow & & \downarrow \\
I_k & \xrightarrow{\text{ram}_s} & \pi_1(\overline{Y}).
\end{array}
$$

If $\text{ram}_{s'} = 0$, then $\text{ram}_s(I_k)$ is finite, hence vanishes as $\pi_1(\overline{Y})$ is torsion free.   □

The converse of Lemma 85 however is unclear.

Lemma 86. *(1) Let s : $\mathrm{Gal}_k \to \pi_1(X)$ be a section. Then ρ factors over the ramification ram_s as follows*

$$\rho : I_k \xrightarrow{\ \mathrm{ram}_s\ } \pi_1(\overline{Y}) \twoheadrightarrow R_{\mathcal{X}/S}.$$

(2) If $\pi_1(X/k)$ admits an unramified section, then $\mathrm{sp} : \pi_1(\overline{X}) \twoheadrightarrow \pi_1(\overline{Y})$ is surjective.

Proof. (1) Obvious by a diagram chase, and (2) follows from (1) because ρ is surjective. □

Next we discuss a stabilized converse to Lemma 86 (2), which justifies our interest in the group $R_{\mathcal{X}/S}$.

Proposition 87. *Let s : $\mathrm{Gal}_k \to \pi_1(X)$ be a section. We assume that for all neighbourhoods $X' \to X$ of s the specialisation map $\mathrm{sp} : \pi_1(\overline{X'}) \twoheadrightarrow \pi_1(\overline{Y'})$ is surjective, where Y' is the reduced special fibre of the normalization $\mathcal{X}'$ of $\mathcal{X}$. Then the section s and its lifts to neighbourhoods are unramified.*

Proof. It clearly suffices to argue for s itself. Let $H \subset \pi_1(\overline{Y})$ be a characteristic open subgroup. Then $\mathrm{sp}^{-1}(H) \subset \pi_1(\overline{X})$ is a normal subgroup in $\pi_1(X)$ and thus $\mathrm{sp}^{-1}(H) \cdot s(\mathrm{Gal}_k)$ defines a neighbourhood $h : X' \to X$ of the section s with $\pi_1(\overline{X'}) = \mathrm{sp}^{-1}(H)$. By the naturality of the ramification map (8.5) and since $\mathrm{sp} : \pi_1(\overline{X'}) \to \pi_1(\overline{Y'})$ is assumed to be surjective, we compute

$$\mathrm{ram}_s(I_k) \subseteq \mathrm{im}\big(\pi_1(\overline{Y'}) \to \pi_1(\overline{Y})\big) = \mathrm{im}\big(\pi_1(\overline{X'}) \to \pi_1(\overline{Y})\big) = \mathrm{sp}(\mathrm{sp}^{-1}(H)) = H.$$

In the limit over all H we find

$$\mathrm{ram}_s(I_k) \subset \bigcap_H H = (1)$$

and ram_s must be the trivial map, so that s is unramified. □

A more careful examination of the construction in the proof of Proposition 87 yields a commutative diagram

$$
\begin{array}{ccc}
\pi_1(\overline{Y'}) & \longrightarrow & R_{\mathcal{X}'/S} \\
{\scriptstyle \mathrm{ram}_{s'}} \nearrow \quad \downarrow & & \downarrow \\
I_k \xrightarrow[\mathrm{ram}_s]{} \ \pi_1(\overline{Y}) & \longrightarrow & \pi_1(\overline{Y})/H.
\end{array}
$$

In the limit over all H we obtain a factorisation

$$\mathrm{ram}_s : I_k \twoheadrightarrow \varprojlim_H R_{\mathcal{X}/S} \to \pi_1(\overline{Y}).$$

σ-**specialisation of sections.** The inertia group I_k is the semi-direct product of its p-Sylow group V_k and the tame inertia quotient group

$$I_k^{\text{tame}} = \hat{\mathbb{Z}}'(1) = \prod_{\ell \neq p} \mathbb{Z}_\ell(1).$$

The quotient map $I_k \twoheadrightarrow I_k^{\text{tame}}$ splits, and the images of splittings are called *tame inertia complements*.

Theorem 88 (Melnikov, Tavgen [MeTa85]). *The following extension splits:*

$$1 \to I_k \to \text{Gal}_k \to \text{Gal}_{\mathbb{F}} \to 1.$$

Proof. The crucial part is the splitting of the extension

$$1 \to V_k \to \text{Gal}_k \to \text{Gal}_k^{\text{tame}} \to 1$$

which Kuhlmann attributes to Pank, see [KPR86] Theorem 2.2. Independently, the result has been obtained by Melnikov and Tavgen [MeTa85].   □

A choice of splitting $\sigma : \text{Gal}_{\mathbb{F}} \to \text{Gal}_k$ leads to a specialisation map

$$\text{sp}_\sigma : \mathscr{S}_{\pi_1(X/k)} \to \mathscr{S}_{\pi_1(Y/\mathbb{F})} \tag{8.6}$$
$$\text{sp}_\sigma(s) = \text{sp} \circ s \circ \sigma.$$

At least for the unramified sections s the specialisation $\text{sp}_\sigma(s)$ is independent of the choice of σ.

Corollary 89. *If* $\pi_1(X/k)$ *splits, then* $\pi_1(Y/\mathbb{F})$ *splits as well.*   □

8.3   Bounds for the Cokernel of Specialisation

Lemma 86 allows to reprove the finiteness assertion of Lemma 80 as follows. After a finite extension k'/k we have $a' \in X(k')$ and $\pi_1(X_{k'}/k')$ has an unramified section. Let $\mathscr{X}'$ be the normalization $\mathscr{X}'$ of $\mathscr{X}$ in $X_{k'}$, and let Y' be the reduced special fibre of $\mathscr{X}'$. By Lemma 86 (2), the specialisation map

$$\text{sp}_{\mathscr{X}'} : \pi_1(\overline{X}) \to \pi_1(\overline{Y'})$$

is surjective. The map $\pi_1(\overline{Y'}) \to \pi_1(\overline{Y})$ has an image of finite index as the map of fibres is dominant. Thus the composite

$$\text{sp}_{\mathscr{X}} : \pi_1(\overline{X}) \to \pi_1(\overline{Y'}) \to \pi_1(\overline{Y})$$

has a finite cokernel $R_{\mathcal{X}/S}$. This argument for the finiteness of $R_{\mathcal{X}/S}$ proves the following more precise bound, see also [SGA1] Exposé X Corollary 2.4 for (3).

Proposition 90. *(1) Let k'/k be a finite extension with $X(k') \neq \emptyset$. Then the extension $k_{\mathcal{X}/S}/k^{\mathrm{nr}}$ corresponding to $\rho : I_k \twoheadrightarrow R_{\mathcal{X}/S}$ is contained in $k^{\mathrm{nr}}k'$.*
(2) The order $\#R_{X/S}$ divides $\gcd(e_\alpha)$, where the e_α are the multiplicities of the irreducible components Y_α of the special fibre Y.
(3) If $\gcd(e_\alpha) = 1$, then the specialisation map $\mathrm{sp} : \pi_1(\overline{X}) \to \pi_1(\overline{Y})$ is surjective.

Proof. (1) Let S' be the normalization of S in k', let $\mathcal{X}'/S'$ be the normalization of $\mathcal{X}$ in $X_{k'}$, and let Y' be the reduced special fibre of $\mathcal{X}'$. We have a commutative diagram

$$
\begin{array}{ccc}
I_{k'} & \xrightarrow{\;\rho_{\mathcal{X}'/S'}\;} & R_{\mathcal{X}'/S'} \\
\downarrow & & \downarrow \\
I_k & \xrightarrow{\;\rho_{\mathcal{X}/S}\;} & R_{\mathcal{X}/S}
\end{array}
$$

in which $\rho_{\mathcal{X}'/S'} = 0$ by Lemma 86. The inclusion $I_{k'} \subset \ker(\rho_{\mathcal{X}/S})$ translates into

$$k_{\mathcal{X}/S} \subset k^{\mathrm{nr}}k'.$$

(2) The degree $[k^{\mathrm{nr}}k' : k^{\mathrm{nr}}]$ equals the ramification index $e(k'/k)$. In order to deduce (2) from (1) it therefore suffices to construct for every irreducible component Y_α of Y a field extension k'/k with $e(k'/k) = e_\alpha$ and $X(k') \neq \emptyset$. We construct such a point by a transversal construction.

As $\mathcal{X}$ is normal and $\mathbb{F}$ is perfect, the locus where both $\mathcal{X}$ is regular and $Y/\mathbb{F}$ is smooth meets every component of Y nontrivially. Let $y \in Y_\alpha$ be a closed point such that $\mathcal{X}$ is regular in y and $Y_\alpha/\mathbb{F}$ is smooth in y. Then choose an sequence of parameters $t, f_1, \ldots, f_n$ for $\mathcal{O}_{\mathcal{X},y}$ with Y being described by $t = 0$ near y. The vanishing locus of $f_1, \ldots, f_n$ describes a closed subscheme

$$S' \subset \mathrm{Spec}(\mathcal{O}_{\mathcal{X},y})$$

which is regular and finite flat over S and thus describes also a closed subscheme $S' \subset \mathcal{X}$. The extension of discrete valuation rings corresponding to S'/S has

$$e = e_\alpha = (S' \bullet Y_\alpha),$$

and this is what we were looking for.

(3) is an immediate consequence of (2). $\qquad\square$

The work of Raynaud [Ra70] contains more detailed information on $R_{\mathscr{X}/S}$. By [Ra70] Proposition 6.3.5(iii), we have a short exact sequence of finite groups

$$1 \to R_{\mathscr{X}/S,p} \to R_{\mathscr{X}/S} \to \mu_{d'}(\overline{\mathbb{F}}) \to 1$$

where $R_{\mathscr{X}/S,p}$ is the p-Sylow group of $R_{\mathscr{X}/S}$ and where d' is the maximal prime to p integer such that $\frac{1}{d'}Y$ is a Cartier-divisor on $\mathscr{X}$. In the regular case this is nothing but the prime to p-part of $\gcd(e_\alpha)$ for the multiplicities e_α of the components of the special fibre Y. In particular, in the regular case the bound provided by Proposition 90 (2) is sharp with respect to the prime to p part.

The maximal abelian quotient $R^{\mathrm{ab}}_{\mathscr{X}/S}$ of $R_{\mathscr{X}/S}$ finds a description in [Ra70] Proposition 6.3.5(iv) as the Cartier dual to the maximal multiplicative subgroup of the special fibre $E_{\mathbb{F}}$ of the closure E of $0 \in \underline{\mathrm{Pic}}_{\mathscr{X}/S}$.

In [Ra70] §9.6, Raynaud constructs examples where $R_{\mathscr{X}/S}$ is a non-commutative p-group or a non-cyclic abelian p-group.

8.4   Specialisation for Curves

We turn our attention to the case of curves, the relevant case for the section conjecture, and exploit a special property of the geometric fundamental group of curves: the maximal abelian quotient of an open subgroup is always torsion free.

No tame ramification on curves. We first address tame ramification and show that in the proper case sections cannot be tamely ramified essentially due to weight arguments.

Proposition 91. *In addition to the above we assume that* $f : \mathscr{X} \to S$ *is a proper curve and that the residue field* $\mathbb{F}$ *is finitely generated over its prime field.*

Let $s : \mathrm{Gal}_k \to \pi_1(X)$ *be a section. Then* $\mathrm{ram}_s(I_k)$ *is a free pro-p group and* $R_{\mathscr{X}/S}$ *is a finite p-group.*

Proof. We argue by contradiction and assume that $\ell \neq p$ divides the order of $\mathrm{ram}_s(I_k)$. We replace Y, and thus X and $\mathscr{X}$ by a suitable finite étale cover corresponding to an open subgroup H with

$$\mathrm{ram}_s(I_k) \subset H \subset \pi_1(Y)$$

in such a way, that ram_s induces a nontrivial $\mathrm{Gal}_{\mathbb{F}}$-equivariant map

$$\mathrm{ram}_s^{\mathrm{ab},\ell} : \mathbb{Z}_\ell(1) \to \pi_1^{\mathrm{ab}}(\overline{Y}) \otimes \mathbb{Z}_\ell. \tag{8.7}$$

The group $\pi_1^{\mathrm{ab}}(\overline{Y})$ is a free $\mathbb{Z}_\ell$-module of finite rank. If $\mathbb{F}$ is finite, then $\mathrm{ram}_s^{\mathrm{ab},\ell}$ being nontrivial contradicts the Frobenius weights in étale cohomology of proper varieties over finite fields, because the weights on π_1^{ab} are ≥ -1 while $\mathbb{Z}_\ell(1)$ has weight -2.

If $\mathbb{F}$ is finitely generated over its prime field, then by the usual spreading out and specialisation technique we can also apply the weight argument above in this case.

If $p = 0$, then we have shown that $\mathrm{ram}_s = 0$. Otherwise, if $p > 0$, then the p-cohomological dimension of $\pi_1(\overline{Y})$ is ≤ 1. Hence we have

$$\mathrm{cd}_p(\mathrm{ram}_s(I_k)) \leq 1$$

and $\mathrm{ram}_s(I_k)$ is a free pro-p group. $\qquad\qquad\qquad\qquad\qquad\qquad\qquad\qquad$ □

We introduce some notation. With the characteristic quotient

$$\mathrm{GI}_k^{(p)} = \mathrm{Gal}_k / \ker\left(I_k \twoheadrightarrow I_k^{\mathrm{pro}\text{-}p}\right)$$

we have the maximal inertia-pro-p extension

$$1 \to I_k^{\mathrm{pro}\text{-}p} \to \mathrm{GI}_k^{(p)} \to \mathrm{Gal}_{\mathbb{F}} \to 1.$$

Similarly, with the characteristic quotient

$$\mathrm{GV}_k^{(p)} = \mathrm{Gal}_k / \ker\left(V_k \twoheadrightarrow I_k^{\mathrm{pro}\text{-}p}\right)$$

we have the extension

$$1 \to I_k^{\mathrm{pro}\text{-}p} \to \mathrm{GV}_k^{(p)} \to \mathrm{Gal}_k^{\mathrm{tame}} \to 1.$$

Corollary 92. *We keep the assumptions of Proposition 91. The σ-specialisation*

$$\mathrm{sp}_\sigma : \mathscr{S}_{\pi_1(X/k)} \to \mathscr{S}_{\pi_1(Y/\mathbb{F})}$$

for a $\sigma : \mathrm{Gal}_{\mathbb{F}} \to \mathrm{Gal}_k$ depends only on the induced section $\bar{\sigma} : \mathrm{Gal}_{\mathbb{F}} \to \mathrm{GI}_k^{(p)}$.

Lemma 93. *Any section $\bar{\sigma} : \mathrm{Gal}_{\mathbb{F}} \to \mathrm{GI}_k^{(p)}$ is induced by a composite section*

$$\sigma : \mathrm{Gal}_{\mathbb{F}} \xrightarrow{\sigma_1} \mathrm{Gal}_k^{\mathrm{tame}} \xrightarrow{\sigma_2} \mathrm{Gal}_k$$

and depends only on σ_2.

Proof. A diagram chase in the commutative diagram

$$
\begin{array}{ccccccccc}
1 & \longrightarrow & I_k^{\mathrm{pro}\text{-}p} & \longrightarrow & \mathrm{GV}_k^{(p)} & \longrightarrow & \mathrm{Gal}_k^{\mathrm{tame}} & \longrightarrow & 1 \\
 & & \| & & \downarrow & & \downarrow & & \\
1 & \longrightarrow & I_k^{\mathrm{pro}\text{-}p} & \longrightarrow & \mathrm{GI}_k^{(p)} & \longrightarrow & \mathrm{Gal}_{\mathbb{F}} & \longrightarrow & 1.
\end{array}
$$

using the fact that

$$I_k^{\text{tame}} = \ker\left(\text{Gal}_k^{\text{tame}} \to \text{Gal}_{\mathbb{F}}\right)$$

is prime to p yields that a section $\bar{\sigma}$ of $\text{GI}_k^{(p)} \to \text{Gal}_{\mathbb{F}}$ corresponds uniquely to a section $\tilde{\sigma}$ of $\text{GV}_k^{(p)} \to \text{Gal}_k^{\text{tame}}$. As

$$\text{cd}_p(\text{Gal}_k^{\text{tame}}) = 1,$$

any section $\tilde{\sigma}$ of $\text{GV}_k^{(p)} \to \text{Gal}_k^{\text{tame}}$ lifts to a section σ_2 of $\text{Gal}_k^{\text{tame}} \to \text{Gal}_k$. With respect to any section $\sigma_1 : \text{Gal}_{\mathbb{F}} \to \text{Gal}_k^{\text{tame}}$ we find that $\sigma_2 \circ \sigma_1$ induces $\bar{\sigma}$.   □

Wild ramification on curves: Abelian approach. We restrict our attention to the case where $R = \mathfrak{o}_k$ is the ring of integers in a finite extension $K/\mathbb{Q}_p$. Unfortunately, wild ramfication of a section, i.e., that $\text{ram}_s(I_k)$ is a non-trivial p-group, cannot be excluded so far. Nevertheless, we would like to present two structures which originate from the ramification map.

First, we consider the abelian approach. If ram_s does not vanish, then after possibly replacing $\mathscr{X}$ by a finite étale cover $\mathscr{X}' \to \mathscr{X}$, whose generic fibre is a neighbourhood of the section s, we find a nontrivial $\text{Gal}_{\mathbb{F}}$-equivariant *abelian ramification* map

$$\text{ram}_s^{\text{ab},p} \; : \; I_k^{\text{pro-}p,\text{ab}} \to \pi_1^{\text{ab}}(\overline{Y}) \otimes \mathbb{Z}_p.$$

Via class field theory and the p-adic logarithm we find that $I_k^{\text{pro-}p,\text{ab}}$ contains a submodule isomorphic to $\mathfrak{o}_k[[\text{Gal}_{\mathbb{F}}]]$, see [NSW08] Lemma 7.4.4, such that in the exact sequence

$$0 \to \mathfrak{o}_k[[\text{Gal}_{\mathbb{F}}]] \to I_k^{\text{pro-}p,\text{ab}} \to T \to 0 \tag{8.8}$$

the cokernel T is killed by a power of p that depends only on the absolute ramification $e(k/\mathbb{Q}_p)$. Since

$$\text{Hom}_{\text{Gal}_{\mathbb{F}}}\left(T, \pi_1^{\text{ab}}(\overline{Y})\right) = 0$$

we find from the $\text{Ext}_{\text{Gal}_{\mathbb{F}}}(-, \pi_1^{\text{ab}}(\overline{Y}))$-sequence applied to (8.8)

$$0 \to \text{Hom}_{\text{Gal}_{\mathbb{F}}}\left(I_k^{\text{pro-}p,\text{ab}}, \pi_1^{\text{ab}}(\overline{Y})\right) \to \text{Hom}\left(\mathfrak{o}_k, \pi_1^{\text{ab}}(\overline{Y})\right) \to \text{Ext}_{\text{Gal}_{\mathbb{F}}}^1\left(T, \pi_1^{\text{ab}}(\overline{Y})\right)$$

that $\text{Hom}_{\text{Gal}_{\mathbb{F}}}\left(I_k^{\text{pro-}p,\text{ab}}, \pi_1^{\text{ab}}(\overline{Y})\right)$ is of finite index in $\text{Hom}\left(\mathfrak{o}_k, \pi_1^{\text{ab}}(\overline{Y})\right)$ and hence a free $\mathbb{Z}_p$-module of finite rank. In particular, the section s induces a $\mathbb{Z}_p$-linear map

$$\mathfrak{o}_k \to \pi_1^{\text{ab}}(\overline{Y})$$

which behaves well under unramified base change and geometrically connected covers $X' \to X$, which are neighbourhoods of s. Its vanishing in all cases is equivalent to the vanishing of the ramification of a section in all cases and also to the surjectivity of the specialisation map in case that a section exists.

Mimicking the argument of the tame case fails badly. Although there is some restricting information known on the $\mathrm{Gal}_{\mathbb{F}}$-action on $\pi_1^{\mathrm{ab}}(\overline{Y})$, namely the action is with weights -1 and 0 and corresponds to the slope 0 part of crystalline cohomology, the hope is dashed by the fact that $I_k^{\mathrm{pro}\text{-}p,\mathrm{ab}}$ provides any irreducible finite rank $\mathbb{Z}_p[\mathrm{Gal}_{\mathbb{F}}]$-module among its quotients.

Wild ramification on curves: Non-abelian approach. By Proposition 91, the ramification of a section factorizes through the map r in

$$
\begin{array}{ccccccccc}
1 & \longrightarrow & I_k^{\mathrm{pro}\text{-}p} & \longrightarrow & \mathrm{GI}_k^{(p)} & \longrightarrow & \mathrm{Gal}_{\mathbb{F}} & \longrightarrow & 1 \\
& & \downarrow{\scriptstyle r} & & \downarrow & & \downarrow & & \\
1 & \longrightarrow & \pi_1(\overline{Y}) & \longrightarrow & \pi_1(Y) & \longrightarrow & \mathrm{Gal}_{\mathbb{F}} & \longrightarrow & 1.
\end{array}
$$

The group $I_k^{\mathrm{pro}\text{-}p}$ is a free pro-p group. Therefore the existence of a nontrivial map r as in the diagram yields no immediate constraint.

If we push further to the maximal pro-p quotient of $\pi_1(Y)$ and if we assume, as is anyway true after a finite unramified extension of k, that the action of $\mathrm{Gal}_{\mathbb{F}}$ on $\pi_1^{\mathrm{pro}\text{-}p}(\overline{Y})$ is through a pro-p group, then we obtain a diagram

$$
\begin{array}{ccccccccc}
1 & \longrightarrow & I_{k(p)/k} & \longrightarrow & \mathrm{Gal}_k^{\mathrm{pro}\text{-}p} & \longrightarrow & \mathbb{Z}_p & \longrightarrow & 1 \\
& & \downarrow{\scriptstyle r} & & \downarrow & & \downarrow & & \\
1 & \longrightarrow & \pi_1^{\mathrm{pro}\text{-}p}(\overline{Y}) & \longrightarrow & \pi_1^{\mathrm{pro}\text{-}p}(Y) & \longrightarrow & \mathbb{Z}_p & \longrightarrow & 1
\end{array}
$$

where $I_{k(p)/k}$ is the inertia group in the maximal p extension $k(p)/k$ of k.

If we assume that $\mu_p \subset k$, which is legitimate if we want to prove that sections are unramified by Lemma 85, then $\mathrm{Gal}_k^{\mathrm{pro}\text{-}p}$ is a Demuškin group, and hence has a known description

$$
\mathrm{Gal}_k^{\mathrm{pro}\text{-}p} = \langle X_0, X_1, \ldots, X_n, X_{n+1} \mid X_0^q (X_0, X_1)(X_2, X_3) \ldots (X_{2n}, X_{2n+1}) \rangle
$$

in terms of generators and relations, see [NSW08] Theorem 7.5.12. We ignore the case $p = 2$ for the time being.

Let $a_i \in \mathbb{Z}_p$ be the image of X_i under the map

$$
\mathrm{Gal}_k^{\mathrm{pro}\text{-}p} \twoheadrightarrow \mathrm{Gal}_{\mathbb{F}}^{\mathrm{pro}\text{-}p} = \mathbb{Z}_p.
$$

We necessarily have $a_0 = 0$. Let v be the p-adic valuation of k with the convention that $v(0) = \infty$. We set for $i = 1, \ldots, n$

$$\varepsilon_i = \begin{cases} a_{2i+1}/a_{2i} \in \mathbb{Z}_p & \text{if } a_{2i} \neq 0 \text{ and } v(a_{2i}) \leq v(a_{2i+1}) \\ a_{2i}/a_{2i+1} \in \mathbb{Z}_p & \text{if } v(a_{2i}) > v(a_{2i+1}) \\ 1 & \text{if } a_{2i} = a_{2i+1} = 0 \end{cases}$$

and construct the following free pro-p intermediate quotient F

$$\mathrm{Gal}_k^{\mathrm{pro}\text{-}p} \xrightarrow{\ \varphi\ } F = \langle Y_1, \dots, Y_n \rangle \xrightarrow{\ \psi\ } \mathrm{Gal}_{\mathbb{F}}^{\mathrm{pro}\text{-}p}$$

where

$$\varphi(X_{2i}), \varphi(X_{2i+1}) = \begin{cases} 0, Y_1 & \text{if } i = 0 \\ Y_i, Y_i^{\varepsilon_i} & \text{if } v(a_{2i}) \leq v(a_{2i+1}) \\ Y_i^{\varepsilon_i}, Y_i & \text{if } v(a_{2i}) > v(a_{2i+1}) \text{ and } i > 0 \end{cases}$$

and

$$\psi(Y_i) = \begin{cases} a_1 & \text{if } i = 0 \\ a_{2i} & \text{if } v(a_{2i}) \leq v(a_{2i+1}) \\ a_{2i+1} & \text{if } v(a_{2i}) > v(a_{2i+1}) \text{ and } i > 0. \end{cases}$$

Unfortunately, this free quotient F maps nontrivially in many ways to $\pi_1(Y)$, even augmented to $\mathbb{Z}_p = \mathrm{Gal}_{\mathbb{F}}^{\mathrm{pro}\text{-}p}$, and again we fail in our attempt to contradict a nontrivial wild part of the ramification map associated to a section.

8.5   Specialisation for Sections Associated to Points

Another viewpoint on specialisation reveals a limitation of variance of sections associated to rational points for some characteristic quotient. We remind the reader that $S = \mathrm{Spec}(R)$ is the spectrum of an excellent henselian discrete valuation ring with generic point $\mathrm{Spec}(k)$ and closed point $\mathrm{Spec}(\mathbb{F})$ with $\mathbb{F}$ a perfect field of characteristic $p \geq 0$.

Theorem 94. *Let $\mathscr{X} \to S$ be a smooth and separated map with geometrically connected generic fibre $X = \mathscr{X}_k$. Let $b, x \in \mathscr{X}(R)$ be integral points, and let b_k, x_k denote their restriction in $X(k)$. With the characteristic map*

$$\varphi : \pi_1(\overline{X}, b) \twoheadrightarrow \Delta(\overline{X}, b) := \left(\pi_1(\overline{X}, b) \right)^{\mathrm{pro}\text{-}p'}$$

to the maximal quotient of order prime to p, the quotient path torsor

$$\Delta(\overline{X}; b, x) := \varphi_*\left(\pi_1(\overline{X}; b, x) \right) \in \mathrm{H}^1\left(k, \Delta(\overline{X}, b) \right)$$

of the path torsor $\pi_1(\overline{X}; b, x)$ *is the trivial element if* b *and* x *have the same reduction in the special fibre* $\mathscr{X}_{\mathbb{F}}$.

Remark 95. (1) In the case of curves and limited to the pro-solvable prime to p quotient, Theorem 94 was obtained by Kim and Tamagawa in [KiTa08] Theorem 0.1. The above version is modeled on their result.

(2) In the case of good reduction, i.e., if $\mathscr{X} \to S$ is proper and smooth, then Theorem 94 follows from Lemma 84, because then the specialisation map induces an isomorphism

$$\mathrm{sp} : \pi_1(\overline{X}, b_k)^{\mathrm{pro}\text{-}p'} \xrightarrow{\sim} \pi_1(\overline{\mathscr{X}_{\mathbb{F}}}, b_{\mathbb{F}})^{\mathrm{pro}\text{-}p'}$$

of maximal prime to p quotients.

(3) In light of the pro-p injectivity of Theorem 76, the above Theorem 94 emphasizes the arithmetic content of the p-part versus the prime to p-part.

Proof. Let $b_{\mathbb{F}} = y = x_{\mathbb{F}}$ be the common reduction of b, x in the special fibre $\mathscr{X}_{\mathbb{F}}$ of $\mathscr{X}/S$. In rigid analytic terms this means that b_k and x_k both lie in the residue polydisc $D =]y[_{\mathscr{X}}$ of rigid analytic points that specialize to y. The map

$$f : D \subset \mathscr{X}^{\mathrm{rig}} \to X$$

induces a map of (quotients of) étale fundamental groups

$$\pi_1(\overline{D}, b) \xrightarrow{f_*} \pi_1(\overline{X}, b) \xrightarrow{\varphi} \Delta(\overline{X}, b)$$

and a corresponding Gal_k-equivariant map of path spaces

$$\pi_1(\overline{D}; b, x) \to \pi_1(\overline{X}; b, x) \to \Delta(\overline{X}; b, x),$$

from rigid analytic paths on the geometric polydisc $\overline{D} = D \times \bar{k}$ to prime to p and prime to p algebraic paths on $\overline{X}$. Since the triviality of $\Delta(\overline{X}; b, x)$ is equivalent to the existence of a Gal_k-invariant path, it suffices to show that the composition $\varphi \circ f_*$ is the trivial map.

Let $h : \overline{X}' \to \overline{X}$ be a connected, finite étale Galois cover of degree prime to p. Then, by Abhyankar's Lemma, after a suitable tamely ramified extension of S we may assume that there is a finite étale cover $\mathscr{X}' \to \mathscr{X}$ which induces h, and such that the fibre above y is completely decomposed. Then, the pullback to D

$$D \times_{\mathscr{X}} \mathscr{X}' = \bigsqcup_{y' \mapsto y}]y'[_{\mathscr{X}'}$$

consists of $\deg(h)$ many residue polydiscs. Hence, for all y' above y, the polydisc $]y'[_{\mathscr{X}'}$ projects isomorphically onto D, so that the cover $\overline{D} \times_{\mathscr{X}} \mathscr{X}' \to \overline{D}$ splits completely. As h is arbitrary, this splitting shows that $\varphi \circ f_*$ is the trivial map. $\square$

Chapter 9
The Space of Sections in the Arithmetic Case and the Section Conjecture in Covers

We resume the discussion of the space of sections from Chap. 4 under arithmetic assumptions. The space

$$\mathscr{S}_{\pi_1}(X/k)$$

turns out to be a compact profinite space if X/k is a smooth projective curve over an algebraic number field, see Proposition 97 or for any hyperbolic curve if k is a finite extension of $\mathbb{Q}_p$. We also recall the known relation between a weak form of the section conjecture, see Conjecture 100, and the genuine version, see Conjecture 2, and between the claim for affine versus proper curves.

In the arithmetic situation, the centraliser of a section is always trivial, see Proposition 104. It follows that, just like rational points, sections obey Galois descent, see Corollary 107. Furthermore, we examine *going up* and *going down* for the section conjecture with respect to a finite étale map. The discussion relies on Chap. 3.

9.1 Compactness of the Space of Sections in the Arithmetic Case

We consider the space of sections in the case where k is either a number field, or a p-adic local field or $\mathbb{R}$.

Proposition 96. *Let X be a geometrically connected variety over a p-adic or an archimedean local field k. Then the space of sections $\mathscr{S}_{\pi_1}(X/k)$ is a profinite topological space and in particular compact.*

Proof. As Gal_k is topologically finitely generated by [Ja88] Theorem 5.1(c), see also [NSW08] Theorem 7.4.1, we can apply Corollary 45. ☐

J. Stix, *Rational Points and Arithmetic of Fundamental Groups*, Lecture Notes in Mathematics 2054, DOI 10.1007/978-3-642-30674-7_9,
© Springer-Verlag Berlin Heidelberg 2013

Proposition 97. *Let X be a proper, geometrically connected variety over an algebraic number field k. Then the space of sections $\mathscr{S}_{\pi_1(X/k)}$ is a profinite topological space and in particular compact.*

Proof. By Lemma 44 we need to show the finiteness of the image of

$$\mathscr{S}_{\pi_1(X/k)} \to \mathscr{S}_{Q_n(\pi_1(X/k))}$$

for each n. Let

$$B_n \subseteq \mathrm{Spec}(\mathfrak{o}_k[\tfrac{1}{n!}])$$

be an open subset such that the extension

$$Q_n(\pi_1(X/k))$$

is induced from an extension of $\pi_1(B_n)$ by $Q_n(\pi_1(\overline{X}))$. Any section of $Q_n(\pi_1(X/k))$ coming from a section of $\pi_1(X/k)$ will be unramified at places in B_n by Proposition 91, because the order of $Q_n(\pi_1(\overline{X}))$ is only divisible by primes $\leq n$. Any two such sections thus differ by a non-abelian cohomology class in

$$H^1\left(B_n, Q_n(\pi_1(\overline{X}))\right),$$

which is a finite set by Hermite's Theorem: the finiteness of the set of field extensions of an algebraic number field with bounded degree and places of ramification, see Lemma 146 below. □

Remark 98. (1) Unlike a countable profinite group, a countable profinite space need not be finite. In fact, the space

$$M = \{1/n \; ; \; n \in \mathbb{N}\} \cup \{0\}$$

with the topology inherited as a subspace $M \subset \mathbb{R}$ is a countable profinite set. Hence, contrary to a long term belief, the topological result of Proposition 97 does not reprove the Faltings–Mordell Theorem stating the finiteness of $X(k)$ for a smooth, projective curve X of genus at least 2 over a number field k once the section conjecture is known for X/k.

(2) In the number field case, the compactness in Proposition 97 had been obtained independently by Deligne, see his letter to Thakur [Sx11] Appendix A. Deligne observes that the space

$$S = \kappa(X(k)) \subseteq \mathscr{S}_{\pi_1(X/k)}$$

of sections associated to rational points is *equicontinuous* as a subspace of the continuous group homomorphisms

$$\mathrm{Hom}(\mathrm{Gal}_k, \pi_1(X)).$$

Indeed, for every finite quotient $\varphi : \pi_1(X) \twoheadrightarrow G$ there is a finite extension k'/k such that the map $\varphi \circ s|_{\mathrm{Gal}_{k'}}$ is independent of $s \in S$. Our proof of Proposition 97 implies that even the full space $\mathscr{S}_{\pi_1(X/k)}$ of all sections is equicontinuous.

However, the result of Deligne in [Sx11] Appendix A holds more generally for a smooth geometrically connected variety over an algebraic number field.

9.2 Weak Versus Strong: The Arithmetic Case

We discuss the content of Proposition 54 in case the base field k is arithmetic.

Corollary 99 (Tamagawa). *Let X/k be a geometrically connected variety. If for every geometrically connected finite étale cover $h : X' \to X$ the set of rational points $X'(k)$ can be endowed with a compact topology such that the maps*

$$h : X'(k) \to X(k)$$

are continuous, then

$$\kappa : X(k) \to \mathscr{S}_{\pi_1(X/k)}$$

is surjective if and only if for every neighbourhood $X' \to X$ of a section s of $\pi_1(X/k)$ we have $X'(k) \neq \emptyset$.

Proof. Let $X_s = (X')$ be the decomposition tower of a section s. The claim follows because

$$X_s(k) = \varprojlim X'(k)$$

and by [Bo98] I §9.6 Proposition 8 a filtered projective limit of compact non empty sets is non empty. We then apply Proposition 54 (3) to deduce the surjectivity of the profinite Kummer map of X/k. $\qquad\square$

The *weak section conjecture* over a field k claims the following (and is meant to be applied to algebraic number fields k or finite extensions of $\mathbb{Q}_p$).

Conjecture 100. *A hyperbolic curve X/k has a rational point if $\pi_1(X/k)$ splits.*

The link between the weak section conjecture and the section conjecture itself is described by the following corollary.

Corollary 101 (Tamagawa). *Let k be an algebraic number field or a finite extension of $\mathbb{Q}_p$. Then the section conjecture holds for a smooth, projective curve X over k of genus at least 2, i.e., the profinite Kummer map*

$$\kappa : X(k) \to \mathscr{S}_{\pi_1(X/k)}$$

is bijective, if and only if the weak section conjecture holds for neighbourhoods of sections s of $\pi_1(X/k)$, i.e., for every neighbourhood $X' \to X$ of such a section s we have $X'(k) \neq \emptyset$.

Proof. The map κ is injective by Corollary 74, and κ is surjective, because we assume that for every section s of $\pi_1(X/k)$ the assumptions of Corollary 99 are satisfied. Indeed, if k is an algebraic number field, then for every neighbourhood X' the set $X'(k)$ is compact with the discrete topology by the Faltings–Mordell Theorem [Fa83]. And if k is a finite extension of $\mathbb{Q}_p$, then the set $X'(k)$ is compact with respect to the p-adic topology. □

Corollary 102. *In order to prove the section conjecture for all hyperbolic curves it suffices to treat the case of curves without rational points.*

Proof. This is simply a reformulation of the claim proved in Corollary 101 that the weak section conjecture implies the genuine section conjecture. □

9.3 Affine Versus Proper

Reducing the section conjecture for smooth projective curves of genus ≥ 2 to the case of affine hyperbolic curves requires to overcome serious obstructions. We have to lift a section $s : \mathrm{Gal}_k \to \pi_1(X)$ of a smooth projective curve X to a section of $\pi_1(U/k)$ for some nontrivial open subset $U \subset X$. This lifting, known as the cuspidalization problem, seems to be considerably obstructed with obstructions that carry substantial arithmetic information. If lifting to open subcurves is always possible, then the section conjecture for hyperbolic curves over algebraic number fields can be reduced to the case of

$$\mathbb{P}^1 - \{0, 1, \infty\}$$

over all algebraic number fields, see Esnault and Hai [EsHa08] Proposition 7.9. That the dévissage to the case of $\mathbb{P}^1 - \{0, 1, \infty\}$ requires varying the base field over all algebraic number fields has its source in the same variation occurring in the results of Sect. 9.5 on going up and down along étale covers. For further details we refer to Sect. 18.4.

The converse problem of deducing the section conjecture in the affine case from the section conjecture in the proper case is settled by the following proposition.

Proposition 103. *Let k be a number field or a finite extension of $\mathbb{Q}_p$. If the section conjecture is true for smooth proper geometrically connected curves X/k of genus at least 2, then the section conjecture is also true for affine hyperbolic curves U/k.*

Proof. Let U/k be an affine hyperbolic curve and let s be a section of $\pi_1(U/k)$. It suffices to work with suitable neighbourhoods of s. If s is different from any cuspidal section associated to a rational point in the boundary of a smooth projective completion, see Chap. 18, then, by the known injectivity of the profinite Kummer map and Proposition 54 (1), we find a neighbourhood $U' \to U$ of s such that U' has genus ≥ 2 and the boundary or U' has no k-rational point any more. In particular, now $U'(k)$ is compact with the discrete topology if k is an algebraic number field or the p-adic topology for a finite extension $k/\mathbb{Q}_p$. Now the argument of Corollary 99 applies and shows that $s = s_a$ for some $a \in X(k)$, provided we can show that for every finer neighbourhood

$$U'' \to U' \to U$$

of s the set $U''(k)$ is not empty. But the lift of s to a section s'' of $\pi_1(U''/k)$ induces a section $j_*(s'')$ of the smooth projective completion $j : U'' \subset X''$ of U'' to which we can apply the assumption. Hence the section $j(s'')$ forces $X''(k) \neq \emptyset$, and by construction of U' we have $U''(k) = X''(k)$. This proves the proposition. $\qquad\square$

9.4 Centralisers of Sections in the Arithmetic Case

It is a necessity for the validity of the section conjecture that sections have trivial centraliser in the sense of Sect. 3.3. Indeed, the injectivity part of the section conjecture implies that a restricted version of assertion (c) of Proposition 36 holds for sections s_a associated to rational points $a \in X(k)$. For $1 \neq g \in G$ acting faithfully on X/k we have $a \neq g(a)$ and thus by the injectivity of the profinite Kummer map

$$s_a \neq s_{g(a)} = g_*(s_a).$$

The same argument as in the proof of Proposition 36 shows that sections of the form s_a have trivial centraliser. Now the surjectivity part of the section conjecture claims that the above argument covers all sections of $\pi_1(X/k)$. Consequently, we better have a priori control on the centraliser of arbitrary sections.

Proposition 104. *Let k be either a number field or a finite extension of $\mathbb{Q}_p$. For a smooth and geometrically connected curve U/k the image $s(\mathrm{Gal}_k)$ of any section $s : \mathrm{Gal}_k \to \pi_1(U)$ has trivial centraliser in $\pi_1(U)$.*

Proof. We argue by contradiction. As Gal_k is known to have trivial center, we may assume that $s(\mathrm{Gal}_k)$ centralises a nontrivial $\gamma \in \pi_1(\overline{U})$. Let $H \trianglelefteq \pi_1(\overline{U})$ be a characteristic open subgroup with $\gamma \notin H$. Then

$$\langle H, \gamma \rangle s(\mathrm{Gal}_k) \subseteq \pi_1(U)$$

is an open subgroup corresponding to a finite étale cover $U' \to U$ with U' geometrically connected over k. The element γ has nontrivial image in

$$\pi_1^{\mathrm{ab}}(\overline{U}') = \langle H, \gamma \rangle^{\mathrm{ab}},$$

as can be seen in the cyclic quotient $\langle H, \gamma \rangle / H$. Consequently, $\pi_1^{\mathrm{ab}}(\overline{U}')$ contains a nontrivial Gal_k-invariant subspace $\mathrm{H}^0\left(k, \pi_1^{\mathrm{ab}}(\overline{U}')\right)$.

For ease of notation we replace U' by U and let X be a smooth projective completion of U with reduced divisor $Y = X \setminus U$. For every $n \in \mathbb{N}$ the sequence

$$0 \to \mathrm{H}^1(\overline{X}, \mu_n) \to \mathrm{H}^1(\overline{U}, \mu_n) \to \mathrm{H}^2_Y(\overline{X}, \mu_n) \to \mathrm{H}^2(\overline{X}, \mu_n)$$

is exact and yields in the projective limit over n of the $\mathrm{Hom}(-, \mathbb{Q}/\mathbb{Z}(1))$-duals the exact sequence of Gal_k-modules

$$\hat{\mathbb{Z}}(1) \xrightarrow{\mathrm{diag}} \hat{\mathbb{Z}}(1)[Y(\bar{k})] \to \pi_1^{\mathrm{ab}}(\overline{U}) \to \pi_1^{\mathrm{ab}}(\overline{X}) \to 0$$

where *diag* is the diagonal map. Let A be the Albanese variety of X/k. Then $\pi_1^{\mathrm{ab}}(\overline{U})$ is an extension of the Tate-module

$$\mathrm{T}(A) = \varprojlim_n A[n](\bar{k}) = \pi_1^{\mathrm{ab}}(\overline{X})$$

with the Tate-module $\mathrm{T}(\mathbb{T})$ of the torus with character group

$$\mathbb{Z}[Y(\bar{k})]/\mathrm{diag}(\mathbb{Z}).$$

It follows from Lemma 105 below that $\mathrm{H}^0\left(k, \pi_1^{\mathrm{ab}}(\overline{U})\right) = 0$, which achieves a contradiction and completes the proof. □

Lemma 105. *Let k be either a number field or a finite extension of $\mathbb{Q}_p$. For a semiabelian variety B over k we have $\mathrm{H}^0\left(k, \mathrm{T}(B)\right) = 0$.*

Proof. We compute

$$\mathrm{H}^0\left(k, \mathrm{T}(B)\right) = \varprojlim_n B[n](k) = \varprojlim_n B(k)[n] = 0,$$

because the torsion group of $B(k)$ is finite if k is a number field or a p-adic local field. Indeed, for a torus B we enlarge k so that B splits and reduce to the case $B = \mathbb{G}_m$ which is obvious. For an abelian variety B we rely on the Mordell–Weil Theorem for k a number field and the description of Mattuck–Tate [Ma55] of $B(k)$ as a topologically finitely generated abelian profinite group. The general case follows by dévissage. □

Remark 106. The proof of Proposition 104 is limited to smooth curves for the following reason. If for a smooth geometrically connected variety U/k we knew that every neighbourhood U' of the section $s \in \mathscr{S}_{\pi_1(U/k)}$ has torsion free $\pi_1^{\mathrm{ab}}(\overline{U'})$, then we could at least in the number field case argue with weights with respect to primes of good reduction. Indeed, $\pi_1^{\mathrm{ab}}(\overline{U})$ is dual to $\mathrm{H}^1_{\mathrm{\acute{e}t}}(\overline{U}, \hat{\mathbb{Z}})$ and thus of mixed weights -1 and -2 with no room for a nontrivial $\mathrm{H}^0\left(k, \pi_1^{\mathrm{ab}}(\overline{U})\right)$.

The assumption of torsion free quotients $\pi_1^{\mathrm{ab}}(\overline{U'})$ is known for curves and semiabelian varieties, but I am not aware of any other fundamentally different example.

Corollary 107. *Let k be either a number field or a finite extension of $\mathbb{Q}_p$, and let X/k be a smooth and geometrically connected curve. The map*

$$E \mapsto \mathscr{S}_{\pi_1(X/k)}(E)$$

is a sheaf of sets on $\mathrm{Spec}(k)_{\text{ét}}$. *In other words the space of sections satisfies Galois descent, i.e., for a Galois extension E/F of finite extensions of k the natural map*

$$\mathscr{S}_{\pi_1(X/k)}(F) \to \mathscr{S}_{\pi_1(X/k)}(E)$$

is injective and has the $\mathrm{Gal}(E/F)$-*invariants as its image.*

Proof. This follows immediately from Propositions 28 and 104. □

Corollary 108. *In order to prove the section conjecture for all hyperbolic curves it suffices to treat the case of curves which have at least one rational point.*

Proof. If s is a section of $\pi_1(X/k)$, then for a finite scalar extension k'/k such that $X(k') \neq \emptyset$, we then know that $s \otimes k' = s_a$ for some $a \in X(k')$. By Galois descent for sections as in Corollary 107 we know that $a \in X(k)$ and $s = s_a$. □

9.5 Scalar Extensions and Geometric Covers

Varieties X_1, X_2 over a field k are k-*forms* of the same variety if and only if there is a finite extension k'/k and an isomorphism of k'-varieties

$$X_1 \times_k k' \cong X_2 \times_k k'.$$

Proposition 109. *Let k be an algebraic number field or a finite extension of $\mathbb{Q}_p$. If the section conjecture holds for a smooth geometrically connected curve X/k with respect to all finite extensions k'/k, i.e., the map*

$$\kappa_X : X(-) \to \mathscr{S}_{\pi_1(X/k)}(-)$$

is an isomorphism of sheaves on $\mathrm{Spec}(k)_{\text{ét}}$, *then the same holds for all k-forms of X.*

Proof. Let Y/k be a k-form of X. By Corollary 107, the relevant spaces of sections indeed form sheaves. Therefore the map

$$\kappa_Y : Y(-) \to \mathscr{S}_{\pi_1(Y/k)}(-)$$

is an isomorphism of sheaves of sets if and only if it is so locally. But étale locally the maps κ_Y and κ_X can be identified with each other by means of an isomorphism $X_{k'} \cong Y_{k'}$, and so κ_Y is also bijective. □

Now we examine *going up* and *going down* for the section conjecture with respect to a finite étale map.

Proposition 110 (Going up an étale map). *Let $h : Y \to X$ be a finite étale cover of the geometrically connected variety X over the field k. Let us assume that all sections of $\pi_1(X/k)$ have trivial centraliser.*

If the section conjecture holds for X over k, i.e., the profinite Kummer map κ_X is bijective, then the section conjecture holds also for Y over k, i.e., the profinite Kummer map κ_Y is bijective.

Proof. Let $s : \mathrm{Gal}_k \to \pi_1(Y)$ be a section. The section $h_*(s)$ of $\pi_1(X/k)$ comes by assumption from a unique k-rational point $a \in X(k)$. Hence $s \in h_*^{-1}(s_a)$ and by Proposition 33 and Corollary 34 we find a unique

$$b \in h^{-1}(a)(k) = \underline{h^{-1}(s_a)}(k) \xrightarrow{\sim} h_*^{-1}(s_a)$$

with $s_b = s$. This establishes an inverse to the profinite Kummer map κ_Y.   □

Proposition 111 (Going down an étale map). *Let $h : Y \to X$ be a finite étale cover of the geometrically connected variety X over the field k. Let us assume that for all finite extensions k'/k all sections of $\pi_1(X_{k'}/k')$ have trivial centraliser.*

Then the section conjecture for Y over all finite extensions k'/k implies the section conjecture for X over all finite extensions k'/k.

Proof. As $\pi_1(Y) \subset \pi_1(X)$ is a subgroup, the hypothesis of trivial centralisers of sections also applies to the sections of $\pi_1(Y_{k'}/k')$. We obtain the following diagram of sheaves, see Proposition 28,

$$
\begin{array}{ccc}
Y(-) & \xrightarrow{\;\;h\;\;} & X(-) \\
{\scriptstyle \kappa_Y}\downarrow & & \downarrow{\scriptstyle \kappa_X} \\
\mathscr{S}_{\pi_1(Y/k)}(-) & \xrightarrow{\;\;h_*\;\;} & \mathscr{S}_{\pi_1(X/k)}(-)
\end{array}
$$

where the horizontal maps are surjective as maps of sheaves on $\mathrm{Spec}(k)_{\text{ét}}$. As κ_Y is bijective by assumption, we find that κ_X is surjective. And κ_X is injective, if for $a \in X(k)$ the map of sheaf-fibres

$$\underline{h^{-1}(a)} \to h_*^{-1}(s_a)$$

is surjective. The latter follows from Proposition 33 (1).   □

Remark 112. (1) By Proposition 104 the conclusion of Proposition 110 and Proposition 111 hold for k an algebraic number field or a finite extension of $\mathbb{Q}_p$.

(2) Let us assume that k is a number field and $h : Y \to X$ is a finite étale cover of proper geometrically connected k-varieties X. Then the field extension k'/k necessary to deduce the section conjecture for X over k as in Proposition 111 from the section conjecture for Y over k' is bounded by ramification and degree constraints. Thus by the Hermite–Minkowski Theorem it is enough to choose a fixed specific finite extension k' of k. We omit the details, as we do not see a specific interest in this refinement yet.

Part III
On the Passage from Local to Global

Chapter 10
Local Obstructions at a p-adic Place

Let X be a geometrically connected variety over a number field k, and let k_v denote the completion of k in a place v. Base change $s \mapsto (s \otimes k_v)$ as in Sect. 3.2 induces a localisation map

$$\mathscr{S}_{\pi_1}(X/k) \to \prod_v \mathscr{S}_{\pi_1}(X \times_k k_v / k_v)$$

where v ranges over all places of k. We will address the section conjecture from this local to global point of view. More of the purely local problem will be addressed in Chap. 16, and the classical obstructions against the passage from local to global form the topic of Chap. 11.

If the local curve $X_v = X \times_k k_v$ does not admit a section of $\pi_1(X_v/k_v)$, then we say that there is a *local obstruction against sections* for $\pi_1(X/k)$ at the place v. Any consequence of the existence of the section for a curve over a local field can be turned around to produce an example of a curve over a number field that satisfies the section conjecture: the example has neither a section nor a rational point, because sections are obstructed locally due to the arithmetic consequence not being true. Examples of this kind by considering period and index of p-adic curves with a section were first constructed in [Sx10b] §7 and are recalled below in Theorem 114.

If a p-adic curve with a section fails to have index 1, then we can construct a $\mathbb{Q}_p$-linear form on $\mathrm{Lie}(\mathrm{Pic}^0_X)$ for which no natural explanation exists and which we therefore consider bizarre (in the hope that this case does not exist in compliance with the local section conjecture).

10.1   Period and Index

The following definitions work for arbitrary base fields K.

Definition 113. Let X/K be a geometrically integral variety.

J. Stix, *Rational Points and Arithmetic of Fundamental Groups*, Lecture Notes in Mathematics 2054, DOI 10.1007/978-3-642-30674-7_10,
© Springer-Verlag Berlin Heidelberg 2013

(1) The *index* of X is the positive number

$$\text{index}(X) = \gcd\{\deg(a) = [\kappa(a) : K] \; ; \; a \text{ is a closed point of } X\}.$$

(2) The *period* of X is the order period(X) of the universal Albanese torsor, see [Wi08], as a principal homogeneous space under the Albanese variety of X.
(3) The *relative Brauer group* of X/K is the kernel $\text{Br}(X/K)$ of the pullback map

$$\text{Br}(K) \to \text{Br}(X).$$

The period–index result. We now put ourselves in a p-adic local framework and let K be a finite extension of $\mathbb{Q}_p$. In the case of curves we can prove the following.

Theorem 114. *Let K be a finite extension of $\mathbb{Q}_p$ and let X/K be a smooth, projective geometrically connected curve of genus > 0, such that $\pi_1(X/K)$ admits a section.*

(1) For p odd, we must have

$$\text{period}(X) = \text{index}(X),$$

and both are powers of p.
(2) For $p = 2$, both period(X) and index(X) are powers of 2. If we moreover assume that we have an even degree finite étale cover $X \to X_0$, then also here

$$\text{period}(X) = \text{index}(X).$$

Proof. This was proved in [Sx10b] Theorem 15+16. Another proof depending on the cycle class of the section was given later in [EsWi09] Corollary 3.6. □

It is the goal of the following Sect. 10.2 to connect Theorem 114 along the proof following [EsWi09] Corollary 3.6 to another interesting construction associated to a section s of $\pi_1(X/K)$.

Example 115. (1) Let k be a number field with a place $v \mid p$ and choose $A \in \text{Br}(k)$ with $\text{inv}_v(A)$ not of p-primary order. Let A by abuse of notation be an Azumaya algebra over k representing $A \in \text{Br}(k)$. Any smooth projective geometrically connected curve X/k with a map

$$f : X \to \text{BS}_A \qquad\qquad (10.1)$$

to the Brauer-Severi variety BS_A of A has non split $\pi_1(X/k)$, since the following argument obstructs sections locally at v. The map f yields that $A \otimes k_v$ lies in $\text{Br}(X_v/k_v)$. By a theorem of Tate, Roquette and Lichtenbaum, see [Li69] Theorem 3, we have

$$\text{index}(X_v) = \#\,\text{Br}(X_v/k_v),$$

and by construction here this order is not a power of p. This yields the local obstruction to sections due to Theorem 114. For details see [Sx10b] §7.

(2) An explicit example following (1) is as follows. Let $p \equiv 3 \mod 4$ and X be the plane curve over $\mathbb{Q}$ which is given by the following homogeneous equation

$$U^{2p} + V^{2p} = pW^{2p}.$$

The curve X maps to the Brauer–Severi variety of the division algebra with local invariant $\mathrm{inv}_p(A) = \frac{1}{2}$

$$A = \left(\frac{p, -1}{\mathbb{Q}} \right).$$

Hence $\pi_1(X/\mathbb{Q})$ has no section for a local obstruction at p.

(3) The local obstruction to sections due to Theorem 114 survives a finite extension k'/k of the base field such that one of the local extensions $k'_{v'}/k_v$ has degree prime to p. So the example in (1) provides actually a curve X/k such that for infinitely many finite extensions k'/k we have $\mathscr{S}_{\pi_1(X/k)}(k') = \emptyset$. More explicitly, we can take the curve $X/\mathbb{Q}$ from (2) and consider the maximal ℓ-cyclotomic extension $\mathbb{Q}(\mu_{\ell^\infty})/\mathbb{Q}$ for a prime ℓ such that $p \nmid \ell - 1$.

Remark 116. The conclusion of Theorem 114 does not hold under the weaker assumption that $\pi_1^{\mathrm{ab}}(X/K)$ splits. We construct an example as follows. Let $K/\mathbb{Q}_p$ be a finite extension with $p > 2$ and let $\mathrm{BS}_{1/2}$ be the Brauer–Severi variety for the quaternion algebra A of invariant $1/2$. As A splits under a quadratic extension, we have a closed point $y \in \mathrm{BS}_{1/2}$ of $\deg(y) = 2$. By Riemann–Roch, we find a nonconstant

$$f \in \mathrm{H}^0\left(\mathrm{BS}_{1/2}, \mathcal{O}(y) \right)$$

and another closed point $z \in \mathrm{BS}_{1/2}$ with $\mathrm{div}(f) = z - y$. We consider the branched cover

$$h : X \to \mathrm{BS}_{1/2}$$

given by $\sqrt[3]{f}$ which is a μ_3-torsor, that is totally branched above exactly z, y. By Riemann–Hurwitz, the curve X has genus 2. Lichtenbaum's congruences, see [Li69], show that $\mathrm{period}(X) = 1$ and $\mathrm{index}(X) = 2$. In particular $\pi_1(X/K)$ does not split as the conclusion of Theorem 114 does not hold. However, $\mathrm{period}(X) = 1$ implies that $\mathrm{Pic}_X^1(K)$ is nonempty and $\pi_1^{\mathrm{ab}}(X/K) = \pi_1(\mathrm{Pic}_X^1/K)$ splits.

Even or odd Jacobian. Following Poonen and Stoll [PoSt99] §3, a polarization $\lambda \in \mathrm{H}^0(k, \mathrm{NS}_A)$ of an abelian variety A over a number field k defines a non-degenerate Cassels–Tate pairing

$$\langle \alpha, \beta \rangle_\lambda = \langle \alpha, \lambda(\beta) \rangle \in \mathbb{Q}/\mathbb{Z}$$

on the Tate–Shafarevich group $\text{III}(A/k)$. The pairing $\langle -, - \rangle_\lambda$ is antisymmetric by [PoSt99] Corollary 6. This defines a special element $c \in \text{III}(A/k)$ in the 2-torsion inducing the linear map

$$\alpha \mapsto \langle \alpha, \alpha \rangle_\lambda = \langle \alpha, c \rangle_\lambda \in \frac{1}{2}\mathbb{Z}/\mathbb{Z}$$

for all $\alpha \in \text{III}(A/k)$. We consider the well defined endomorphism of $\text{III}(A/k)$

$$\alpha \mapsto \alpha^c = \alpha + \langle \alpha, c \rangle c$$

where we consider $\langle \alpha, c \rangle$ taking values in $\mathbb{Z}/2\mathbb{Z}$. The modified pairing

$$\langle \alpha, \beta \rangle^c = \langle \alpha, \beta^c \rangle_\lambda$$

is alternating and, by [PoSt99] Theorem 8, non-degenerate if and only if $\langle c, c \rangle = 0$. If $\langle c, c \rangle = 1/2$, then $\text{III}(A/k)$ is the direct sum of $C = \{0, c\}$ and the orthogonal complement $C^\perp$ with both pairings inducing the same non-degenerate alternating paring on $C^\perp$.

Definition 117. The polarized abelian variety is called *even* if $\langle c, c \rangle = 0$ and it is called *odd* if $\langle c, c \rangle = \frac{1}{2}$.

By [PoSt99] Corollary 12, for the Jacobian Pic_X^0 of a smooth projective curve X of genus g endowed with its canonical polarization, the class c is given by the class

$$c = [\text{Pic}_X^{g-1}] \in \text{III}(\text{Pic}_X^0/k)$$

and

$$\langle c, c \rangle = N/2 \in \mathbb{Q}/\mathbb{Z}$$

where N is the number of *deficient* places v of k, namely those v such that $\text{Pic}^{g-1}(X_v) = \emptyset$, so

$$\text{index}(X_v) \nmid g - 1.$$

By Lichtenbaum's congruences, see [Li69] second theorem in the introduction, as $\text{period}(X_v) \mid g - 1$, this is equivalent to

$$\text{index}(X_v)/\text{period}(X_v) = 2 \nmid \frac{g-1}{\text{period}(X_v)}$$

which itself is equivalent by loc. cit. to $\text{index}(X_v) = 2 \, \text{period}(X_v)$.

Corollary 118. *Let X/k be a geometrically connected smooth projective curve over a number field k. If $\pi_1(X/k)$ splits, and for all places $v \mid 2$ we either have a rational divisor on $X_v = X \times_k k_v$ of degree $g - 1$ or there is an étale map $X \to X_0$ of even degree, then X has even Jacobian Pic_X^0.*

Proof. By Theorem 114 and the special assumption at places $v \mid 2$, the deficient places can only be real places. For a real place $k_v = \mathbb{R}$, the proven real analogue of the section conjecture, see Sect. 16.1, shows that we have a real point on X_v and so X_v has index 1, which means that v is also not deficient. Hence the number of deficient places is 0, and the Jacobian Pic_X^0 is indeed even. $\qquad\square$

The section conjecture predicts that a section of $\pi_1(X/k)$ gives a rational point and so no place is deficient and Pic_X^0 is even.

10.2 A Linear Form on the Lie-Algebra

We keep the hypothesis that $K/\mathbb{Q}_p$ is a finite extension and that X/K is a smooth, projective variety. In addition we fix an algebraic closure $\bar{K}$ of K and assume that X is a $\text{K}(\pi, 1)$-space in degree ≤ 2 so that we have a natural isomorphism

$$\text{H}^2\left(\pi_1(X), \hat{\mathbb{Z}}(1)\right) \xrightarrow{\sim} \text{H}^2\left(X, \hat{\mathbb{Z}}(1)\right).$$

Continuous Kummer sequence. The Kummer sequence with continuous coefficients à la Jannsen, see [Ja88] (3.27), gives rise to a short exact sequence

$$0 \to \varprojlim_n \text{Pic}(X)/n\,\text{Pic}(X) \xrightarrow{c_1} \text{H}^2\left(X, \hat{\mathbb{Z}}(1)\right) \to \text{Hom}(\mathbb{Q}/\mathbb{Z}, \text{Br}(X)) \to 0, \quad (10.2)$$

because, with the cohomological Brauer group $\text{Br}(X) = \text{H}^2(X, \mathbb{G}_m)_{\text{tors}}$,

$$0 \to \text{Hom}(\mathbb{Q}/\mathbb{Z}, \text{Br}(X)) \to \varprojlim\left(\text{H}^2(X, \mathbb{G}_m), n \cdot\right) \to \text{H}^2(X, \mathbb{G}_m)$$

is exact, and also, see [Ja88] §4,

$$\text{Pic}(X) \to \varprojlim_n \text{Pic}(X)/n\,\text{Pic}(X) \to \varprojlim{}^1\left(\text{Pic}(X), n \cdot\right) \to 0$$

is exact. The section s of $\pi_1(X/K)$ yields an evaluation map

$$s^* : \text{H}^2\left(X, \hat{\mathbb{Z}}(1)\right) = \text{H}^2\left(\pi_1(X), \hat{\mathbb{Z}}(1)\right) \to \text{H}^2\left(K, \hat{\mathbb{Z}}(1)\right) = \text{Hom}(\mathbb{Q}/\mathbb{Z}, \text{Br}(K)),$$

which is the degree 2 analogue of the evaluation map of Chap. 5.

Galois equivariant evaluation. We set $X' = X \times_K K'$ for a finite extension K'/K. There is a restriction map for which (10.2) is natural. Moreover, the evaluation with respect to s and to the restriction $s' = s|_{\text{Gal}_{k'}}$ are compatible with restriction

$$\mathrm{H}^2\left(X, \hat{\mathbb{Z}}(1)\right) \xrightarrow{\ s^* \ } \mathrm{Hom}(\mathbb{Q}/\mathbb{Z}, \mathrm{Br}(K))$$

$$\downarrow \qquad\qquad\qquad\qquad \downarrow$$

$$\mathrm{H}^2\left(X', \hat{\mathbb{Z}}(1)\right) \xrightarrow{\ s'^* \ } \mathrm{Hom}(\mathbb{Q}/\mathbb{Z}, \mathrm{Br}(K')).$$

By transport of structure we obtain in the limit a Gal_k-equivariant evaluation map

$$s^* \ : \ \varinjlim_{K'/K} \mathrm{H}^2\left(X \otimes_K K', \hat{\mathbb{Z}}(1)\right) \to \varinjlim_{K'/K} \mathrm{Hom}(\mathbb{Q}/\mathbb{Z}, \mathrm{Br}(K')).$$

We analyse more closely the induced map

$$s^* \circ c_1 \ : \ \varinjlim_{K'/K} \left(\varprojlim_{n} \mathrm{Pic}(X')/n\,\mathrm{Pic}(X') \right) \to \varinjlim_{K'/K} \mathrm{Hom}(\mathbb{Q}/\mathbb{Z}, \mathrm{Br}(K')). \qquad (10.3)$$

The kernel of $\mathrm{Br}(K) \to \mathrm{Br}(K')$ is of bounded exponent and so the transfer maps in the system

$$\varprojlim_{K'} \mathrm{Hom}(\mathbb{Q}/\mathbb{Z}, \mathrm{Br}(K'))$$

are injective. So vanishing of the finite level s^* on the image of c_1 can be checked after scalar extension, or even in the limit over all K'. For K' large enough X acquires a rational point and $\mathrm{Pic}(X') = \mathrm{Pic}_X(K')$ so that naturally

$$\varprojlim_{n} \mathrm{Pic}(X')/n\,\mathrm{Pic}(X') = \widehat{\mathrm{Pic}_X(K')} \cong \mathrm{Pic}_X^0(K') \times \hat{\mathbb{Z}}$$

with the factor $\hat{\mathbb{Z}}$ given by the degree. Since under the identification $\mathrm{Br}(K') = \mathbb{Q}/\mathbb{Z}$ by the local invariant map restriction corresponds to multiplication by the degree, we obtain a Galois equivariant map

$$s^* \circ c_1\,|_{\mathrm{Pic}_X^0} : \mathrm{Pic}_X^0(\bar{K}) \to \hat{\mathbb{Z}} \otimes_{\mathbb{Z}} \mathbb{Q}. \qquad (10.4)$$

Abelian varieties, p-adic logarithm and traces. More generally, we study a Gal_K-map like (10.4) for an abelian variety A/K. Let the homomorphism

$$\varphi : A(\bar{K}) \to \hat{\mathbb{Z}} \otimes \mathbb{Q}$$

be a Gal_K-equivariant. For a finite extension K'/K, the p-adic logarithm

$$\log_A \ : \ A(K') \to \mathrm{Lie}(A) \otimes_K K'$$

is a local homeomorphism and bijective onto its image when restricted to a subgroup of $A(K')$ which is of cotorsion in all of $A(K')$. Hence, the map φ restricted to $A(K')$ factors through $\log_A$. In the limit over all K' we obtain a factorization

$$\varphi \;:\; A(\bar{K}) \xrightarrow{\;\log_A\;} \mathrm{Lie}(A) \otimes_K \bar{K} \xrightarrow{\;\bar{\psi}\;} \hat{\mathbb{Z}} \otimes \mathbb{Q},$$

with $\log_A$ now surjective. The Gal_K module $\mathrm{Lie}(A) \otimes_K \bar{K}$ is the union of all the $\mathrm{Lie}(A) \otimes_K K'$ for K'/K Galois, which by the normal basis theorem are induced modules from trivial $\mathrm{Gal}_{K'}$ modules. The map $\bar{\psi}$ restricted to $\mathrm{Lie}(A) \otimes_K K'$ thus factors through the trace $\mathrm{id} \otimes \mathrm{tr}_{K'/K}$ which for compatibility reasons must be normalized as

$$\tau_{K'/K} := \frac{1}{[K' : K]} \, \mathrm{id} \otimes \mathrm{tr}_{K'/K}$$

that leads to a map

$$\tau = \bigcup_{K'/K} \tau_{K'/K} : \mathrm{Lie}(A) \otimes_K \bar{K} \to \mathrm{Lie}(A)$$

and which describes the maximal Gal_K-invariant quotient. Hence we arrive at a further factorization

$$\varphi \;:\; A(\bar{K}) \xrightarrow{\;\log_A\;} \mathrm{Lie}(A) \otimes_K \bar{K} \xrightarrow{\;\tau\;} \mathrm{Lie}(A) \xrightarrow{\;\psi\;} \hat{\mathbb{Z}} \otimes \mathbb{Q}.$$

The space $\mathrm{Lie}(A)$ is a finite dimensional $\mathbb{Q}_p$ vector space which maps therefore only trivially to the ℓ-component $\mathbb{Q}_\ell$ for $\ell \neq p$ of the finite $\mathbb{Q}$-adeles $\hat{\mathbb{Z}} \otimes \mathbb{Q}$. So the only potentially nontrivial part of ψ is the p-part, namely a map

$$\psi_p \;:\; \mathrm{Lie}(A) \to \mathbb{Q}_p. \tag{10.5}$$

The linear form associated to a section. Let s be a section of $\pi_1(X/K)$ for a smooth, projective variety which is an algebraic $\mathrm{K}(\pi, 1)$ in degree ≤ 2. The above construction yields a $\mathbb{Q}_p$-linear form

$$\psi_p(s) : \mathrm{H}^1(X, \mathcal{O}_X) = \mathrm{Lie}\left(\mathrm{Pic}_X^0\right) \to \mathbb{Q}_p. \tag{10.6}$$

Proposition 119. *If $s = s_a$ comes from $a \in X(K)$, then $\psi_p(s) = 0$.*

Proof. If $s = s_a$, then $s^* \circ c_1$ by naturality factors over

$$\varprojlim_n \mathrm{Pic}(\mathrm{Spec}(K))/n\,\mathrm{Pic}(\mathrm{Spec}(K)),$$

which obviously vanishes.   □

Question 120. Does the conclusion of Proposition 119 hold for an arbitrary section?

In the special case that $K = \mathbb{Q}_p$ and X is a smooth projective curve, then $\psi_p(s)$ by Serre-duality corresponds to a global differential $\omega(s) \in H^0(X, \Omega_X)$. The map $s \mapsto \omega(s)$ is continuous and yields a map

$$\omega : \mathscr{S}_{\pi_1(X/\mathbb{Q}_p)} \to H^0(X, \Omega_X).$$

The author finds it extremely bizarre, that sections of $\pi_1(X/\mathbb{Q}_p)$ should single out a certain compact subset of the space of global differentials, and thus conjectures the map ω to vanish identically.

Moving up in neighbourhoods. Let $h : Y \to X$ be a neighbourhood of the section s of $\pi_1(X/K)$, with the section t of $\pi_1(Y/K)$ mapping to s. Then $s^* = t^* \circ h^*$ which in the limit over all neighbourhoods Y yields a map

$$s^* : \varinjlim_{(Y,t)} \mathrm{Pic}^0(Y) \to \mathrm{Hom}\left(\mathbb{Q}_p/\mathbb{Z}_p, \mathrm{Br}(K)\right) = \mathbb{Z}_p \qquad (10.7)$$

where we only note the p-part, as we have learned above that the p'-part vanishes. We recall the Brauer obstruction map b, occurring in the exact sequence

$$0 \to \mathrm{Pic}(X) \to \mathrm{Pic}_X(K) \xrightarrow{b} \mathrm{Br}(K) \to \mathrm{Br}(X)$$

that controls whether a rational points of Pic_X correspond to an actual line bundle, see for example [Sx10b] §2.

Definition 121. The Pic^0-*part of the relative Brauer group* of the smooth, projective curve X/K is by definition the image $\mathrm{Br}^0(X/K) = b(\mathrm{Pic}_X^0(K))$.

It follows that we have an exact sequence

$$0 \to \mathrm{Pic}^0(X) \to \mathrm{Pic}_X^0(K) \to \mathrm{Br}^0(X/K) \to 0.$$

By the usual cohomological machinery, $\mathrm{Pic}^0(X)$ becomes divisible by n in $\mathrm{Pic}_Y^0(K)$ for a suitably fine neighbourhood $Y \to X$ of s. Consequently, in the extension

$$0 \to \varinjlim_{(Y,t)} \mathrm{Pic}^0(Y) \to \varinjlim_{(Y,t)} \mathrm{Pic}_Y^0(K) \to \varinjlim_{(Y,t)} \mathrm{Br}^0(Y/K) \to 0 \qquad (10.8)$$

where (Y, t) ranges over all neighbourhoods of the section s, the elements of the left hand group become divisible in the middle.

Period and index revisited. In view of the preparation above we can now relate the map Ψ_p to the index of neighbourhoods.

Proposition 122. *Let X/K be a smooth projective curve and let s be a section of $\pi_1(X/K)$. Then the following are equivalent.*

(a) $\lim_{\substack{\longrightarrow \\ (Y,t)}} \mathrm{Br}^0(Y/K) = 0$.

(b) $\lim_{\substack{\longrightarrow \\ (Y,t)}} \mathrm{Br}^0(Y/K)$ *has finite exponent.*

(c) The extension (10.8) splits.

(d) The pushout of the extension (10.8) by s^* from (10.7) splits.

(e) The map s^* from (10.7) vanishes.

(f) For every neighbourhood (Y,t) of s the class cl_t is orthogonal to $c_1(\mathrm{Pic}^0(Y))$.

(g) Every neighbourhood (Y,t) of the section s has index 1.

(d') The map $\psi_p(t)$ vanishes for every neighbourhood (Y,t) of the section s.

If the equivalent conditions hold, then (f) can be improved to $\mathrm{cl}(t) = c_1(\mathscr{L})$ for some line bundle on Y.

Proof. Obviously (a) implies (b). If (b) holds, then the group

$$\lim_{\substack{\longrightarrow \\ (Y,t)}} \mathrm{Pic}^0(Y)$$

is itself divisible and the extension (10.8) splits, hence (c) holds. Again, obviously (c) implies (d). If (d) holds, then s^* extends to a map

$$\lim_{\substack{\longrightarrow \\ (Y,t)}} \mathrm{Pic}_Y^0(K) \to \mathbb{Z}_p$$

from which we find that the image of s^* is divisible in $\mathbb{Z}_p$, hence vanishes and (e) holds. Property (f) is an immediate consequence of (e) and the formula

$$s^*(c_1(\mathscr{L})) = c_1(\mathscr{L}) \cup \mathrm{cl}_s \in \mathrm{H}^4\left(X, \hat{\mathbb{Z}}(2)\right) = \hat{\mathbb{Z}}.$$

Now we assume (f) and use that by Lichtenbaum duality [Li69] the orthogonal complement of $c_1(\mathrm{Pic}^0(Y))$ is

$$c_1(\mathrm{Pic}(Y)) + \mathrm{pr}^*(\mathrm{H}^2(K, \hat{\mathbb{Z}}(1)).$$

Hence there is a $\mathscr{L} \in \mathrm{Pic}(Y)$ and a constant class $\gamma \in \mathrm{H}^2(K, \hat{\mathbb{Z}}(1))$ such that

$$\mathrm{cl}_t = c_1(\mathscr{L}) + \mathrm{pr}^* \gamma.$$

In particular,

$$\deg(\mathscr{L}) = \deg(\mathrm{cl}_t - \mathrm{pr}^* \gamma) = \deg(\mathrm{cl}_t) = 1$$

and $\mathrm{index}(Y) = 1$ showing (g). To conclude (a) from (g) is obvious. Properties (d) and (d') are immediately equivalent. In order to show $\mathrm{cl}_t = c_1(\mathscr{L})$ for a suitable $\mathscr{L} \in \mathrm{Pic}(Y)$ we resort to [EsWi09] Lemma 3.5 which states that

$$\mathrm{cl}_t \cup \left(\mathrm{cl}_t + c_1(\omega_Y)\right) = 0$$

where ω_Y is the dualizing sheaf on Y. With $\mathrm{cl}_t = c_1(\mathscr{L}) + \mathrm{pr}^* \gamma$ we thus compute

$$0 = \left(c_1(\mathscr{L}) + \mathrm{pr}^* \gamma\right) \cup \left(c_1(\mathscr{L} \otimes \omega_Y) + \mathrm{pr}^* \gamma\right)$$
$$= \left(\deg(\mathscr{L}) + \deg(\mathscr{L} \otimes \omega_Y)\right) \cdot \gamma = 2g \cdot \gamma$$

so that $\gamma = 0$ and thus $\mathrm{cl}_t = c_1(\mathscr{L})$ follows.   □

The proof above could be split into ℓ-primary parts, in which case for $\ell \neq p$ condition (f) is a consequence of the factorisation of $s^* \circ c_1$ over $\psi_p(s)$ as in (10.6). This reasoning supplies an alternative proof of Theorem 114.

The bizarre case. Let us assume that the equivalent conditions of Proposition 122 do not hold, i.e., we are in the bizarre case that $\psi_p(s)$ does not vanish. Then

$$\varinjlim_{(Y,t)} \mathrm{Br}^0(Y/K) = \mathbb{Q}_p/\mathbb{Z}_p$$

and the pushout of the extension (10.8) by s^* gives a nontrivial element

$$\beta_{X/K}(s) \in \mathrm{Ext}^1_{\mathbb{Z}_p}(\mathbb{Q}_p/\mathbb{Z}_p, \mathbb{Z}_p) = \mathbb{Z}_p.$$

Obviously, for a neighbourhood (Y, t) of the section s we have

$$\beta_{Y/K}(t) = \beta_{X/K}(s).$$

Extending scalars by K'/K scales $\mathbb{Q}_p/\mathbb{Z}_p$ by $[K' : K]$ coming from the restriction map of Brauer groups. The argument $\mathbb{Z}_p$ is scaled as in the following diagram:

$$
\begin{array}{ccc}
\varinjlim_{(Y,t)} \mathrm{Pic}^0(Y) & \xrightarrow{\ s^*\ } & \mathbb{Z}_p \\
\downarrow & & \downarrow {\scriptstyle [K':K]\cdot} \\
\varinjlim_{(Y',t')} \mathrm{Pic}^0(Y') & \xrightarrow{\ (s')^*\ } & \mathbb{Z}_p.
\end{array}
$$

Here (Y', t') is the scalar extension by K'/K applied to the neighbourhood (Y, t). Consequently, we have

$$\beta_{X_{K'}/K'}(s') = \beta_{X/K}(s)$$

where $s' = s|_{\mathrm{Gal}_K}$. Hence, both natural operations, namely passing to neighbourhoods or passing to scalar extensions, are available but do not help.

10.3   More Examples Exploiting the Relative Brauer Group

In Example 115, curves X over a number field k were constructed such that $\pi_1(X/k)$ does not split. The arithmetic of curves over p-adic fields played a crucial role. Here we present a method to construct such examples over almost arbitrary fields.

We start with $0 \neq A \in \mathrm{Br}(k)$ of order $n \in k^*$ and an auxiliary smooth, projective curve X_0/k of genus at least 2 with $A \in \mathrm{Br}(X_0/k)$. Such curves exist in abundance, see [Sx10b] Proposition 8. From the Kummer sequence we obtain the mod n version of (10.2), namely

$$0 \to \mathrm{Pic}(X_0)/n\,\mathrm{Pic}(X_0) \xrightarrow{c_1} \mathrm{H}^2(\pi_1(X_0), \mu_n) \to \mathrm{Br}(X_0)[n] \to 0. \qquad (10.9)$$

The pullback of A by the map

$$\mathrm{Br}(k)[n] = \mathrm{H}^2(k, \mu_n) \to \mathrm{H}^2(X_0, \mu_n)$$

has by assumption an image $\alpha = c_1(\mathcal{L})$ for some $\mathcal{L} = \mathcal{O}(D) \in \mathrm{Pic}(X_0)$ with some divisor D on X_0. By Riemann–Roch, there exists a principal divisor

$$\mathrm{div}(f) = R + \sum_{P \in D} P$$

where the sum is over the points in the support of D and R is disjoint from D. The μ_n-torsor $h : X \to X_0$ obtained by extracting an nth root of f is geometrically connected as along the support of D we are taking an nth root of a uniformizer. Moreover, h^*D becomes divisible by n in the divisor group and thus there is $\mathcal{M} \in \mathrm{Pic}(X)$ with $h^*\mathcal{L} = \mathcal{M}^{\otimes n}$. By the naturality of (10.9) with respect to h we see that

$$A \in \ker\left(\mathrm{H}^2(k, \mu_n) \to \mathrm{H}^2(\pi_1(X), \mu_n)\right).$$

So $\pi_1(X/k)$ does not split, because the retraction by pullback s^* with a section s forces this kernel to be trivial.

We note that in these examples, even if k is a finite extension of $\mathbb{Q}_p$, the index can be a power of p. Namely, we start with $\mathrm{index}(X_0)$ a power of p as witnessed by the divisor E_0 on X_0 of degree this power of p. Then, in constructing the branched cover $h : X \to X_0$ we also take care that h is totally ramified above the points of E_0. Thus $h^*E_0 = nE$ with $\deg(E) = \deg(E_0)$ and $\mathrm{index}(X)$ is still a power of p.

Chapter 11
Brauer–Manin and Descent Obstructions

In Chap. 10 we have discussed obstructions to sections arising from arithmetic at p-adic places. Later in Chap. 16 we will discuss what is known about the local analogues of the section conjecture over real and p-adic local fields. The present Chapter concerns the usual next step, when the local problem is considered settled (which it is not for the section conjecture). In order to possibly arise from a common global rational point, the tuple of local solutions must survive known obstructions from arithmetic duality: the Brauer–Manin obstruction and the descent obstruction. We develop here the analogous obstructions to a collection of local sections against being the restriction of a common global section.

The Brauer–Manin obstruction for adelic sections was developed in [Sx11]. The descent obstruction is actually the older sibling of the Brauer–Manin obstruction. The technique of descent goes back to Fermat. Descent using torsors under tori was developed and studied in detail by Colliot-Thélène and Sansuc, while later Harari and Skorobogatov [HaSk02] analysed descent obstructions coming from non-abelian groups. Descent obstructions under finite groups have been thoroughly analysed by Stoll [St07]. The transfer of the descent obstruction to spaces of sections was essentially worked out in [HaSx12].

Unlike a priori for adelic points, for adelic sections the constant finite descent obstruction turns out to be the only obstruction to globalisation, see Theorem 144. This gives a link between the section conjecture and strong approximation that either yields interesting applications to the section conjecture, see Corollary 158, or, if one is pessimistic, opens up an approach to disproving the section conjecture eventually.

11.1 The Brauer–Manin Obstruction for Rational Points

The Brauer–Manin obstruction was introduced by Manin in [Ma71] with the goal to explain the failure of the local–global principle, see [Sk01] §5.2.

J. Stix, *Rational Points and Arithmetic of Fundamental Groups*, Lecture Notes in Mathematics 2054, DOI 10.1007/978-3-642-30674-7_11,
© Springer-Verlag Berlin Heidelberg 2013

Let $\mathbb{A}_k = \prod'_v k_v$ be the adeles of an algebraic number field k, and let X/k be a smooth, geometrically connected variety. The space of *modified adelic points* of X is the quotient

$$X(\mathbb{A}_k)_\bullet \tag{11.1}$$

of the set of adelic points $X(\mathbb{A}_k)$ where the archimedean components are replaced by their respective set of connected components.

For a global Brauer class $A \in \mathrm{Br}(X)$ we consider the function

$$\langle A, - \rangle : X(\mathbb{A}_k)_\bullet \to \mathbb{Q}/\mathbb{Z},$$

$$x = (x_v)_v \mapsto \sum_v \mathrm{inv}_v(A(x_v))$$

where $A(x_v)$ is the evaluation of A in x_v and

$$\mathrm{inv}_v : \mathrm{Br}(k_v) \to \mathbb{Q}/\mathbb{Z}$$

is the local invariant map. The Hasse–Brauer–Noether local global principle for Brauer groups, see [NSW08] Theorem 8.1.17, i.e., the exactness of

$$0 \to \mathrm{Br}(k) \to \bigoplus_v \mathrm{Br}(k_v) \xrightarrow{\sum_v \mathrm{inv}_v} \mathbb{Q}/\mathbb{Z} \to 0, \tag{11.2}$$

shows that the global points $X(k)$ lie in the Brauer kernel

$$X(\mathbb{A}_k)_\bullet^{\mathrm{Br}} := \{x \in X(\mathbb{A}_k)_\bullet \, ; \, \langle A, x \rangle = 0 \text{ for all } A \in \mathrm{H}^2(X, \mathbb{G}_\mathrm{m})\}. \tag{11.3}$$

If $X(\mathbb{A}_k) \neq \emptyset$ but $X(\mathbb{A}_k)_\bullet^{\mathrm{Br}}$ is empty, then $X(k) = \emptyset$ and the failure of the local–global principle for X is explained by a Brauer–Manin obstruction.

There are examples due to Skorobogatov [Sk99], e.g., some elliptic surfaces, where the Brauer-Manin obstruction fails to explain the failure of a local–global principle. However, it is a recent widespread conjecture due among others to Scharaschkin [Sch98] see also [ErSch06], Poonen [Po06] and Stoll [St07] Conjecture 9.1, that for smooth curves the Brauer–Manin obstruction will suffice to decide the existence of a rational point, see Sect. 11.9.

11.2   The Brauer–Manin Obstruction for Sections

We now aim to establish a Brauer–Manin obstruction on adelic spaces of sections.

Definition 123. The space of *adelic sections* of $\pi_1(X/k)$ is the subset

$$\mathscr{S}_{\pi_1(X/k)}(\mathbb{A}_k) \subseteq \prod_v \mathscr{S}_{\pi_1(X/k)}(k_v)$$

of all tuples (s_v) in the product over all places v of k, such that for every continuous homomorphism $\varphi : \pi_1(X) \to G$ with G finite the composites $\varphi \circ s_v : \mathrm{Gal}_{k_v} \to G$ are unramified for almost all v.

Proposition 124. *The natural maps yield a commutative diagram*

$$
\begin{array}{ccc}
X(k) & \longrightarrow & X(\mathbb{A}_k)_\bullet \\
\downarrow & & \downarrow \\
\mathscr{S}_{\pi_1(X/k)} & \longrightarrow & \mathscr{S}_{\pi_1(X/k)}(\mathbb{A}_k).
\end{array}
$$

Proof. For a section s of $\pi_1(X/k)$ and a homomorphism $\varphi : \pi_1(X) \to G$ with G finite, the composite

$$\varphi \circ s : \mathrm{Gal}_k \to G$$

describes a G-torsor over k which therefore is unramified almost everywhere. Hence the tuple $(s \otimes k_v)_v$ is an adelic section.

Secondly, for an adelic point $(x_v) \in X(\mathbb{A}_k)_\bullet$ the tuple of associated sections s_{x_v} is adelic. Indeed, the G-torsor over X corresponding to a homomorphism $\varphi : \pi_1(X) \to G$ with G finite extends to a G-torsor over some flat model $\mathscr{X}/B$ for a nonempty open

$$B \subset \mathrm{Spec}(\mathfrak{o}_k)$$

for the integers $\mathfrak{o}_k$ in k. By the adelic condition, for almost all places v the point $x_v \in X(k_v)$ actually is a point

$$x_v : \mathrm{Spec}(\mathfrak{o}_v) \to \mathscr{X}.$$

This means that $\varphi \circ s_{x_v}$ factors as in the following diagram

$$
\begin{array}{ccccccc}
\mathrm{Gal}_{k_v} & \xrightarrow{\ s_{x_v}\ } & \pi_1(X \times_k k_v) & \subseteq & \pi_1(X) & \xrightarrow{\ \varphi\ } & G \\
\downarrow & & \downarrow & & \downarrow & \nearrow & \\
\pi_1(\mathrm{Spec}(\mathfrak{o}_v)) & \xrightarrow{\ \pi_1(x_v)\ } & \pi_1(\mathscr{X} \times_B \mathfrak{o}_v) & \longrightarrow & \pi_1(\mathscr{X}) & &
\end{array}
$$

which shows that for such v the composite $\varphi \circ s_{x_v}$ is unramified. □

Proposition 125. *Let X/k be a proper geometrically connected variety. Then all tuples in*

$$\prod_v \mathscr{S}_{\pi_1(X/k)}(k_v)$$

are adelic sections of $\pi_1(X/k)$.

Proof. Let $\varphi : \pi_1(X) \to G$ be a homomorphism with G finite and (s_v) a tuple of local sections. Let $\mathscr{X}/B$ be a flat proper model such that the G-torsor corresponding to φ has good reduction over $\mathscr{X}$ and such that $\#G$ is invertible on B. For all places $v \in B$ the restriction of $\varphi \circ s_v$ to the inertia group I_{k_v} factors over the ramification ρ_{s_v} that by Proposition 91 has a pro-p group as its image and therefore dies in G. Hence for all such places v the corresponding $\varphi \circ s_v$ is unramified.   □

Let $\alpha \in H^2(\pi_1(X), \mu_n)$ be represented by the extension

$$1 \to \mu_n \to E_\alpha \to \pi_1(X) \to 1.$$

The evaluation of α in a section s of $\pi_1(X/k)$, respectively in s_v, is the Brauer class $s^*(\alpha) \in H^2(k, \mu_n)$, respectively $s_v^*(\alpha) \in H^2(k_v, \mu_n)$, that is represented by the pullback via s, respectively s_v, of the extension E_α.

Proposition 126. *Let (s_v) be an adelic section of $\pi_1(X/k)$. Then $s_v^*(\alpha)$ vanishes for all but finitely many places v.*

Proof. The extension E_α comes by inflation via a homomorphism $\varphi : \pi_1(X) \to G$ with G finite from a class in $H^2(G, \mu_n)$ represented by an extension

$$1 \to \mu_n \to E \to G \to 1.$$

The class $s_v^*(\alpha)$ vanishes if and only if the composite $\varphi \circ s_v$ lifts to E. The adelic condition on (s_v) yields that for almost all v the map $\varphi \circ s_v$ factors over $\mathrm{Gal}_{\kappa(v)} \cong \hat{\mathbb{Z}}$ and thus lifts.   □

Proposition 126 allows the following definition of a function on adelic sections.

$$\langle \alpha, - \rangle \; : \; \mathscr{S}_{\pi_1(X/k)}(\mathbb{A}_k) \to \mathbb{Q}/\mathbb{Z},$$

$$(s_v) \mapsto \sum_v \mathrm{inv}_v\left(s_v^*(\alpha)\right)$$

Theorem 127. *The image of the natural map $\mathscr{S}_{\pi_1(X/k)} \to \mathscr{S}_{\pi_1(X/k)}(\mathbb{A}_k)$ lies in the Brauer kernel*

$$\mathscr{S}_{\pi_1(X/k)}(\mathbb{A}_k)^{\mathrm{Br}} := \left\{ (s_v) \in \mathscr{S}_{\pi_1(X/k)}(\mathbb{A}_k) \; ; \; \begin{array}{l} \langle \alpha, (s_v) \rangle = 0 \; for \; all \; n \in \mathbb{N} \\ and \; \alpha \in H^2(\pi_1(X), \mu_n) \end{array} \right\}.$$

Proof. For a section $s \in \mathscr{S}_{\pi_1(X/k)}$ and $\alpha \in H^2(\pi_1(X), \mu_n)$ we have

$$\langle \alpha, (s \otimes k_v)_v \rangle = \sum_v \mathrm{inv}_v\left((s \otimes k_v)^*(\alpha)\right)$$

$$= \sum_v \mathrm{inv}_v\left(s^*(\alpha) \otimes_k k_v\right) = \left(\sum_v \mathrm{inv}_v\right)(s^*(\alpha)) = 0,$$

that vanishes by the local global principle for Brauer groups (11.2).   □

The connection with the Brauer–Manin obstruction for rational points considers the composition

$$b \; : \; \mathrm{H}^2(\pi_1(X), \mu_n) \subseteq \mathrm{H}^2(X, \mu_n) \twoheadrightarrow \mathrm{Br}(X)[n]$$

of the comparison map with the quotient map from the Kummer sequence (10.2) for X. The naturality of the Kummer sequence and

$$\mathrm{H}^2(k_v, \mu_n) = \mathrm{Br}(k_v)[n]$$

imply

$$\langle b(\alpha), (x_v) \rangle = \langle \alpha, (s_{x_v}) \rangle \tag{11.4}$$

and thus a commutative diagram

$$\begin{array}{ccccccc}
X(k) & \longrightarrow & X(\mathbb{A}_k)_\bullet^{\mathrm{Br}} & \subseteq & X(\mathbb{A}_k)_\bullet & & \\
\downarrow{\scriptstyle \kappa} & & \downarrow{\scriptstyle \kappa} & & \downarrow{\scriptstyle \kappa} & \searrow^{\langle b(\alpha), - \rangle} & \\
& & & & & & \mathbb{Q}/\mathbb{Z}. \\
\mathscr{S}_{\pi_1}(X/k) & \longrightarrow & \mathscr{S}_{\pi_1(X/k)}(\mathbb{A}_k)^{\mathrm{Br}} & \subseteq & \mathscr{S}_{\pi_1(X/k)}(\mathbb{A}_k) & \nearrow_{\langle \alpha, - \rangle} &
\end{array}$$

$$\tag{11.5}$$

Remark 128. It is by no means clear, albeit predicted by the local section conjecture, that the function $\langle \alpha, (s_v) \rangle$ on adelic sections only depends on $b(\alpha) \in \mathrm{Br}(X)$. Our choice of notation $\mathscr{S}_{\pi_1(X/k)}(\mathbb{A}_k)^{\mathrm{Br}}$ is short and suggestive and hopefully that sufficiently justifies its use.

Proposition 129. *Let X/k be a smooth, geometrically connected curve. Then the set $X(\mathbb{A}_k)_\bullet^{\mathrm{Br}}$ is the intersection of $X(\mathbb{A}_k)_\bullet$ and $\mathscr{S}_{\pi_1(X/k)}(\mathbb{A}_k)^{\mathrm{Br}}$ inside $\mathscr{S}_{\pi_1(X/k)}(\mathbb{A}_k)$.*

Proof. It suffices to prove that $X(\mathbb{A}_k)_\bullet$ is empty or that the map

$$b : \mathrm{H}^2(\pi_1(X), \mu_n) \to \mathrm{Br}(X)[n]$$

is surjective for every $n \geq 1$. Unless X is a form of $\mathbb{P}_k^1$, a smooth curve X has

$$\mathrm{H}^2(\pi_1(X), \mu_n) = \mathrm{H}^2(X, \mu_n),$$

see [Sx02] Appendix A.4. Thus the second condition follows from (10.2).

A form of $\mathbb{P}_k^1$ with local points everywhere is isomorphic to $\mathbb{P}_k^1$ by the Hasse local–global principle. So the non-trivial forms X have $X(\mathbb{A}_k)_\bullet = \emptyset$. It remains to discuss $X = \mathbb{P}_k^1$. But then the composite

$$\mathrm{H}^2(k, \mu_n) = \mathrm{H}^2(\pi_1(X), \mu_n) \subset \mathrm{H}^2(X, \mu_n) \xrightarrow{b} \mathrm{Br}(X)[n]$$

is an isomorphism, because $c_1(\mathcal{O}(1))$ kills the part of $H^2(X, \mu_n)$ which does not come from $H^2(\pi_1(X), \mu_n)$.   □

Remark 130. (1) That a section s which locally belongs to points $x_v \in X(k_v)$, namely $s_{x_v} = s \otimes k_v$ leads to an adelic point $(x_v) \in X(\mathbb{A}_k)_\bullet$ which survives all Brauer–Manin obstruction was also observed by Wittenberg, and probably others.

(2) It was finally proven in [HaSx12] Corollary 3.1, that for a smooth, projective geometrically connected curve X/k, a section s of $\pi_1^{ab}(X/k)$ such that $s \otimes k_v = s_{x_v}$ for an adelic point $(x_v) \in X(\mathbb{A}_k)$, then (x_v) already lies in $X(\mathbb{A}_k)_\bullet^{Br}$.

(3) The result (2) implicitly comments on whether the Brauer–Manin obstruction goes *beyond the abelian quotient* into perspective, see [Sx11] §6. As an argument in favour of the *non-abelian* content of the Brauer–Manin obstruction [Sx11] cites a curve found in [BLM84], see [Sk99] p. 128, which has index 1 and adelic points but no global point. It turns out that none of the sections of $\pi_1^{ab}(X/k)$ actually come from points locally everywhere. The non-abelian content of the Brauer–Manin obstruction to rational points here is hidden in the constraint that a global section of $\pi_1^{ab}(X/k)$ shall be adelic with respect to points on X.

(4) There are examples of curves with adelic points for which the absence of sections could be verified by establishing a Brauer–Manin obstruction. These are (arithmetic twists of) the affine Reichardt–Lind curve over $\mathbb{Q}$, see [Sx11] §5 and originally [Li40] [Re42], and as an improvement Schinzel's curve

$$2(Y^2 + 4Z^2)^2 = X^4 - 17Z^4,$$

still over $\mathbb{Q}$, see [Wi12] Theorem 1.1.

(5) It would be interesting to take a known example of a smooth, projective surface S/k whose absence of rational points is due to a Brauer–Manin obstruction and to show that there are also no sections of $\pi_1(S/k)$ due to again a (probably the same) Brauer–Manin obstruction. Possible candidates are the Cassels–Guy cubic surface

$$5X^3 + 12Y^3 + 9Z^3 + 10W^3 = 0$$

in $\mathbb{P}^3_{\mathbb{Q}}$, or the cubic surface used by Swinnerton–Dyer, see [Ma71] §5,

$$T(T + X)(2T + X) = \prod (X + \theta Y + 2\theta Z)$$

again in $\mathbb{P}^3_{\mathbb{Q}}$ with the product ranging over roots θ of $\theta^3 - 7\theta^2 + 14\theta - 7 = 0$.

Any curve with a nontrivial map $X \rightarrow S$ for example the rational fibres of a fibration $S \rightarrow B$ into curves would yield an example of a curve with neither sections nor points, hence giving rise to families of examples for the section conjecture. By more complicated means following an argument of Elkies we establish in [Sx11] Proposition 33, that the rational fibres of an isotrivial family of twists of the affine Reichardt–Lind curve all neither have points nor sections.

11.3   The Descent Obstruction for Rational Points

Having dealt with the Brauer–Manin obstruction above, we now turn our attention to the descent obstruction. To an algebraic group G/k and a right G-torsor $h : Y \to X$ for the étale topology on X representing an element $[h] \in \mathrm{H}^1(X, G)$ is associated the descent obstruction

$$X(\mathbb{A}_k)_\bullet^h = \left\{ (x_v) \in X(\mathbb{A}_k)_\bullet \; ; \; (x_v^*([h]))_v \in \mathrm{diag}\left(\mathrm{H}^1(k, G) \to \prod_v \mathrm{H}^1(k_v, G) \right) \right\}$$

where diag is the diagonal restriction map from global to local cohomology, see [Sk01] §5.3. There is an equivalent description

$$X(\mathbb{A}_k)_\bullet^h = \bigcup_{\tau \in \mathrm{H}^1(k,G)} {}^\tau h\left({}^\tau Y(\mathbb{A}_k)_\bullet \right) \qquad (11.6)$$

exploiting twists

$$ {}^\tau h : {}^\tau Y \to X $$

of the torsor by a cocycle representing $\tau \in \mathrm{H}^1(k, G)$. Clearly, global points satisfy the descent obstruction posed by any torsor

$$X(k) \subseteq X(\mathbb{A}_k)_\bullet^h.$$

The descent obstruction is defined as the intersection

$$X(\mathbb{A}_k)_\bullet^{\mathrm{descent}} = \bigcap_{G,h} X(\mathbb{A}_k)_\bullet^h \qquad (11.7)$$

over all torsors for all algebraic groups. The descent obstruction succeeds to show absence of rational points if $X(\mathbb{A}_k)_\bullet^h = \emptyset$, namely if a suitable torsor $h : Y \to X$ together with all its twists lacks local points at some place which might depend on the twist.

The connection between the descent obstruction and the Brauer–Manin obstructions was clarified recently by work of Demarche [De09] and Skorobogatov [Sk09] and is summarized by the chain of inclusions

$$X(k) \subseteq X(\mathbb{A}_k)_\bullet^{\mathrm{descent}} = X(\mathbb{A}_k)_\bullet^{\text{ét-Br}} \subseteq X(\mathbb{A}_k)_\bullet^{\mathrm{Br}} \subseteq X(\mathbb{A}_k)_\bullet. \qquad (11.8)$$

Here $X(\mathbb{A}_k)_\bullet^{\text{ét-Br}}$ is a finite étale stabilized version of the Brauer obstruction set. All inclusions are known to be potentially strict, see [Sk99] and [Po08].

11.4   Torsors Under Finite Étale Groups in Terms of Fundamental Groups

The descent obstruction has been studied in relation to sections in [HaSx12]. Here we would like to give an alternative description in terms of fundamental groups and sections, in order to establish the descent obstruction also on the space of adelic sections.

Groups and torsors. A finite étale group G/k will be treated as the finite group $G(\bar{k})$ together with its action by Gal_k. In the sequel we will indistinguishably use the notation G for a finite Gal_k-group and for the corresponding finite étale group G/k. The semidirect product $G \rtimes \mathrm{Gal}_k$ shall be constructed by means of the Gal_k-action given by the structure of G as a Gal_k-group.

Definition 131. Let G be a finite Gal_k-group, and let X/k be a geometrically connected variety with a geometric point $\bar{x} \in X$.

(1) A *right G-torsor on X (in terms of π_1-data)* is defined as a $\pi_1(X, \bar{x})$-set M together with a $\pi_1(X, \bar{x})$-equivariant transitive free right action by G.
(2) A *pointed right G-torsor on X (in terms of π_1-data)* is a right G-torsor M together with a marked element $m \in M$.

Lemma 132. *There is a natural bijection between the following three sets of objects up to isomorphism.*

(a) *Pointed G-torsors in terms of π_1-data (M, m).*
(b) *G-torsors $h : Y \to X$ together with a point $\bar{y} \in h^{-1}(\bar{x})$.*
(c) *Continuous group homomorphisms*

$$\varphi : \pi_1(X, \bar{x}) \to G \rtimes \mathrm{Gal}_k$$

that are compatible with the projection to Gal_k.

If we forget the pointing, respectively consider φ up to conjugation by an element of G, then all three isomorphism classifications are naturally in bijection with

$$\mathrm{H}^1(\pi_1(X, \bar{x}), G) = \mathrm{H}^1(X, G).$$

Proof. The equivalence of (a) and (b) follows from the fact that finite étale covers and finite continuous $\pi_1(X, \bar{x})$-sets are equivalent by taking the fibre above $\bar{x}$, and that this fibre functor is compatible with products. The latter transforms the G-action $Y \times G \to Y$ over X to a right G-action on $M = h^{-1}(\bar{x})$.

The equivalence of (a) and (c) is constructed as follows. Let s_0 be the standard section $\mathrm{Gal}_k \to G \rtimes \mathrm{Gal}_k$. Then we map

$$\varphi : \pi_1(X, \bar{x}) \to G \rtimes \mathrm{Gal}_k$$

to the set

$$M_\varphi = G \rtimes \mathrm{Gal}_k / s_0(\mathrm{Gal}_k)$$

with left action by $\pi_1(X, \bar{x})$ via φ and left translation, and with right G-action via the preferred representatives

$$G = G \rtimes \mathrm{Gal}_k / s_0(\mathrm{Gal}_k)$$

and right translation. The marked element of M_φ shall be $1 \in G \rtimes \mathrm{Gal}_k / s_0(\mathrm{Gal}_k)$. That $(M_\varphi, 1)$ is indeed endowed with a $\pi_1(X, \bar{x})$-equivariant transitive free right action by G follows from the following computation. For $\gamma \in \pi_1(X, \bar{x})$, $g \in G$ and $m \in M_\varphi$ given as the preferred representative in G, we have

$$\gamma.(m.g) = \varphi(\gamma)(mg)$$

$$= \Big(\varphi(\gamma)m s_0(\mathrm{pr}_*(\gamma))^{-1}\Big)\Big(s_0(\mathrm{pr}_*(\gamma))g s_0(\mathrm{pr}_*(\gamma))^{-1}\Big) = (\gamma.m).(\gamma.g)$$

where $\mathrm{pr}_* : \pi_1(X, \bar{x}) \to \mathrm{Gal}_k$ is the canonical map.

For the converse, we assign to a pointed G-torsor (M, m) the map

$$\varphi : \pi_1(X, \bar{x}) \to G \rtimes \mathrm{Gal}_k$$

$$\gamma \mapsto \big(\tau_\gamma, \mathrm{pr}_*(\gamma)\big)$$

with $\gamma.m = m.(\tau_\gamma)$. The map φ is a group homomorphism if and only if $\gamma \mapsto \tau_\gamma$ is a 1-cocycle on $\pi_1(X, \bar{x})$ with values in G, which follows from

$$m.\tau_{\gamma\delta} = (\gamma\delta).m = \gamma.(m.\tau_\delta) = (m.\tau_\gamma).(\gamma.\tau_\delta) = m.\big(\tau_\gamma \gamma(\tau_\delta)\big).$$

The isomorphism classification of the unpointed structures by $\mathrm{H}^1(\pi_1(X, \bar{x}), G)$ is clear.   □

Lemma 133. *Let* $\varphi : \pi_1(X, \bar{x}) \to G \rtimes \mathrm{Gal}_k$ *be a pointed right G-torsor that corresponds to* $h : Y \to X$ *with* $\bar{y} \in Y$. *Let* Y^0 *be the connected component of Y that contains* $\bar{y}$. *Then* $\pi_1(h)$ *identifies* $\pi_1(Y^0, \bar{y})$ *with* $\varphi^{-1}(s_0(\mathrm{Gal}_k))$.

Proof. The subgroup $\pi_1(h)\big(\pi_1(Y, \bar{y})\big)$ is the stabilizer of $\bar{y}$ in $h^{-1}(\bar{x})$ which is the stabilizer of $1 \in M_\varphi$ and clearly agrees with $\varphi^{-1}(s_0(\mathrm{Gal}_k))$.   □

Connectedness of torsors. The base change to $\bar{k}$ of a pointed G-torsor on X is described by the restriction to $\pi_1(\overline{X}, \bar{x})$ of any of the equivalent structures of Lemma 132, in particular by a homomorphism

$$\overline{\varphi} : \pi_1(\overline{X}, \bar{x}) \to G.$$

Without the pointing, the homomorphism $\overline{\varphi}$ is only well defined up to G-conjugation, namely as an element

$$[\overline{\varphi}] \in H^1(\overline{X}, G) = H^1\left(\pi_1(\overline{X}, \bar{x}), G\right) = \mathrm{Hom}_{\mathrm{out}}\left(\pi_1(\overline{X}), G\right)$$

of the group of outer homomorphism.

Lemma 134. *Let M be a pointed G-torsor in π_1-data corresponding to the G-torsor $h : Y \to X$, and let $\varphi : \pi_1(X, \bar{x}) \to G \rtimes \mathrm{Gal}_k$ be the corresponding homomorphism.*

(1) The following are equivalent.

    (a) M is geometrically connected, i.e.,

$$\overline{\varphi} = \varphi|_{\pi_1(\overline{X}, \bar{x})} : \pi_1(\overline{X}, \bar{x}) \to G$$

       is surjective.
    (b) $\pi_1(\overline{X}, \bar{x})$ acts transitively on M.
    (c) φ is surjective.
    (d) Y/k is geometrically connected.

(2) The following are equivalent.

    (e) M is connected, i.e., $\pi_1(X, \bar{x})$ acts transitively on M.
    (f) The product set $\varphi(\pi_1(X, \bar{x})) \cdot s_0(\mathrm{Gal}_k)$ is all of $G \rtimes \mathrm{Gal}_k$.
    (g) Y is connected.

In particular, if M is geometrically connected, then it is also connected.

Proof. (1) It follows from the theory of Galois categories that (a), (b) and (d) are equivalent. The equivalence of (a) and (c) follows from an easy diagram chase.

    (2) The equivalence of (e) and (g) follows from the theory of Galois categories. The equivalence of (e) and (f) is a consequence of $M = G \rtimes \mathrm{Gal}_k / s_0(\mathrm{Gal}_k)$. $\square$

11.5 Twisting

We are going to explain twisting of torsors in terms of fundamental group data.

Twisting of groups. Let τ be a 1-cocycle of Gal_k with values in the finite Gal_k-group G. The *inner twist* $^\tau G$ of G by τ has the same underlying finite group, but the Gal_k-action is modified as follows. The twisted action $\sigma._\tau g$ of $\sigma \in \mathrm{Gal}_k$ on $g \in G$ is given by

$$\sigma._\tau g = \tau_\sigma(\sigma.g)(\tau_\sigma)^{-1}. \tag{11.9}$$

That $^{\tau}G$ is again a Gal_k-group by (11.9) can for example be seen by Proposition 8.
The original action ρ in the diagram

$$
\begin{array}{ccccccccc}
 & & & & & & \mathrm{Gal}_k & & \\
 & & & & {\scriptstyle\rho}\nearrow & & \downarrow{\scriptstyle\bar\rho} & & \\
1 & \longrightarrow & \mathrm{Inn}(G) & \longrightarrow & \mathrm{Aut}(G) & \longrightarrow & \mathrm{Out}(G) & \longrightarrow & 1
\end{array}
$$

was twisted by the cocycle $\mathrm{int}(\tau)$ which is the image of τ under the natural map

$$\mathrm{H}^1(k, G) \to \mathrm{H}^1(k, \mathrm{Inn}(G))$$

to the new action $\mathrm{int}(\tau)\rho$. The isomorphism class of the twist $^{\tau}G$ does only depend
on the class $\tau \in \mathrm{H}^1(k, G)$ and more precisely on

$$\mathrm{int}(\tau) \in \mathrm{H}^1(k, \mathrm{Inn}(G)).$$

The twist bitorsor. Let τ be a 1-cocycle of Gal_k with values in the finite Gal_k-
group G. Then we denote by $P(\tau)$ the pointed Gal_k-equivariant $(G, {}^{\tau}G)$-bitorsor
with underlying set $P(\tau) = G$, distinguished point $\mathbf{1} \in G$, action by $(g, h) \in$
$G \times {}^{\tau}G$ on $p \in P(\tau) = G$ by

$$g.(p.h) = gph = (g.p).h$$

and Gal_k-action of $\sigma \in \mathrm{Gal}_k$ on $p \in P(\tau) = G$ by

$$\sigma._{\tau} p = \sigma.p(\tau_\sigma)^{-1}.$$

Indeed, the bitorsor structure is equivariant:

$$\sigma._{\tau}(gph) = (\sigma.g)(\sigma.p)(\sigma.h)(\tau_\sigma)^{-1}$$

$$= (\sigma.g)\Big((\sigma.p)(\tau_\sigma)^{-1}\Big)\tau_\sigma(\sigma.h)(\tau_\sigma)^{-1} = (\sigma.g)(\sigma._{\tau} p)(\sigma._{\tau} h).$$

Twisting of torsors. We keep the notation of the preceding section. Let X/k be
a geometrically connected variety with a geometric point $\bar x \in X$. The *twist* of a
pointed G-torsor (M, m_1) on X is the pointed $^{\tau}G$-torsor $(^{\tau}M, {}^{\tau}m_1)$ given by

$$^{\tau}M = M \times_G P(\tau)$$

which is the quotient of $M \times P(\tau)$ by the equivalence relation

$$(mg, p) \sim (m, gp)$$

for all $g \in G, m \in M$ and $p \in P(\tau)$, with distinguished point ${}^\tau m_1 = (m_1, \mathbf{1})$. Here the $\pi_1(X, \bar{x})$-action on $P(\tau)$ is through the natural projection $\pi_1(X, \bar{x}) \twoheadrightarrow \mathrm{Gal}_k$. The isomorphism type of the twist does only depend on the class $\tau \in \mathrm{H}^1(k, G)$ which therefore induces a map

$$\mathrm{H}^1(X, G) \to \mathrm{H}^1(X, {}^\tau G).$$

Following Lemma 132 the G-torsors on X occur in three incarnations. Twisting has a description in each, and the natural identifications are compatible with twisting. The twist of the corresponding G-torsor $h : Y \to X$ is the ${}^\tau G$-torsor

$$ {}^\tau h \,:\, {}^\tau Y = T \times_G P(\tau) \to X$$

which is the quotient of

$$Y \times P(\tau) = \coprod_{p \in P(\tau)} Y$$

by the antidiagonal G-action, and where $P(\tau)$ is the corresponding finite étale k-scheme. The pointing is treated similarly to the case of ${}^\tau M$.

The corresponding group homomorphism

$$\varphi : \pi_1(X, \bar{x}) \to G \rtimes \mathrm{Gal}_k$$

is twisted to

$$ {}^\tau\varphi : \pi_1(X, \bar{x}) \to {}^\tau G \rtimes \mathrm{Gal}_k$$

in the following way. The projection $G \rtimes \mathrm{Gal}_k \to \mathrm{Gal}_k$ has a standard section s_0 and, according to Proposition 8, a twisted section ${}^\tau s_0$. Using this twisted section as the standard section we obtain an isomorphism

$$G \rtimes \mathrm{Gal}_k \xrightarrow{\sim} {}^\tau G \rtimes \mathrm{Gal}_k$$

given by

$$(g, \sigma) \mapsto (g\tau_\sigma, \sigma).$$

Another way of seeing this isomorphism is as the identity of $G \rtimes \mathrm{Gal}_k$, but shifting the semi-direct product structure. This view will be adopted in the sequel, because it is most convenient for the purpose of the section conjecture. Now ${}^\tau\varphi = \varphi$, but with the target group equipped by a new standard section.

Lemma 135. *With the notation as above we have a natural identification*

$$M_{\tau\varphi} = {}^\tau(M_\varphi)$$

of pointed ${}^\tau G$-torsors on X.

Proof. We have a bijection

$$f : {}^{\tau}M_{\varphi} = G \rtimes \mathrm{Gal}_k / s_0(\mathrm{Gal}_k) \times_G P(\tau) \xrightarrow{\sim} M_{\tau\varphi} = G \rtimes \mathrm{Gal}_k / {}^{\tau}s_0(\mathrm{Gal}_k)$$

by $(g, p) \mapsto gp$ for representatives $g, p \in G$. For $\gamma \in \pi_1(X, \bar{x})$ with $\sigma = \mathrm{pr}_*(\gamma)$ we compute, essentially in $G \rtimes \mathrm{Gal}_k$,

$$\gamma . f(g, p) = {}^{\tau}\varphi(\gamma) g p \big({}^{\tau}s_0(\sigma)^{-1} \big) = \varphi(\gamma) g p (\tau_\sigma s_0(\sigma))^{-1}$$

$$= \varphi(\gamma) g \big(p (s_0(\sigma))^{-1} \big) (\tau_\sigma)^{-1} = \varphi(\gamma) g (s_0(\sigma))^{-1} \sigma . p(\tau_\sigma)^{-1}$$

$$= (\gamma . g)(\sigma ._\tau p) = f(\gamma .(g, p)),$$

thus f is $\pi_1(X, \bar{x})$-equivariant. As f is clearly right ${}^{\tau}G$-equivariant, the lemma is proven. □

Pullback of torsors. Let $f : X' \to X$ be a map of geometrically connected k-varieties with compatible geometric points $\bar{x} = f(\bar{x}')$. Then composing with

$$f_* : \pi_1(X', \bar{x}') \to \pi_1(X, \bar{x})$$

defines a map

$$f^* : \mathrm{H}^1(X, G) \to \mathrm{H}^1(X', G)$$

$$\varphi \mapsto \varphi \circ f_*$$

that is compatible with twisting by $\tau \in \mathrm{H}^1(k, G)$, namely ${}^{\tau}\varphi \circ f_* = {}^{\tau}(\varphi \circ f_*)$.

Fibres. The fibre of a right G-torsor $\varphi : \pi_1(X, \bar{x}) \to G \rtimes \mathrm{Gal}_k$ in a section s is the pullback

$$s^*\varphi = \varphi \circ s : \mathrm{Gal}_k \to G \rtimes \mathrm{Gal}_k$$

which represents an element $s^*\varphi \in \mathrm{H}^1(k, G)$. Changing s by a $\pi_1(\overline{X}, \bar{x})$-conjugate, leads to a G-conjugate of the section $s^*\varphi$, hence to the same class $s^*\varphi$. We therefore find a well defined evaluation map

$$\mathscr{S}_{\pi_1(X/k)} \to \mathrm{H}^1(k, G)$$

$$s \mapsto s^*\varphi.$$

Lemma 136. *The following are equivalent.*

(i) *The fibre $s^*\varphi$ is the trivial torsor.*
(ii) *For a suitable G-conjugate of the standard section $s_0 : \mathrm{Gal}_k \to G \rtimes \mathrm{Gal}_k$ we have $\varphi \circ s = s_0$.*
(iii) *The section s lifts to $\pi_1(Y^0, \bar{y})$ where Y^0 is the connected component of the pointing $\bar{y} \in Y$ in the corresponding G-torsor $h : Y \to X$.*

Proof. This is an obvious consequence of Lemma 133. □

11.6   The Descent Obstruction for Sections

With twisting formulated in the language of fundamental groups at our disposal, we can now address the descent obstruction for sections.

Definition 137. Let X/k be a geometrically connected variety over an algebraic number field k with geometric point $\bar{x} \in X$, and let G be a finite Gal_k-group. Let

$$\varphi : \pi_1(X, \bar{x}) \to G \rtimes \mathrm{Gal}_k$$

be a right G-torsor on X. For each place v of k with completion k_v the pullback of φ to $X_v = X \times_k k_v$ is denoted by φ_v. The *descent obstruction posed by φ on adelic sections* of $\pi_1(X/k)$ is the subset

$$\left(\mathscr{S}_{\pi_1(X/k)}(\mathbb{A}_k) \right)^{\varphi} = \left\{ (s_v) \; ; \; (s_v^* \varphi_v)_v \in \mathrm{diag}\big(\mathrm{H}^1(k, G) \to \prod_v \mathrm{H}^1(k_v, G) \big) \right\}$$

of $\mathscr{S}_{\pi_1(X/k)}(\mathbb{A}_k)$.

Theorem 138. *The image of the natural map $\mathscr{S}_{\pi_1(X/k)} \to \mathscr{S}_{\pi_1(X/k)}(\mathbb{A}_k)$ lies in the finite descent obstruction set on adelic sections*

$$\mathscr{S}_{\pi_1(X/k)}(\mathbb{A}_k)^{\text{f-descent}} := \bigcap_{G, \varphi} \left(\mathscr{S}_{\pi_1(X/k)}(\mathbb{A}_k) \right)^{\varphi},$$

where G ranges over all finite Gal_k-groups and φ over all right G-torsors on X.

Proof. Clearly, for a global section s we find $s_v^* \varphi_v = (s^* \varphi)_v$ and thus global sections satisfy the descent obstruction posed by any φ.   □

Compatibility. Let $f : G \to G'$ be a group homomorphism of finite Gal_k-groups. Using f we can push a right G-torsor

$$\varphi : \pi_1(X, \bar{x}) \to G \rtimes \mathrm{Gal}_k$$

to a right G'-torsor $f_* \varphi$ defined by

$$f_* \varphi = (f, \mathrm{id}) \circ \varphi \; : \; \pi_1(X, \bar{x}) \to G' \rtimes \mathrm{Gal}_k \, .$$

Pushing obviously commutes with pullback, hence also with taking fibres in sections. The following lemma is an immediate consequence.

Lemma 139. *(1) There is an inclusion of descent obstructions*

$$\left(\mathscr{S}_{\pi_1(X/k)}(\mathbb{A}_k) \right)^{\varphi} \subseteq \left(\mathscr{S}_{\pi_1(X/k)}(\mathbb{A}_k) \right)^{f_* \varphi}.$$

(2) *Let τ be a 1-cocycle representing a class in $\mathrm{H}^1(k, G)$. Then the descent obstruction sets*

$$\left(\mathscr{S}_{\pi_1(X/k)}(\mathbb{A}_k)\right)^{\tau_\varphi} = \left(\mathscr{S}_{\pi_1(X/k)}(\mathbb{A}_k)\right)^\varphi$$

agree.

Proof. (1) Let (s_v) be an adelic section that survives the descent obstruction posed by φ. Then there is a class $\alpha \in \mathrm{H}^1(k, G)$ with image

$$\alpha_v = s_v^* \varphi_v \in \mathrm{H}^1(k_v, G).$$

The class $f_* \alpha \in \mathrm{H}^1(k, G')$ provides the interpolation for

$$(f_* \alpha)_v = f_* \alpha_v = f_*(s_v^* \varphi_v) = s_v^*(f_* \varphi_v) = s_v^*(f_* \varphi)_v$$

and (s_v) also survives the descent obstruction posed by $f_* \varphi$.

(2) The description of the twist $^\tau\varphi$ shows that conversely also φ is a twist of $^\tau\varphi$. By symmetry, it suffices therefore to show an inclusion

$$\left(\mathscr{S}_{\pi_1(X/k)}(\mathbb{A}_k)\right)^\varphi \subseteq \left(\mathscr{S}_{\pi_1(X/k)}(\mathbb{A}_k)\right)^{\tau_\varphi}.$$

With (s_v) and α as above, now the class $^\tau\alpha \in \mathrm{H}^1(k, {}^\tau G)$ provides the interpolation for

$$(^\tau\alpha)_v = {}^\tau(\alpha_v) = {}^\tau(s_v^*\varphi_v) = s_v^*({}^\tau(\varphi_v)) = s_v^*(({}^\tau\varphi)_v)$$

and (s_v) also survives the descent obstruction posed by $^\tau\varphi$.   □

Obstructions from connected torsors. Without passing to multi-Galois categories and anabelioids, the alternative description (11.9) of the descent obstruction set in the classical case of rational points is available only for torsors φ such that all twists $^\tau\varphi$ are geometrically connected. However, an ostensibly stronger form is the following.

Lemma 140. *Let $\varphi : \pi_1(X, \bar{x}) \to G \rtimes \mathrm{Gal}_k$ be a pointed G-torsor corresponding to $h : Y \to X$. Then the natural map $\mathscr{S}_{\pi_1(X/k)} \to \mathscr{S}_{\pi_1(X/k)}(\mathbb{A}_k)$ factors over the subset*

$$\bigcup_{\tau \in \mathrm{H}^1(k,G)} {}^\tau h_*(\mathscr{S}_{\pi_1(^\tau Y^0/k)}(\mathbb{A}_k)) \subseteq \left(\mathscr{S}_{\pi_1(X/k)}(\mathbb{A}_k)\right)^\varphi,$$

with τ ranging over those classes such that $^\tau Y^0/k$ is geometrically connected.

Proof. It suffices to treat the case of trivial τ with Y^0/k geometrically connected. Let (t_v) be an adelic section of $\pi_1(Y^0/k)$, and let $(s_v) = (h_*(t_v))$ be its image. Then

$$s_v^*(\varphi) = \varphi \circ s_v = \varphi \circ h_* \circ t_v$$

takes values in

$$\varphi(h_*(\pi_1(Y^0))) \subseteq G \rtimes \text{Gal}_k,$$

hence in $s_0(\text{Gal}_k)$ according to Lemma 133. It follows that $\varphi \circ s_v = s_0|_{\text{Gal}_{k_v}}$ and thus (s_v) survives the descent obstruction posed by φ.

If $s : \text{Gal}_k \to \pi_1(X, \bar{x})$ is a section, then $\varphi \circ s$ is a section of $G \rtimes \text{Gal}_k \twoheadrightarrow \text{Gal}_k$. Replacing φ by an appropriate twist, we may assume that $\varphi \circ s = s_0$. But then s lifts to a section t of $\pi_1(Y^0/k)$ because by Lemma 133 the subgroup

$$\pi_1(Y^0, \bar{y}) \subset \pi_1(X, \bar{x})$$

is nothing but the preimage of $s_0(\text{Gal}_k)$. In particular, this Y^0 must be geometrically connected over k. The local components of t map to the local components of s under h_* and thus the proof of the lemma is complete.   □

Remark 141. It follows from Theorem 144 below that the intersection over all possible (G, φ) of both sides of the inclusion in Lemma 140 actually agrees, see Corollary 150.

The following obstruction to sections works over an arbitrary field k of characteristic 0. For a number field k we could deduce the result from Lemma 140.

Theorem 142. *Let* $\varphi : \pi_1(X, \bar{x}) \to G \rtimes \text{Gal}_k$ *be a pointed G-torsor corresponding to* $h : Y \to X$, *such that all twists are connected. If there is a twist* $^\tau Y$ *which is not geometrically connected over k, then* $\pi_1(X/k)$ *does not split*

Proof. The effect of twisting the torsor by a 1-cocycle τ representing a class in $H^1(k, G)$ disappears after base change to $\bar{k}$. Hence, if a single twist $^\tau Y$ is not geometrically connected, then none of them is

We argue by contradiction. If s splits $\pi_1(X/k)$, then there is a 1-cocycle τ with $\varphi \circ s = \tau.s_0$. After replacing φ by $^\tau \varphi$ we may assume that $\varphi \circ s = s_0$. By Lemma 134 (e) and (f) we deduce that the connectedness of Y implies that φ is surjective, hence by (c) and (a) we may deduce that Y is even geometrically connected.   □

Strong approximation. We are going to prove a strong approximation result for sections.

Definition 143. The *constant finite descent obstruction set* on adelic sections is the set

$$\mathscr{S}_{\pi_1(X/k)}(\mathbb{A}_k)^{\text{cf-descent}} = \bigcap_{G,\varphi} \left(\mathscr{S}_{\pi_1(X/k)}(\mathbb{A}_k) \right)^\varphi,$$

where G ranges over all finite constant Gal_k-groups, i.e., the Galois action on G is trivial, and φ ranges over all right G-torsors on X.

Clearly, a priori the constant descent obstruction is weaker than the descent obstruction imposed by all torsors. But there is the following important result.

Theorem 144. *Let X/k be a geometrically connected variety over an algebraic number field.*

(1) The natural map

$$\mathscr{S}_{\pi_1(X/k)} \twoheadrightarrow \mathscr{S}_{\pi_1(X/k)}(\mathbb{A}_k)^{\text{cf-descent}}$$

is surjective.

(2) In particular, the constant descent obstruction is as strong as the descent obstruction for sections, i.e.,

$$\mathscr{S}_{\pi_1(X/k)}(\mathbb{A}_k)^{\text{f-descent}} = \mathscr{S}_{\pi_1(X/k)}(\mathbb{A}_k)^{\text{cf-descent}}.$$

Remark 145. (1) In [HaSx12] Theorem 2.1 the assertion of Theorem 144 was spelled out only for Selmer sections, i.e., sections that locally belong to rational points. The proof here is a straight forward extract of the proofs in [HaSx12].

(2) The map $\mathscr{S}_{\pi_1(X/k)} \to \mathscr{S}_{\pi_1(X/k)}(\mathbb{A}_k)$ would be part of a non-abelian profinite Tate–Poitou sequence, which currently exists only in fragments, see Chap. 12. Nevertheless, Theorem 144 already answers the question of the image of this map.

Before we start the proof, we recall the following finiteness result.

Lemma 146 (Minkowski–Hermite). *Let G be a finite Gal_k-group over an algebraic number field. Then the localisation map has finite fibres:*

$$\text{res} : H^1(k, G) \to \prod_v H^1(k_v, G).$$

Proof. A class $\tau \in H^1(k, G)$ corresponds to a finite étale k-algebra A/k, which is a G-torsor. A class $\tau' \in \text{res}^{-1}(\text{res}(\tau))$ corresponds to a finite étale k-algebra A'/k, which is locally isomorphic. Hence A' is unramified at the same places as A. Because ramification (resp. the degree) of the number fields which are factors of A' is bounded to a fixed finite set (resp. by $\#G$), the classical theorem of Minkowski–Hermite [Ne92] Theorem 2.13 shows that there are only finitely many possible such number fields up to isomorphism, and a fortiori only finitely many algebras A'. For the structure of a G-torsor on such an algebra A there are also only finitely many possibilities. This proves the lemma.   □

Proof (of Theorem 144). Clearly (2) follows from (1). To prove (1) we take an adelic section (s_v) that survives all torsors under constant groups. For a finite quotient

$$\alpha : \pi_1(X, \bar{x}) \twoheadrightarrow G_\alpha$$

we consider G_α as a constant k-group and further consider the torsor

$$\varphi_\alpha = (\alpha, \text{pr}) : \pi_1(X) \to G_\alpha \times \text{Gal}_k .$$

The set
$$T_\alpha = \{t : \mathrm{Gal}_k \to G_\alpha \; ; \; t|_{\mathrm{Gal}_{k_v}} \text{ is } G_\alpha\text{-conjugate to } \alpha \circ s_v\}$$

is non-empty because (s_v) survives the torsor φ_α, i.e., the class $\alpha \circ s_v = s_v^* \varphi_\alpha$ is in the image of
$$\mathrm{H}^1(k, G_\alpha) \to \prod_v \mathrm{H}^1(k_v, G_\alpha).$$

Moreover, by Lemma 146 the set T_α is finite. By [Bo98] I §9.6 Proposition 8 the projective limit $\varprojlim_\alpha T_\alpha$ is also nonempty. An element
$$t \in \varprojlim_\alpha T_\alpha$$

is nothing but a homomorphism $t : \mathrm{Gal}_k \to \pi_1(X)$ with the following interpolation property. For any α and v the set
$$C_{v,\alpha} = \{g \in G_\alpha \; ; \; g(-)g^{-1} \circ t|_{\mathrm{Gal}_{k_v}} = \alpha \circ s_v\}$$

is finite and nonempty. Thus there exist
$$c_v \in \varprojlim_\alpha C_{v,\alpha} \subseteq \pi_1(X, \bar{x})$$

which provides the equation
$$c_v(-)c_v^{-1} \circ t|_{\mathrm{Gal}_{k_v}} = s_v. \qquad (11.10)$$

So far, the homomorphism t need not be a section and also the interpolation of the s_v is only up to conjugation from $\pi_1(X, \bar{x})$. It turns out that these two defects are easily remedied. We apply Lemma 147 below to the composition $u = \mathrm{pr}_* \circ t$, which therefore is an isomorphism. The homomorphism $s = t \circ u^{-1}$ satisfies
$$\mathrm{pr}_* \circ s = \mathrm{pr}_* \circ (t \circ u^{-1}) = (\mathrm{pr}_* \circ t) \circ u^{-1} = u \circ u^{-1} = \mathrm{id}$$

and thus is a section of $\pi_1(X/k)$. Moreover, applying the projection to (11.10), we find with $\gamma_v = \mathrm{pr}_*(c_v)$ that
$$\gamma_v(-)\gamma_v^{-1} \circ u|_{\mathrm{Gal}_{k_v}} = \mathrm{id}_{\mathrm{Gal}_{k_v}}$$

or
$$u^{-1}|_{\mathrm{Gal}_{k_v}} = \gamma_v(-)\gamma_v^{-1}$$

so that
$$s|_{\mathrm{Gal}_{k_v}} = (t \circ u^{-1})|_{\mathrm{Gal}_{k_v}} = t \circ \left(\gamma_v(-)\gamma_v^{-1}\right)|_{\mathrm{Gal}_{k_v}}$$
$$= t(\gamma_v)(-)t(\gamma_v)^{-1} \circ t|_{\mathrm{Gal}_{k_v}} = t(\gamma_v)(-)t(\gamma_v)^{-1} \circ c_v^{-1}(-)c_v \circ s_v.$$

Since $t(\gamma_v)c_v^{-1}$ is in $\pi_1(\overline{X}, \bar{x})$, the section s interpolates the adelic section (s_v), i.e., the local components $s|_{\text{Gal}_{k_v}}$ are $\pi_1(\overline{X}, \bar{x})$-conjugate to s_v for all v. $\qquad\qquad\square$

Lemma 147. *Let u be an endomorphism of Gal_k such that $u|_{\text{Gal}_{k_v}} : \text{Gal}_{k_v} \to \text{Gal}_k$ is conjugation by an element of Gal_k for all places v. Then u is an automorphism.*

Proof. For every $\sigma \in \bigcup_v \bigcup_{g \in \text{Gal}_k} g\, \text{Gal}_{k_v}\, g^{-1}$ the image $u(\sigma)$ is conjugate to σ in Gal_k. By the Chebotarev density theorem the set $\bigcup_v \bigcup_{g \in \text{Gal}_k} g\, \text{Gal}_{k_v}\, g^{-1}$ is dense in Gal_k, and so u preserves every conjugacy class of Gal_k by continuity and compactness of Gal_k. In particular u is injective.

In every finite quotient $\text{Gal}_k \twoheadrightarrow G$ the image of $u(\text{Gal}_k)$ is a subgroup $H \leq G$ such that the union of the conjugates of H covers G. An old argument that goes back to at least Camille Jordan [Jo1872], namely the estimate

$$|G| = \Big| \bigcup_{g \in G/H} gHg^{-1} \Big| \leq (G : H) \cdot (|H| - 1) + 1 = |G| - (G : H) + 1 \leq |G|,$$

shows that necessarily $H = G$. Thus u is also surjective. $\qquad\qquad\square$

A part of the proof of Theorem 144 applied in the case when the auxiliary isomorphism u is actually the identity to start with, shows the following corollary, for which we give an immediate alternative proof.

Corollary 148. *Let X/k be a geometrically connected variety over an algebraic number field. Let (s_v) be an adelic section of $\pi_1(X/k)$ and let $s : \text{Gal}_k \to \pi_1(X, \bar{x})$ be a section that interpolates (s_v) up to conjugation by elements $\gamma_v \in \pi_1(X \otimes k_v, \bar{x})$. Then s actually interpolates (s_v) even up to conjugation by elements from $\pi_1(\overline{X}, \bar{x})$.*

Proof. The local Galois groups Gal_{k_v} for finite places v have trivial center. Hence γ_v must lie in $\pi_1(\overline{X}, \bar{x})$. On the other hand, for real infinite places v, the local Galois group is abelian. Thus we may alter γ_v to by $\gamma_v \cdot \big(s(\text{pr}_*(\gamma_v))\big)^{-1}$. $\qquad\qquad\square$

11.7 Consequences of Strong Approximation for Sections

The strong approximation result Theorem 144 has a couple of interesting consequences.

Comparison of finite descent with the Brauer–Manin obstruction. The strong approximation result Theorem 144 makes it easy to compare the finite descent obstruction with the Brauer–Manin obstruction, and to obtain an analogue of the result of Demarche [De09] and Skorobogatov [Sk09].

Definition 149. We define the *étale Brauer–Manin obstruction to adelic sections* as the subset

$$\mathscr{S}_{\pi_1(X/k)}(\mathbb{A}_k)^{\text{ét-Br}} = \bigcap_{G,h:Y\to X} \bigcup_\tau {}^\tau h_*\left(\mathscr{S}_{\pi_1(({}^\tau Y)^0/k)}(\mathbb{A}_k)^{\text{Br}}\right)$$

where G ranges over all finite Gal_k-groups, $h : Y \to X$ over all pointed G-torsors, and τ ranges over 1-cocycles representing classes in $\mathrm{H}^1(k, G)$, such that the k-variety $({}^\tau Y)^0$, which is the connected component of the distinguished point in the twist ${}^\tau Y$, is geometrically connected.

Corollary 150 (Analogue of Demarche–Skorobogatov). *Let X/k be a geometrically connected variety over an algebraic number field.*

(1) The Brauer–Manin obstruction to adelic sections is coarser than the finite descent obstruction, i.e.,

$$\mathscr{S}_{\pi_1(X/k)}(\mathbb{A}_k)^{\text{f-descent}} \subseteq \mathscr{S}_{\pi_1(X/k)}(\mathbb{A}_k)^{\text{Br}}.$$

(2) The étale Brauer–Manin obstruction is as strong as the finite descent obstruction, i.e.,

$$\mathscr{S}_{\pi_1(X/k)}(\mathbb{A}_k)^{\text{f-descent}} = \mathscr{S}_{\pi_1(X/k)}(\mathbb{A}_k)^{\text{ét-Br}}.$$

Proof. (1) As obviously the étale Brauer–Manin obstruction to adelic sections is contained in the Brauer kernel, we see that (2) implies (1).

(2) For a given torsor $h : Y \to X$, the natural map $\mathscr{S}_{\pi_1(X/k)} \to \mathscr{S}_{\pi_1(X/k)}(\mathbb{A}_k)$ factors by Lemma 140 over

$$\bigcup_\tau {}^\tau h_*\left(\mathscr{S}_{\pi_1(({}^\tau Y)^0/k)}(\mathbb{A}_k)\right)$$

and by Theorem 127 over the union of Brauer kernels

$$\bigcup_\tau {}^\tau h_*\left(\mathscr{S}_{\pi_1(({}^\tau Y)^0/k)}(\mathbb{A}_k)^{\text{Br}}\right)$$

hence the intersection $\mathscr{S}_{\pi_1(X/k)}(\mathbb{A}_k)^{\text{ét-Br}}$ contains its image, described by Theorem 144 as $\mathscr{S}_{\pi_1(X/k)}(\mathbb{A}_k)^{\text{f-descent}}$. The other inclusion follows from Lemma 140. □

Remark 151. (1) Because for sections we only deal with finite descent obstructions, actually both the analogue of Skorobogatov's part (with the obvious definition of the étale stabilized finite descent obstruction set)

$$\mathscr{S}_{\pi_1(X/k)}(\mathbb{A}_k)^{\text{ét-f-descent}} = \mathscr{S}_{\pi_1(X/k)}(\mathbb{A}_k)^{\text{f-descent}}$$

and the analogue of Demarche's part of Corollary 150

$$\mathscr{S}_{\pi_1(X/k)}(\mathbb{A}_k)^{\text{ét-Br}} \subseteq \mathscr{S}_{\pi_1(X/k)}(\mathbb{A}_k)^{\text{f-descent}}$$

are almost immediate consequences of the definitions. The nontrivial part provided by Theorem 144 consists in

$$\mathscr{S}_{\pi_1(X/k)}(\mathbb{A}_k)^{\text{ét-f-descent}} \subseteq \mathscr{S}_{\pi_1(X/k)}(\mathbb{A}_k)^{\text{ét-Br}}.$$

(2) Corollary 150 leaves us with a hierarchy of obstructions on adelic sections, namely the subsets of $\mathscr{S}_{\pi_1(X/k)}(\mathbb{A}_k)$ given by

$$\mathscr{S}_{\pi_1(X/k)} \twoheadrightarrow \mathscr{S}_{\pi_1(X/k)}(\mathbb{A}_k)^{\text{f-descent}} = \mathscr{S}_{\pi_1(X/k)}(\mathbb{A}_k)^{\text{ét-Br}} \subseteq \mathscr{S}_{\pi_1(X/k)}(\mathbb{A}_k)^{\text{Br}},$$

which is the analogue of (11.8).

Consequences for strong approximation of rational points. Even without the section conjecture being established we can draw conclusions by means of sections.

Corollary 152. *Let X/k be a geometrically connected variety over an algebraic number field.*

(1) We have equality

$$X(\mathbb{A}_k)_\bullet^{\text{f-descent}} = X(\mathbb{A}_k)_\bullet^{\text{cf-descent}}$$

of the analogues of $X(\mathbb{A}_k)_\bullet^{\text{descent}}$ where we intersect only the descent obstruction sets $X(\mathbb{A}_k)_\bullet^h$ corresponding to G-torsors $h : Y \to X$ with G finite (resp. finite and constant as a k-group).

(2) If X/k is an algebraic $K(\pi,1)$ space, then the Brauer–Manin obstruction is coarser than the finite descent obstruction, i.e, we have an inclusion

$$X(\mathbb{A}_k)_\bullet^{\text{f-descent}} \subseteq X(\mathbb{A}_k)_\bullet^{\text{Br}}.$$

Proof. An adelic point $(x_v) \in X(\mathbb{A}_k)$ survives the descent obstruction posed by a G-torsor under a finite group G/k if and only if the associated adelic section (s_{x_v}) survives the same torsor. Hence, in the following diagram

$$
\begin{array}{ccccc}
X(\mathbb{A}_k)_\bullet^{\text{f-descent}} & \subseteq & X(\mathbb{A}_k)_\bullet^{\text{cf-descent}} & \subseteq & X(\mathbb{A}_k)_\bullet \\
\downarrow & & \downarrow & & \downarrow \\
\mathscr{S}_{\pi_1(X/k)}(\mathbb{A}_k)^{\text{f-descent}} & \subseteq & \mathscr{S}_{\pi_1(X/k)}(\mathbb{A}_k)^{\text{cf-descent}} & \subseteq & \mathscr{S}_{\pi_1(X/k)}(\mathbb{A}_k)
\end{array}
$$

both squares are cartesian. Part (1) follows at once from Theorem 144 (2).

Assertion (2) is a direct consequence of Corollary 150 (1) and the fact that for an adelic point $(x_v) \in X(\mathbb{A}_k)_\bullet$ the associated adelic section (s_{x_v}) decides whether (x_v) survives the Brauer–Manin obstruction, see Proposition 129, respectively the finite descent obstruction. □

A test for the section conjecture in higher dimensions. The following observation is no danger to the section conjecture in its original form, because for a smooth, projective curve X/k over an algebraic number field of genus at least 2 the set of rational points $X(k)$ is finite and a forteriori closed in $X(\mathbb{A}_k)_\bullet$.

Corollary 153. *Let X/k be a geometrically connected variety over an algebraic number field, such that X embeds into its Albanese variety and $X(k)$ is not closed in $X(\mathbb{A}_k)_\bullet$. Then the profinite Kummer map is injective but not surjective. In particular, the analogue of the section conjecture for X/k fails.*

Proof. The assumption of an injective Albanese map shows that the profinite Kummer map over k and over the completions k_v are injective, see Proposition 73, where the smoothness assumption is superfluous for this application.

For any G-torsor $h : Y \to X$ the obstruction set $X(\mathbb{A}_k)_\bullet^h$ is closed in $X(\mathbb{A}_k)_\bullet$. Hence, with $X(k)$ also the elements of the closure $\overline{X(k)}$ satisfy all descent obstructions. If the profinite Kummer map for X were surjective, then by Theorem 144 the map

$$X(k) \xrightarrow{\kappa} \mathscr{S}_{\pi_1(X/k)} \twoheadrightarrow \mathscr{S}_{\pi_1(X/k)}(\mathbb{A}_k)^{\text{f-descent}}$$

were surjective. But on the other hand, this map agrees with the injection

$$X(k) \subsetneq \overline{X(k)} \subseteq X(\mathbb{A}_k)_\bullet^{\text{f-descent}} \subseteq \mathscr{S}_{\pi_1(X/k)}(\mathbb{A}_k)^{\text{f-descent}}.$$

By assumption, $X(k) \subsetneq \overline{X(k)}$ is not surjective. This provides a contradiction. $\square$

Torsor structure on the connected component of a torsor. Another application of Theorem 144 gives a more precise form of [De09] Lemma 4, which essentially describes a key result of [St07] Lemma 5.7. The strengthening consist in our morphism $G^0 \to G$ being an inclusion of a subgroup.

Corollary 154. *Let $h : Y \to X$ be a right G-torsor with G/k finite. If there is an adelic section (s_v) of $\pi_1(X/k)$ that survives every finite constant descent obstruction, e.g., if there is an adelic point $(x_v) \in X(\mathbb{A}_k)^{\text{f-descent}}$, then there exists*

  (i) a 1-cocycle τ representing a class in $\mathrm{H}^1(k, G)$,
  (ii) an inclusion of Gal_k-groups $\iota : G^0 \hookrightarrow {}^\tau G$ and a geometrically connected G^0-torsor $h^0 : Y^0 \to X$,
  (iii) with a G^0-equivariant map $Y^0 \hookrightarrow {}^\tau Y$ above X, i.e., an identification of ${}^\tau G$-torsors $\iota_* Y^0 = {}^\tau Y$,
  (iv) such that (s_v) lifts to an adelic section on Y^0 (resp. (x_v) lifts to $Y^0(\mathbb{A}_k)$).

Proof. By Theorem 144 there is a section s of $\pi_1(X/k)$ interpolating the adelic section (s_v). Let

$$\varphi : \pi_1(X, \bar{x}) \to G \rtimes \mathrm{Gal}_k$$

be a description of the G-torsor $h : Y \to X$ together with a distinguished point $\bar{y} \in Y$ above $\bar{x}$. Then $\varphi \circ s = \tau . s_0$ for a 1-cocycle τ as in (i). We may replace φ by $^{\tau}\varphi$ and thus assume that $\varphi \circ s = s_0$. Then $\Phi = \mathrm{im}(\varphi)$ agrees with the product set

$$\varphi(\pi_1(X, \bar{x})) \cdot s_0(\mathrm{Gal}_k).$$

With $\iota : G^0 = \Phi \cap G \hookrightarrow G$ we find $\Phi = G^0 \rtimes \mathrm{Gal}_k$ and so φ factors over a G^0-torsor

$$\varphi^0 : \pi_1(X, \bar{x}) \to G^0 \rtimes \mathrm{Gal}_k$$

corresponding to $h^0 : Y^0 \to X$. Clearly $\iota_* \varphi^0 = \varphi$, hence (iii) holds.

The map φ^0 is surjective by definition, hence the corresponding G^0-torsor is geometrically connected by Lemma 134 (c), which shows (ii). By Lemma 133 we have $\pi_1(Y^0, \bar{y}) = \varphi^{-1}(s_0(\mathrm{Gal}_k))$, so that the section s lifts to a section t of $\pi_1(Y^0/k)$. The corresponding adelic section (t_v) then lifts (s_v), which shows (iv).

$\square$

Remark 155. In Corollary 154, after twisting, the G^0-torsor $Y^0 \to X$ is the connected component of the distinguished point in the G-torsor $Y \to X$. In general however, the connected component of a G-torsor has no torsor structure, see the Example 156 below. The assumption of an adelic section $(s_v) \in \mathscr{S}_{\pi_1(X/k)}(\mathbb{A}_k)^{\mathrm{cf\text{-}descent}}$ is essential for the conclusion of Corollary 154.

Example 156. Let $p \geq 5$ be a prime number and consider $G = \mathrm{SL}_2(\mathbb{F}_p)$ with the action by $\mathrm{Gal}_{\mathbb{R}} = \langle \sigma \rangle$ via $A \mapsto (A^t)^{-1}$. In $G \rtimes \mathrm{Gal}_{\mathbb{R}}$ we consider the cyclic subgroup

$$\Phi = \left\langle \begin{pmatrix} 1 & 1 \\ & 1 \end{pmatrix} \cdot \sigma \right\rangle.$$

An easy computation shows that $\Phi \cong \mathbb{Z}/12\mathbb{Z}$ and that $\Phi \cup \Phi \cdot \sigma$ is not a subgroup of $G \rtimes \mathrm{Gal}_{\mathbb{R}}$, for example because this set has 24 elements and contains the unipotent element of order p.

We now assume that there is a geometrically connected variety $X/\mathbb{R}$ with a surjective map of extensions

$$
\begin{array}{ccccccccc}
1 & \longrightarrow & \pi_1(\overline{X}, \bar{x}) & \longrightarrow & \pi_1(X, \bar{x}) & \longrightarrow & \mathrm{Gal}_{\mathbb{R}} & \longrightarrow & 1 \\
 & & \downarrow{\overline{\psi}} & & \downarrow{\psi} & & \| & & \\
1 & \longrightarrow & 2\mathbb{Z}/12\mathbb{Z} & \longrightarrow & \mathbb{Z}/12\mathbb{Z} & \longrightarrow & \mathbb{Z}/2\mathbb{Z} & \longrightarrow & 1
\end{array}
$$

The pointed $\mathrm{SL}_2(\mathbb{F}_p)$-torsor $Y \to X$ corresponding to

$$\varphi : \pi_1(X, \bar{x}) \twoheadrightarrow \mathbb{Z}/12\mathbb{Z} = \Phi \subset \mathrm{SL}_2(\mathbb{F}_p) \rtimes \mathrm{Gal}_{\mathbb{R}}$$

has no torsor structure on its connected component of the distinguished point. Otherwise the image of φ would be a subgroup of $G \rtimes \mathrm{Gal}_\mathbb{R}$ and a semi-direct product, which it is obviously not.

Remark 157. Rungtanapirom constructed in his Diplomarbeit [Ru12] for any field k, any finite group G, and any continuous extension

$$1 \to G \to E \to \mathrm{Gal}_k \to 1$$

a generalized Godeaux–Serre variety X/k with $\pi_1(X/k) \cong E$ as extensions of Gal_k. In particular, a variety $X/\mathbb{R}$ with $\pi_1(X) \cong \mathbb{Z}/12\mathbb{Z}$ as used in Example 156 exists.

Consequences of strong approximation in light of the section conjecture. In conjunction with the section conjecture, Theorem 144 makes the following predictions.

Corollary 158. *Let X/k be a smooth projective curve of genus at least 2 over an algebraic number field such that the section conjecture holds for X/k. Then the following holds.*

(1) $X(k) = X(\mathbb{A}_k)^{\mathrm{cf\text{-}descent}}_\bullet$, in particular $X(\mathbb{A}_k)^{\mathrm{cf\text{-}descent}}_\bullet$ is finite.
(2) The map $\mathscr{S}_{\pi_1(X/k)} \to \mathscr{S}_{\pi_1(X/k)}(\mathbb{A}_k)^{\mathrm{cf\text{-}descent}}$ is bijective.
(3) Any local point $x_v \in X(k_v)$ which is the v-component of an adelic point

$$(x_v) \in X(\mathbb{A}_k)^{\mathrm{cf\text{-}descent}}_\bullet$$

   is global, i.e., belongs to $X(k)$.

Proof. In the commutative diagram

$$
\begin{array}{ccc}
X(k) & \hookrightarrow & X(\mathbb{A}_k)^{\mathrm{cf\text{-}descent}}_\bullet \\[2mm]
{\scriptstyle \kappa}\downarrow {\scriptstyle \cong} & & \downarrow \\[2mm]
\mathscr{S}_{\pi_1(X/k)} & \twoheadrightarrow & \mathscr{S}_{\pi_1(X/k)}(\mathbb{A}_k)^{\mathrm{cf\text{-}descent}}
\end{array}
$$

the section conjecture and Theorem 144 imply that the diagonal map is surjective, while Corollary 74 shows that it is injective. Hence all maps are bijective and all statements of the corollary follow. $\square$

Remark 159. It is a nuisance that despite serious efforts I am not able to show that

$$\mathscr{S}_{\pi_1(X/k)} \to \mathscr{S}_{\pi_1(X/k)}(\mathbb{A}_k)^{\mathrm{descent}}$$

is injective for a smooth projective curve X/k of genus at least 2 over an algebraic number field. The map occurs again in the exact sequence of non-abelian

Tate–Poitou duality theory with profinite coefficients as discussed in Sect. 12.3. Due to an ad hoc definition of some terms in the exact sequence, non-abelian Tate–Poitou theory in its current form cannot decide the above injectivity question.

11.8 Generalizing the Descent Obstruction

Let G/k be a group scheme. We fix $i \in \mathbb{N}$ and if $i \geq 2$ we ask G to be commutative. For a class $\alpha \in \mathrm{H}^i \left(\pi_1(X), G(\bar{k}) \right)$ the obstruction set is the subset of $\mathscr{S}_{\pi_1(X/k)}(\mathbb{A}_k)$

$$
\mathscr{S}_{\pi_1(X/k)}(\mathbb{A}_k)^\alpha = \left\{ (s_v) \; ; \; (s_v^*(\alpha)) \in \mathrm{diag}\big(\mathrm{H}^i(k, G) \to \prod_v \mathrm{H}^i(k_v, G)\big) \right\}.
$$

Clearly, the restriction map $s \mapsto (s \otimes k_v)$ factors as

$$
\mathscr{S}_{\pi_1(X/k)} \to \mathscr{S}_{\pi_1(X/k)}(\mathbb{A}_k)^\alpha,
$$

since for a section s of $\pi_1(X/k)$ we have

$$
\mathrm{diag}(s^*(\alpha)) = \big((s \otimes k_v)^*(\alpha) \big).
$$

This cohomological obstruction set generalizes the Brauer–Manin obstruction of Sect. 11.2, here $i = 2$ and $G = \mu_n$, and the finite descent obstruction with respect to torsors under abelian groups of Sect. 11.6, here $i = 1$ and G is finite étale.

When $i \geq 3$, a theorem of Tate, see [NSW08] Theorem 8.6.10, shows that

$$
\mathrm{H}^i(k, G) \to \prod_v \mathrm{H}^i(k_v, G) \to \prod_{v \text{ real}} \mathrm{H}^i(k_v, G)
$$

are isomorphisms. It follows that cohomology classes in degree $i \geq 3$ do not contribute further cohomological constraints on adelic sections.

We next discuss the case of $i = 2$ with G a diagonalizable group scheme. The Tate–Poitou exact sequence extends to coefficients G, see [Mi06] II §4 Theorem 4.6(b) and [HaSz05] Theorem 5.6. For any character $\chi : G \to \mathbb{G}_m$ we therefore obtain a commutative diagram with exact rows

$$
\begin{array}{ccccccc}
\mathrm{H}^2(k, G) & \longrightarrow & \bigoplus_v \mathrm{H}^2(k_v, G) & \longrightarrow & \mathrm{H}^0(k, \mathscr{H}om(G, \mathbb{G}_m))^\vee & \longrightarrow & 0 \\[2mm]
\downarrow{\scriptstyle \chi_*} & & \downarrow{\scriptstyle \chi_*} & & \downarrow{\scriptstyle \mathrm{ev}_\chi} & & \\[2mm]
\mathrm{H}^2(k, \mathbb{G}_m) & \longrightarrow & \bigoplus_v \mathrm{H}^2(k_v, \mathbb{G}_m) & \longrightarrow & \mathbb{Q}/\mathbb{Z} & \longrightarrow & 0
\end{array}
$$

where ev_χ is the evaluation in χ.

For a class $\alpha \in H^2\left(\pi_1(X), G(\bar{k})\right)$ we have $\chi_*(\alpha) \in H^2\left(\pi_1(X), \mathbb{G}_m(\bar{k})\right)$ and the diagram shows

$$\mathscr{S}_{\pi_1(X/k)}(\mathbb{A}_k)^\alpha = \bigcap_\chi \mathscr{S}_{\pi_1(X/k)}(\mathbb{A}_k)^{\chi_*(\alpha)},$$

so that the cohomological constraints originating from arbitrary G are already captured in the special case of $G = \mathbb{G}_m$. Moreover, the group $H^2\left(\pi_1(X), \mathbb{G}_m(\bar{k})\right)$ is torsion and so a class $\alpha \in H^2\left(\pi_1(X), \mathbb{G}_m(\bar{k})\right)$ lifts to a class

$$\beta \in H^2(\pi_1(X), \mu_n) = H^2(X, \mu_n)$$

for a suitable n. Naturality of the obstruction set with respect to maps in the coefficient G shows

$$\mathscr{S}_{\pi_1(X/k)}(\mathbb{A}_k)^\beta \subseteq \mathscr{S}_{\pi_1(X/k)}(\mathbb{A}_k)^\alpha.$$

We conclude that the power of cohomological constraints with $i = 2$ and G a diagonalizable group is exhausted by the Brauer–Manin obstruction for sections.

11.9 The Section Conjecture as an *Only–One* Conjecture

The search for a criterion that decides the existence of a rational solution to any given system of Diophantine equations influenced number theory since its beginnings. A suggestion in this direction can be called an *only–one* conjecture, because the criterion would claim to be the only test that a system of Diophantine equations has to pass in order to admit a rational solution.

When restricted to Diophantine equations that describe smooth curves over an algebraic number field, the section conjecture can be viewed as an attempt towards this direction. Essentially, the section conjecture asks the following, see Corollary 101.

Question 160. For a smooth, projective, geometrically connected curve of genus at least 2 over an algebraic number field k, is the existence of a section of the fundamental group extension the only obstruction to having a k-rational point?

We clarify the relation of the section conjecture with other *only–one* conjectures.

Theorem 161. *Let k be a fixed algebraic number field, and consider the following assertions on smooth projective curves X/k of genus at least 2.*

(SC)   The section conjecture holds for all such X/k.
(BrM)   The Brauer–Manin obstruction is the only one for such X/k, i.e., we have

$$X(\mathbb{A}_k)_\bullet^{\mathrm{Br}} \neq \emptyset \quad \Longrightarrow \quad X(k) \neq \emptyset.$$

(fd)   *The finite descent obstruction is the only one for such X/k, i.e., we have*

$$X(\mathbb{A}_k)_{\bullet}^{\text{f-descent}} \neq \emptyset \qquad \Longrightarrow \qquad X(k) \neq \emptyset.$$

(LSC)   *For all such X/k and all completions k_v of k the local section conjecture holds, i.e., every local section $s_v \in \mathscr{S}_{\pi_1(X/k)}(k_v)$ comes from a k_v-rational point of X.*

Then the following implications hold.

(1)   (BRM) + (LSC) $\Longrightarrow$ (fd) + (LSC) $\Longrightarrow$ (SC)
(2)   (SC) $\Longrightarrow$ (fd).

Proof. (1) By Corollary 152 (2) the first implication is obvious. For the second implication we consider a curve X/k with a section s. By (LSC) we find an adelic point (x_v) with $s \otimes k_v = s_{x_v}$. Then with the adelic section (s_{x_v}) also the adelic point (x_v) satisfies all finite descent obstructions. By (fd) we deduce the existence of a rational point in $X(k)$. Applying the above reasoning for all neighbourhoods of s, Corollary 101 then shows (SC).

(2) Let (x_v) be an adelic point that survives all finite descent obstructions. Then the adelic section (s_{x_v}) lies in $\mathscr{S}_{\pi_1(X/k)}(\mathbb{A}_k)^{\text{f-descent}}$. Thus Theorem 144 shows the existence of a section of $\pi_1(X/k)$ whence by (SC) we are guaranteed a k-rational point of X.   □

Remark 162. (1) Theorem 161 relates the section conjecture to other *only–one* conjectures and thus should broaden the interest in the section conjecture itself.

(2) That the finite descent obstruction is the only one for rational points on smooth projective curves over algebraic number fields has been conjectured by Stoll [St07] Conjecture 9.1.

(3) Educated speculations towards the Brauer–Manin obstruction being the only one for rational points on smooth projective curves over algebraic number fields have been conducted by Scharaschkin [Sch98] see also [ErSch06], Skorobogatov [Sk01] and Poonen [Po06].

(4) In fact, Poonen and Voloch [PoVo10] prove a spectacular result for nonisotrivial curves over function fields in positive characteristic. There the Brauer–Manin obstruction is indeed the only obstruction for having a rational point. Consequently, in these cases the appropriate local version of the section conjecture shows the analogue of the section conjecture itself.

Hilbert's 10th problem. Among the famous list of Hilbert's problems from 1900, the 10th problem addresses Diophantine equations, see [Hi1902]:

> **10. Entscheidung der Lösbarkeit einer diophantischen Gleichung.** Eine diophantische Gleichung mit irgendwelchen Unbekannten und mit ganzen rationalen Zahlkoeffizienten sei vorgelegt: *man soll ein Verfahren angeben, nach welchen sich mittels einer endlichen Anzahl von Operationen entscheiden läßt, ob die Gleichung in ganzen rationalen Zahlen lösbar ist.*

As a result of the effort of Davis, Putnam and Robinson, and culminating in the work of Matjassewitsch we know that the answer to Hilbert's 10th problem is negative.

There is no algorithm which decides whether a given system of Diophantine equations has a rational integer solution or not.

The section conjecture is only mildly affected by this negative answer to Hilbert's 10th problem. First, there is a gap between asking for solutions in algebraic integers or in a given algebraic number field. But this gap is closing in recent research, and it may even be that there is also no algorithm which decides about solutions in $\mathbb{Q}$. More importantly, the claim of the section conjecture only addresses very special Diophantine equations, namely those which lead to smooth, projective curves of genus at least 2. It is conceivable that Hilbert's 10th problem has a positive answer when restricted to this subclass of equations. Only in a worst case scenario, with the section conjecture valid (so not too bad actually), we might end up with a bijection between rational points $X(k)$ and conjugacy classes of sections of $\pi_1(X/k)$, which are finite sets of which we are algorithmically unable to decide if they are empty or not.

Effective diophantine geometry via the section conjecture. The question of decidability for rational points on curves in light of the section conjecture has been studied by Pál [Pa10a], and even with an eye towards effectivity by Kim [Ki12]. The latter also has to assume the Bloch–Kato conjecture on special values of L-functions.

Chapter 12
Fragments of Non-abelian Tate–Poitou Duality

The most sophisticated tool to analyse local–global questions of cohomological nature in number theory is provided by the Tate–Poitou exact sequence and duality. Our interest in the fibres of the map

$$\mathscr{S}_{\pi_1(X/k)} \to \mathscr{S}_{\pi_1(X/k)}(\mathbb{A}_k),$$

which is the central topic of this part, motivates the search for even fragments of a non-abelian generalization of the Tate–Poitou sequence.

12.1   Review of Non-abelian Galois Cohomology

Non-abelian cohomology has an increasing complexity in its coefficients and its definition with increasing cohomological degree. Starting from degree 3 it becomes so complicated that for our purposes non-abelian cohomology ceases to exist. We will proceed along descending cohomological degree, and refer to [EMcL47] for details.

Degree 2. A *k-kernel* or *k-lien* on a finite discrete group $\overline{G}$ is a continuous action $\rho : \mathrm{Gal}_k \to \mathrm{Out}(\overline{G})$ by outer automorphisms. The center $Z(\overline{G})$, as a characteristic abelian subgroup, inherits a Galois module structure.

The non-abelian cohomology set $\mathrm{H}^2(k, (\overline{G}, \rho))$ with coefficients in the k-kernel $(\overline{G}, \rho)$ is the set of equivalence classes of continuous extensions

$$1 \to \overline{G} \to E \to \mathrm{Gal}_k \to 1$$

with associated outer action identical to ρ. Equivalently, $\mathrm{H}^2(k, (\overline{G}, \rho))$ classifies gerbes on $\mathrm{Spec}(k)_{\text{ét}}$ with associated k-lien equal to $(\overline{G}, \rho)$. A gerbe $\mathscr{G}/k$ corresponds here to the extension given by its fundamental group extension $\pi_1(\mathscr{G}/k)$.

J. Stix, *Rational Points and Arithmetic of Fundamental Groups*, Lecture Notes in Mathematics 2054, DOI 10.1007/978-3-642-30674-7_12,
© Springer-Verlag Berlin Heidelberg 2013

The set $\mathrm{H}^2(k, (\overline{G}, \rho))$ may be empty, which would stop our search for further cohomology in lower degrees. There is an obstruction class

$$\eta_{(\overline{G},\rho)} \in \mathrm{H}^3(k, Z(\overline{G}))$$

whose vanishing decides whether $\mathrm{H}^2(k, (\overline{G}, \rho))$ is empty. Moreover, there is a subset of neutral classes

$$\mathrm{H}^2_{\mathrm{nt}}(k, (\overline{G}, \rho)) \subseteq \mathrm{H}^2(k, (\overline{G}, \rho)).$$

Neutral classes correspond to extensions which admit a section, hence semi-direct product extensions, or to gerbes that posses a global section, i.e., neutral gerbes.

Degree 1. We assume that $\eta_{(\overline{G},\rho)} = 0$ so that we can pick a gerbe $\mathscr{G} \in \mathrm{H}^2(k, (\overline{G}, \rho))$. We define the non-abelian cohomology set

$$\mathrm{H}^1(k, \mathscr{G}) = \mathscr{S}_{\pi_1(\mathscr{G}/k)} \tag{12.1}$$

as the set of $\overline{G}$-conjugacy classes of sections of $\pi_1(\mathscr{G}/k)$. Again, the set $\mathrm{H}^1(k, \mathscr{G})$ may be empty, which again would stop our search for further cohomology in lower degrees. Being nonempty is controlled by the associated gerbe $\mathscr{G}$ being neutral, so belonging to the distinguished subset of $\mathrm{H}^2(k, (\overline{G}, \rho))$.

Cocycles for degree 1. We now furthermore assume that $\mathscr{G}$ is neutral so that we can pick a section s_0 of $\pi_1(\mathscr{G}/k)$. The induced conjugation action of Gal_k on $\overline{G}$ lifts the outer action ρ and defines the structure of a finite étale group G/k with $G \times_k \overline{k} = \overline{G}$, i.e, a k-form of $\overline{G}$, and $\mathscr{G}$ becomes identified with the gerbe of torsors under G. By Proposition 8 we obtain a cocycle description via the map $\tau \mapsto \tau.s_0$

$$\mathrm{H}^1(k, G) \xrightarrow{\sim} \mathscr{S}_{\pi_1(\mathscr{G}/k)} = \mathrm{H}^1(k, \mathscr{G}).$$

Degree 0. We define the non-abelian cohomology set

$$\mathrm{H}^0(k, G) = G(k) = \overline{G}^{s_0(\mathrm{Gal}_k)}$$

as the set of invariants under the Gal_k-action enforced by the section s_0. This agrees with the centraliser of the section s_0 intersected with $\overline{G}$.

12.2 The Non-abelian Tate–Poitou Exact Sequence with Finite Coefficients

Let k be an algebraic number field, and let $(\overline{G}, \rho)$ be a k-kernel on a finite group $\overline{G}$ with center $Z = Z(\overline{G})$. The restriction $\rho_v = \rho|_{\mathrm{Gal}_{k_v}}$ induces k_v-kernels for every place v of k. For an algebraic number field k we have

$$H^3(k, Z) = \prod_v H^3(k_v, Z) = \prod_{v:k \hookrightarrow \mathbb{R}} H^3(\mathbb{R}, Z)$$

and thus the obstruction $\eta_{(\overline{G},\rho)}$ vanishes if and only if $\mathrm{res}_v(\eta_{(\overline{G},\rho)}) = \eta_{(\overline{G},\rho_v)}$ vanishes for all (real) places v of k.

Degree ≥ 2. We define artificially the direct sum of the local cohomology sets as the tuples of k_v-gerbes $\mathscr{G}_v$ that differ from a global gerbe at most at finitely many places.

$$\bigoplus_v H^2(k_v, (\overline{G}, \rho_v)) = \left\{ (\mathscr{G}_v) \in \prod_v H^2(k_v, (\overline{G}, \rho_v)) \; ; \quad \begin{array}{l} \text{there is } \mathscr{G} \in H^2(k, (\overline{G}, \rho)) \\ \text{with } \mathscr{G}_v = \mathscr{G} \otimes k_v \\ \text{for almost all } v \end{array} \right\}$$

Let us consider the Cartier dual Gal_k-module

$$Z^D = \mathscr{H}om(Z, \mathbb{G}_\mathrm{m})$$

of the center $Z = Z(\overline{G})$. We assume $\eta_{(\overline{G},\rho)} = 0$ and pick a gerbe $\mathscr{G} \in H^2(k, (\overline{G}, \rho))$. There is a duality pairing

$$\bigoplus_v H^2(k_v, (\overline{G}, \rho_v)) \times H^0(k, Z^D) \to \bigoplus_v H^2(k_v, \mathbb{G}_\mathrm{m}) \xrightarrow{\sum_v \mathrm{inv}_v(-)} \mathbb{Q}/\mathbb{Z} \qquad (12.2)$$

$$(\mathscr{G}_v) \times \chi \mapsto \left(\chi_*(\mathscr{G}_v - \mathscr{G} \otimes k_v) \right)_v \mapsto \sum_v \mathrm{inv}_v\left(\chi_*(\mathscr{G}_v - \mathscr{G} \otimes k_v)\right)$$

where we make use of the fact that $H^2(k_v, (\overline{G}, \rho_v))$ is a free and transitive $H^2(k_v, Z)$-set, so that $\mathscr{G}_v - \mathscr{G} \otimes k_v \in H^2(k_v, Z)$ can be pushed by $\chi : Z \to \mathbb{G}_\mathrm{m}$. By reciprocity, the pairing is in fact independent of the chosen auxiliary global gerbe $\mathscr{G}$, and the difference $\mathscr{G}_v - \mathscr{G} \otimes k_v$ is nontrivial at only finitely many places.

Proposition 163. *The non-abelian Tate–Poitou exact sequence in degrees ≥ 2 is the exact sequence*

$$H^2(k, (\overline{G}, \rho)) \xrightarrow{\mathrm{res}_v} \bigoplus_v H^2(k_v, (\overline{G}, \rho_v)) \to \left(H^0(k, Z^D) \right)^{\vee} \to 0 \qquad (12.3)$$

with maps defined by $\mathrm{res}_v(\mathscr{G}) = \mathscr{G} \otimes k_v$ *and*

$$(\mathscr{G}_v)_v \mapsto \left(\chi \mapsto \sum_v \mathrm{inv}_v\left(\chi_*(\mathscr{G}_v - \mathscr{G} \otimes k_v)\right)\right),$$

and with exactness in the sense of pointed sets with respect to $0 \in \left(H^0(k, Z^D)\right)^{\vee}$.

Proof. The second map is well defined by the duality pairing (12.2). Using the global gerbe $\mathscr{G} \in H^2\left(k, (\overline{G}, \rho)\right)$ we find a commutative diagram

$$
\begin{array}{ccccccc}
\mathrm{H}^2(k,Z)\cdot\mathscr{G} & \xrightarrow{\ \mathrm{res}_v\ } & \bigoplus_v \mathrm{H}^2(k_v,Z)\cdot\mathscr{G}\otimes k_v & \longrightarrow & \left(\mathrm{H}^0(k,Z^D)\right)^{\vee} & \longrightarrow & 0 \\
\downarrow & & \downarrow & & \| & & \\
\mathrm{H}^2(k,(\overline{G},\rho)) & \xrightarrow{\ \mathrm{res}_v\ } & \bigoplus_v \mathrm{H}^2(k_v,(\overline{G},\rho_v)) & \longrightarrow & \left(\mathrm{H}^0(k,Z^D)\right)^{\vee} & \longrightarrow & 0
\end{array}
$$

with bijective vertical arrows. Consequently, the sequence (12.3) becomes an exact homogeneous set version of the exact sequence from Tate–Poitou duality

$$
\mathrm{H}^2(k,Z) \xrightarrow{\ \mathrm{res}_v\ } \bigoplus_v \mathrm{H}^2(k_v,Z) \to \left(\mathrm{H}^0(k,Z^D)\right)^{\vee} \to 0. \qquad \square
$$

Degree ≥ 1. We keep the gerbe $\mathscr{G}$ with kernel $(\overline{G},\rho)$. A section s_v of $\pi_1(\mathscr{G}\otimes k_v/k_v)$ is *unramified* if there is a diagram

$$
\begin{array}{ccccccccc}
1 & \longrightarrow & \overline{G} & \longrightarrow & \pi_1(\mathscr{G}\otimes k_v) & \longrightarrow & \mathrm{Gal}_{k_v} & \longrightarrow & 1 \\
 & & \| & & \downarrow{\scriptstyle \chi} & & \downarrow & & \\
1 & \longrightarrow & \overline{G} & \longrightarrow & E & \longrightarrow & \Gamma & \longrightarrow & 1
\end{array}
$$

with a finite quotient $\mathrm{Gal}_{k_v} \twoheadrightarrow \Gamma$ and such that $\chi\circ s_v : \mathrm{Gal}_{k_v}\to E$ is unramified, see Chap. 8 for a parallel notion. We define the restricted product

$$
{\prod_v}' \mathrm{H}^1(k_v,\mathscr{G}\otimes k_v) = {\prod_v}' \mathscr{S}_{\pi_1(\mathscr{G}/k)}(k_v) = \left\{ (s_v)\ ;\ \begin{array}{l} s_v \text{ is unramified} \\ \text{for almost all } v \end{array} \right\}
$$

as a subset of $\prod_v \mathscr{S}_{\pi_1(\mathscr{G}/k)}(k_v)$. In parallel to Definition 123 we call it the set of *adelic sections* of $\pi_1(\mathscr{G}/k)$ and denote it by

$$
\mathscr{S}_{\pi_1(\mathscr{G}/k)}(\mathbb{A}_k).
$$

Definition 164. The *non-abelian compactly supported cohomology in degree* 2 is the set

$$
\mathrm{H}^2_{\mathrm{c}}\left(k,(\overline{G},\rho)\right) = \left\{ (\mathscr{G}',(s_v'))\ ;\ \mathscr{G}'\in\mathrm{H}^2\left(k,(\overline{G},\rho)\right) \text{ and } (s_v')\in\mathscr{S}_{\pi_1(\mathscr{G}'/k)}(\mathbb{A}_k) \right\}/\!\sim
$$

where $(\mathscr{G}',(s_v')) \sim (\mathscr{G}'',(s_v''))$ if there is a k-isomorphism $f : \mathscr{G}'\cong\mathscr{G}''$ such that $f(s_v') = f(s_v'')$.

Proposition 165. *The non-abelian Tate–Poitou exact sequence in degrees* ≥ 1 *takes the form*

$$\mathrm{H}^1(k,\mathscr{G}) \xrightarrow{\ \mathrm{res}_v\ } \prod'_v \mathrm{H}^1(k_v,\mathscr{G}\otimes k_v) \xrightarrow{\ r_2\ } \mathrm{H}^2_c(k,(\overline{G},\rho)) \;\underset{f_2}{\diagdown}$$

$$\underset{}{\diagup}\; \mathrm{H}^2(k,(\overline{G},\rho)) \xrightarrow{\ \mathrm{res}_v\ } \bigoplus_v \mathrm{H}^2(k_v,(\overline{G},\rho_v)) \longrightarrow \left(\mathrm{H}^0(k,Z^D)\right)^{\vee} \longrightarrow 0$$

with the map $f_2 : \left(\mathscr{G}', (s'_v)\right) \mapsto \mathscr{G}'$ forgetting the adelic section, and the map $r_2((s_v)) = \left(\mathscr{G},(s_v)\right)$ remembering the gerbe of which (s_v) is an adelic section.

Proof. We need to verify exactness at the second to forth entry. At $\mathrm{H}^2\left(k,(\overline{G},\rho)\right)$ exactness means that a k-gerbe $\mathscr{G}'$ is locally neutral if and only if it admits an adelic section. To prove this we choose a diagram

$$
\begin{array}{ccccccccc}
1 & \longrightarrow & \overline{G} & \longrightarrow & \pi_1(\mathscr{G}') & \longrightarrow & \mathrm{Gal}_k & \longrightarrow & 1 \\
 & & \| & & \downarrow & & \downarrow{\scriptstyle\psi} & & \\
1 & \longrightarrow & \overline{G} & \longrightarrow & E & \longrightarrow & \Gamma & \longrightarrow & 1
\end{array}
$$

with a finite group E. The quotient $\psi : \mathrm{Gal}_k \twoheadrightarrow \Gamma$ is unramified for almost all v. When ψ is unramified at v, then $\psi|_{\mathrm{Gal}_{k_v}}$ factors over $\mathrm{Gal}_{\kappa(v)} = \hat{\mathbb{Z}}$ and admits unramified lifts

$$\tilde{\psi}_v : \mathrm{Gal}_{k_v} \twoheadrightarrow \mathrm{Gal}_{\kappa(v)} \to E$$

which correspond to an unramified section of $\pi_1(\mathscr{G}'\otimes k_v/k_v)$.

At $\mathrm{H}^2_c\left(k,(\overline{G},\rho)\right)$ exactness has to be interpreted as $\mathrm{im}(r_2) = f_2^{-1}(\mathscr{G})$, because by the choice of $\mathscr{G}$ the set $\mathrm{H}^2\left(k,(\overline{G},\rho)\right)$ has become a pointed set. This part is tautologically true.

In order to discuss exactness at $\prod'_v \mathrm{H}^1(k_v,\mathscr{G}\otimes k_v)$ we first need to specify a distinguished subset of $\mathrm{H}^2_c\left(k,(\overline{G},\rho)\right)$. The natural choice would be to take all $\left(\mathscr{G},(s_v)\right)$ such that the adelic section (s_v) survives all descent obstructions. Then a trivial generalization of Theorem 144 for gerbes shows exactness here. $\square$

Remark 166. The content of Proposition 165 is fairly meagre for its use of the ad hoc definition of $\mathrm{H}^2_c\left(k,(\overline{G},\rho)\right)$ the maps f_2 and r_2 and furthermore, the distinguished subset of $\mathrm{H}^2_c\left(k,(\overline{G},\rho)\right)$. Only a duality statement with another group or set of interest can make the exact sequence of Proposition 165 useful. Also, exactness at $\mathrm{H}^2_c\left(k,(\overline{G},\rho)\right)$ should make use of the neutral classes in $\mathrm{H}^2\left(k,(\overline{G},\rho)\right)$ rather than just the pointing provided by $\mathscr{G}$.

Degree ≥ 0. We now even assume that the k-gerbe $\mathscr{G}$ is neutral and fix a section s of $\pi_1(\mathscr{G}/k)$. We regard $\overline{G}$ as a Gal_k-group via s and consider the resulting group scheme G/k.

Definition 167. The *non-abelian compactly supported cohomology in degree* 1 is the pointed set

$$H^1_c(k,\mathscr{G}) = \left\{ (s',(g'_v)) \; ; \; \begin{array}{l} s' : \mathrm{Gal}_k \to \pi_1(\mathscr{G}) \text{ a section, and} \\ g'_v \in \overline{G} \text{ for all } v \text{ such that} \\ s' \otimes k_v = g'_v(-)(g'_v)^{-1} \circ (s \otimes k_v) \end{array} \right\} / \sim$$

where $(s',(g'_v)) \sim (s'',(g''_v))$ if there is a $g \in \overline{G}$ such that $s'' = g(-)g^{-1} \circ s'$ and $g''_v = g \cdot g'_v$ for all v. The distinguished point of $H^1_c(k,\mathscr{G})$ is $(s,(1)_v)$.

There is a natural forget support map

$$f_1 : H^1_c(k,\mathscr{G}) \to H^1(k,\mathscr{G}) = \mathscr{S}_{\pi_1(\mathscr{G}/k)}$$

which maps $(s',(g'_v))$ to the $\overline{G}$-conjugacy class of s'. Moreover, we define

$$r_2 : \prod_v G(k_v) \to H^1_c(k,\mathscr{G})$$

as the map $(g_v) \mapsto (s,(g_v))$ which remembers the section used to define G/k.

At this point we have collected all ingredients in our suggestion for a non-abelian Tate–Poitou exact sequence.

Theorem 168. *Let* $(\overline{G},\rho)$ *be a* k-*kernel on the finite discrete group* $\overline{G}$, *let* $\mathscr{G}$ *be a* k-*gerbe with lien* $(\overline{G},\rho)$, *and let* s *be a section of* $\pi_1(\mathscr{G}/k)$ *that leads to the structure of a* k-*group* G/k *on* $\overline{G}$. *The non-abelian Tate–Poitou exact sequence reads*

$$
\begin{array}{ccccccc}
1 & \longrightarrow & G(k) & \xrightarrow{\mathrm{res}^0_v} & \prod_v G(k_v) & \xrightarrow{r_1} & H^1_c(k,\mathscr{G}) \\
\end{array}
\Big)\; f_1
$$

$$
\xrightarrow{\quad} H^1(k,\mathscr{G}) \xrightarrow{\mathrm{res}^1_v} \textstyle\prod'_v H^1(k_v,\mathscr{G} \otimes k_v) \xrightarrow{r_2} H^2_c(k,(\overline{G},\rho)) \Big)\; f_2
$$

$$
\xrightarrow{\quad} H^2(k,(\overline{G},\rho)) \xrightarrow{\mathrm{res}^2_v} \bigoplus_v H^2(k_v,(\overline{G},\rho_v)) \xrightarrow{r_3} \left(H^0(k,Z^D)\right)^\vee \longrightarrow 0
$$

$$(12.4)$$

and is exact in the following senses.

(1) Until $\prod'_v H^1(k_v,\mathscr{G} \otimes k_v)$ *the sequence (12.4) consists of pointed sets and is exact in the sense of pointed sets. The distinguished point is given by* 1, $(1)_v$, $(s,(1)_v)$, s, *and* $(s \otimes k_v)$ *in order of appearance. Moreover, the map* res^0_v *is an injective group homomorphism and the map* r_1 *identifies its image with the quotient set* $G(k)\backslash\left(\prod_v G(k_v)\right)$.

(2) *At $\prod_v' H^1(k_v, \mathcal{G} \otimes k_v)$ exactness means that the image of res_v^1 agrees with the preimage under r_2 of all $(\mathcal{G}, (s_v))$ where the adelic section (s_v) survives every finite descent obstruction.*

(3) *At $H_c^2(k, (\overline{G}, \rho))$ exactness means that the image of r_2 agrees with the preimage under f_2 of the class of the gerbe $\mathcal{G}$.*

(4) *At $H^2(k, (\overline{G}, \rho))$ exactness means that the image of f_2 agrees with the preimage under res_v^2 of the tuples of local gerbes $(\mathcal{G}_v')$ which are all neutral.*

(5) *The bottom line of the sequence (12.4) is homogeneous under an exact sequence of abelian groups as described in Proposition 163.*

Proof. The exactness of until $\prod_v' H^1(k_v, \mathcal{G} \otimes k_v)$ follows at once from the definition of the objects, maps and exactness itself. The exactness of the rest of the sequence was treated in Propositions 165 and 163 □

Remark 169. I feel obliged to stress that the content of Theorem 168 is fairly meagre in that it does not contain the *duality* part of the classical Tat–Poitou exact sequence, namely a description of the terms $H_c^1(k, \mathcal{G})$ and $H_c^2(k, (\overline{G}, \rho))$ via a duality pairing.

12.3 The Non-abelian Tate–Poitou Exact Sequence with Profinite Coefficients

Let X/k be a geometrically connected variety with a section s of $\pi_1(X/k)$. We denote the corresponding Gal_k-group structure on $\pi_1(\overline{X}, \bar{x})$ by $\pi_1(\overline{X}, s)$. The characteristic quotients $\pi_1(\overline{X}, s) \twoheadrightarrow Q_n(\pi_1(\overline{X}, s))$ induce quotient extensions $Q_n(\pi_1(X/k))$, see Chap. 4, that correspond to gerbes $\mathcal{G}_n$ with lien a k-kernel on the finite group $G_n = Q_n(\pi_1(\overline{X}, s))$. The section s induces a section s_n of $\pi_1(\mathcal{G}_n/k)$. If we perform a (delicate) limit over the non-abelian Tate–Poitou exact sequences of Theorem 168 applied to the $\mathcal{G}_n$, then we obtain an exact sequence

$$1 \rightarrow H^0(k, \pi_1(\overline{X}, s)) \rightarrow \prod_v H^0(k_v, \pi_1(\overline{X}, s)) \longrightarrow \varprojlim_n H_c^1(k, \mathcal{G}_n)$$

$$\rightarrow \mathcal{S}_{\pi_1(X/k)} \longrightarrow \mathcal{S}_{\pi_1(X/k)}(\mathbb{A}_k) \longrightarrow \varprojlim_n H_c^2(k, (G_n, \rho))$$

$$\rightarrow H^2(k, \pi_1(\overline{X}, s)) \rightarrow \bigoplus_v H^2(k_v, \pi_1(\overline{X}, s)) \rightarrow \varprojlim_n \left(H^0(k, Z_n^D)\right)^\vee \rightarrow 1$$

where $Z_n = Z(G_n)$ is the center of the k-lien of the gerbe $\mathscr{G}_n$. Now let X/k be a smooth, projective curve of genus at least 2. Then $\varprojlim_n Z_n = Z(\pi_1(\overline{X})) = 1$ is trivial and also the centraliser of s as well as all the s_v are trivial. Thus the sequence degenerates to

$$1 \to \varprojlim_n H_c^1\left(k, \mathscr{G}_n\right) \to \mathscr{S}_{\pi_1(X/k)} \to \mathscr{S}_{\pi_1(X/k)}(\mathbb{A}_k) \to \varprojlim_n H_c^2\left(k, (G_n, \rho)\right) \to 1$$

which unfortunately does not shed a single ray of light onto the problem that we set out to solve by means of a non-abelian Tate–Poitou exact sequence.

Part IV
Analogues of the Section Conjecture

Chapter 13
On the Section Conjecture for Torsors

We explore the weak section conjecture for torsors W under a connected algebraic group G/k: we ask whether a torsor is necessarily trivial if its extension $\pi_1(W/k)$ splits. Continuous Kummer theory, or a direct construction using neighbourhoods, shows that $\pi_1(W/k)$ splits if and only if the torsor W is compatibly divisible by étale isogenies, see Corollary 175. Therefore the weak section conjecture holds for torsors under tori and, if G is an abelian variety, is linked to Bashmakov's problem. As an application, we study the relative Brauer group of hyperbolic curves with completion of genus 0 and show a weak section conjecture for such curves, see Proposition 187.

Torsors under algebraic groups occur in the context of the section conjecture for a smooth projective curve as the universal Albanese torsor $\mathrm{Alb}^1_X = \mathrm{Pic}^1_X$ and via the abelianization of sections, see Definition 26. Zero cycles on a smooth projective variety X/k can be considered as an abelianized version of a rational point. We construct a map that assigns to a zero cycle z of degree 1 on X a section s_z of the abelianized fundamental group extension $\pi_1^{\mathrm{ab}}(X/k)$ that lifts the corresponding section of the rational point of the universal Albanese torsor corresponding to the zero cycle.

We work over a field k of characteristic 0.

13.1 The Fundamental Group of an Algebraic Group

Let G/k be a geometrically connected algebraic group. Since we assume characteristic 0, the Künneth formula holds and the multiplication map $\mu : G \times_k G \to G$ induces a group homomorphism

$$\pi_1(\overline{G}, \bar{e}) \times \pi_1(\overline{G}, \bar{e}) = \pi_1(\overline{G} \times \overline{G}, (\bar{e}, \bar{e})) \xrightarrow{\mu_*} \pi_1(\overline{G}, \bar{e}),$$

where $\bar{e}$ is a geometric point above the neutral element $e \in G$. If follows that $\pi_1(\overline{G}, \bar{e})$ is a group object in the category of profinite groups and hence an abelian profinite group. Even more is true.

J. Stix, *Rational Points and Arithmetic of Fundamental Groups*, Lecture Notes
in Mathematics 2054, DOI 10.1007/978-3-642-30674-7_13,
© Springer-Verlag Berlin Heidelberg 2013

Proposition 170. *Any connected finite étale cover* $h : G' \to G$ *together with a lift* $e' \in G'(k)$ *of* $e \in G(k)$ *has canonically the structure of an étale isogeny with central kernel* $G'[h] = \ker(h)$.

Proof. The argument in [LS57] Theorem 2 deals with the case of an abelian variety but relies only on the validity of the Künneth formula and thus actually proves the general case. It remains to show that $\ker(h)$ is central. The commutator map

$$G' \times_k \ker(h) \to G' \quad (g',a) \mapsto [g',a] = g'a(g')^{-1}a^{-1}$$

takes values in the finite étale $\ker(h)$ and maps $e' \times \ker(h)$ to e'. Thus any connected component of $G' \times_k \ker(h)$ is mapped to e' as well, and so $\ker(h)$ is central. $\quad\square$

Remark 171. Proposition 170 for a connected and locally connected topological group goes back to Schreier [Sch25]. In the case of an abelian variety, this is the familiar theorem of Lang and Serre [LS57] Theorem 2, and for connected linear algebraic groups this is contained in Miyanishi [Mi72]. For more about the prime to p fundamental group of algebraic groups and homogenous spaces see Brion and Szamuely [BrSz12].

A neighbourhood $h : G' \to G$ of the section s_e associated to the neutral element e admits a distinguished lift $e' \in G'(k)$ of e. The projective system

$$(G_h)_h$$

of all neighbourhoods forms a tower of k-forms of the universal pro-étale cover of $\overline{G}$. The continuous *Kummer sequence for* G is the exact sequence of pro-systems, non-abelian but à la Jannsen [Ja88],

$$1 \to \big(\ker(h)\big) \to (G_h) \xrightarrow{(h)} G \to 1$$

indexed by the isomorphy classes of isogenies h, which yields a central extension of G by $\pi_1(\overline{G}, \bar{e})$ with Galois action through s_e via the identification

$$\pi_1(\overline{G}, \bar{e}) = \mathrm{Aut}(\varprojlim G_h / G)^{\mathrm{opp}} = \big(\varprojlim_h \ker(h)\big)^{\mathrm{opp}} = \varprojlim_h \ker(h).$$

An element $a \in \varprojlim_h \ker(h)$ is identified with the left multiplication $g \mapsto \lambda_a(g) = ag$ by a. The interesting part of the corresponding long exact sequence of cohomology reads

$$\varprojlim_h G_h(k) \to G(k) \xrightarrow{\delta_1} \mathrm{H}^1\big(k, \pi_1(\overline{G}, \bar{e})\big) \tag{13.1}$$

$$\to \mathrm{H}^1\big(k, (G_h)\big) \xrightarrow{(h)*} \mathrm{H}^1(k, G) \xrightarrow{\delta_2} \mathrm{H}^2(k, \pi_1(\overline{G}, \bar{e})).$$

Lemma 172. *The map δ_1 in (13.1) agrees with the profinite Kummer map*

$$\kappa : G(k) \to \mathscr{S}_{\pi_1(G/k)}$$

under the identification $\mathrm{H}^1\left(k, \pi_1(\overline{G}, \bar{e})\right) = \mathscr{S}_{\pi_1(G/k)}$ *which maps a cocycle τ to the section $\tau.s_e$.*

Proof. This was proven for an abelian variety in Proposition 70 and Corollary 71. The general case admits the same proof in the light of Proposition 170.   □

13.2   The Fundamental Group of a Torsor

Let W/k be a torsor under the connected algebraic group G/k. The isomorphism

$$(\mathrm{pr}_1, \mu) : W \times_k G \xrightarrow{\sim} W \times_k W$$

that maps $(w, g) \mapsto (w, wg)$ leads via the Künneth formula to the following isomorphism of short exact sequences of Galois modules

$$
\begin{array}{ccccccccc}
1 & \longrightarrow & \pi_1(\overline{G}, \bar{e}) & \longrightarrow & \pi_1(\overline{W} \times \overline{G}, (\bar{w}, \bar{e})) & \longrightarrow & \pi_1(\overline{W}, \bar{w}) & \longrightarrow & 1 \\
 & & \downarrow{\scriptstyle t_{\bar{w},*}} & & \downarrow{\scriptstyle (\mathrm{pr}_1, \mu)_*} & & \downarrow{\scriptstyle \mathrm{Id}} & & \\
1 & \longrightarrow & \pi_1(\overline{W}, \bar{w}) & \longrightarrow & \pi_1(\overline{W} \times \overline{W}, (\bar{w}, \bar{w})) & \longrightarrow & \pi_1(\overline{W}, \bar{w}) & \longrightarrow & 1.
\end{array}
$$

Consequently, translation $t_{\bar{w}} : \overline{G} \to \overline{W}$ induces an isomorphism $\pi_1(\overline{G}, \bar{e}) \cong \pi_1(\overline{W}, \bar{w})$ as Gal_k-modules, which moreover is independent of $\bar{w}$.

Lemma 173. *The total space W'/k of a geometrically connected finite étale cover $f : W' \to W$ is naturally a torsor under an algebraic group G' such that there is a finite étale isogeny $h : G' \to G$ and*

$$W = h_* W' = W' \times_{G'} G.$$

Proof. The subgroup $\pi_1(\overline{W'}) < \pi_1(\overline{W})$ is in fact a Gal_k-submodule. For

$$H = t_{\bar{w},*}^{-1}(\pi_1(\overline{W'})) < \pi_1(\overline{G})$$

we can form the neighbourhood

$$H \rtimes s_e(\mathrm{Gal}_k) < \pi_1(G, \bar{e}),$$

which belongs to an étale isogeny $h : G' \to G$. The action

$$W' \times G' \to W'$$

is constructed as a map of finite étale covers above $W \times G \to W$ using the description as finite sets with action by π_1, namely as the $\pi_1(W \times G, (\bar{w}, \bar{e}))$-equivariant map

$$\pi_1(W)/\pi_1(W') \times \pi_1(G)/\pi_1(G') \to \pi_1(W)/\pi_1(W')$$

induced by $\mu_* : \pi_1(W) \times \pi_1(G) \to \pi_1(W)$. The rest of the lemma is formal. $\square$

Proposition 174. *The map* $\delta_2 : \mathrm{H}^1(k, G) \to \mathrm{H}^2(k, \pi_1(\overline{G}, \bar{e}))$ *in (13.1) is described by*

$$\delta_2([W]) = -[\pi_1(W/k)],$$

where $[W] \in \mathrm{H}^1(k, G)$ *is a class of G-torsors and* $[\pi_1(W/k)] \in \mathrm{H}^2(k, \pi_1(\overline{G}, \bar{e}))$ *is the class of the extension of* Gal_k *by* $\pi_1(\overline{G}, \bar{e})$ *given by* $\pi_1(W/k)$.

Proof. With respect to the point $\bar{w} \in W(\bar{k})$ the class $[W]$ of the G-torsor is represented by the cocycle $a_\sigma \in G(\bar{k})$ with

$$\sigma(\bar{w}g) = \bar{w}a_\sigma\sigma(g). \tag{13.2}$$

Let α_σ be a compatible lift of a_σ to the pro-system $\tilde{G} = (G_h)$ of all neighbourhoods $h : G_h \to G$ of the section $s_e : \mathrm{Gal}_k \to \pi_1(G, \bar{e})$. The class $\delta_2([W])$ is represented by the 2-cocycle, see [Se97] §5.6 definition of Δ,

$$\sigma, \tau \mapsto \left(\alpha_\sigma\sigma(\alpha_\tau)\alpha_{\sigma\tau}^{-1}\right) \in \varprojlim_h \ker(h) = \pi_1(\overline{G}, \bar{e}). \tag{13.3}$$

In order to compute the extension $\pi_1(W/k)$ we fix a universal pro-étale cover $\tilde{W} \to W$. Then we choose a point $\tilde{w} \in \tilde{W}(\bar{k})$ that defines an isomorphism

$$t_{\tilde{w}} : \tilde{G} \times_k \bar{k} \to \tilde{W}$$

by $\tilde{g} \mapsto \tilde{w}\tilde{g}$ of pro-varieties. For any continuous lifts $\tilde{\sigma} \in \mathrm{Aut}(\tilde{W}/W)^{\mathrm{opp}}$ of

$$\mathrm{id}_W \otimes \sigma \in \mathrm{Aut}(\overline{W}/W)^{\mathrm{opp}} = \mathrm{Gal}_k$$

the extension class

$$[\pi_1(W/k)] \in \mathrm{H}^2\left(k, \pi_1(\overline{G}, \bar{e})\right)$$

is represented by the 2-cocycle

$$\sigma, \tau \mapsto t_{\tilde{w}}^{-1} \circ \left(\tilde{\sigma}\tilde{\tau}(\widetilde{\sigma\tau})^{-1}\right) \circ t_{\tilde{w}} \in \mathrm{Aut}(\tilde{G}/\overline{G})^{\mathrm{opp}} = \pi_1(\overline{G}, \bar{e}).$$

With the left translation map $\lambda_a : \overline{G} \to \overline{G}$ defined by $g \mapsto ag$, the formula (13.2) translates for $g : \mathrm{Spec}(\bar{k}) \to \overline{G}$ into

$$(\mathrm{id}_W \otimes \sigma)^{-1} \circ t_{\tilde{w}} \circ g \circ \sigma = t_{\tilde{w}} \circ \lambda_{a_\sigma} \circ (\mathrm{id}_G \otimes \sigma)^{-1} \circ g \circ \sigma$$

which shows that

$$\mathrm{id}_W \otimes \sigma = t_{\tilde{w}} \circ (\mathrm{id}_G \otimes \sigma) \circ \lambda_{a_\sigma^{-1}} \circ t_{\tilde{w}}^{-1}.$$

The lift $\tilde{\sigma}$ can thus be obtained as

$$\tilde{\sigma} = t_{\tilde{w}} \circ (\mathrm{id}_{\tilde{G}} \otimes \sigma) \circ \lambda_{a_\sigma^{-1}} \circ t_{\tilde{w}}^{-1}.$$

The map $a \mapsto \lambda_a$ also identifies

$$\varprojlim_h \ker(h) = \mathrm{Aut}(\tilde{G}/G)$$

so that in $\mathrm{Aut}(\tilde{G}/\overline{G})^{\mathrm{opp}} = \pi_1(\overline{G}, \bar{e})$

$$t_{\tilde{w}}^{-1} \circ \left(\tilde{\sigma}\tilde{\tau}(\widetilde{\sigma\tau})^{-1} \right) \circ t_{\tilde{w}}$$

$$= \left((\mathrm{id}_{\tilde{G}} \otimes \sigma\tau) \circ \lambda_{a_{\sigma\tau}^{-1}} \right)^{-1} \circ \left((\mathrm{id}_{\tilde{G}} \otimes \tau) \circ \lambda_{a_\tau^{-1}} \right) \circ \left((\mathrm{id}_{\tilde{G}} \otimes \sigma) \circ \lambda_{a_\sigma^{-1}} \right)$$

$$= \lambda_{a_{\sigma\tau}} \circ \sigma \left(\lambda_{a_\tau^{-1}} \right) \circ \lambda_{a_\sigma^{-1}} = \lambda_{a_{\sigma\tau}\sigma(a_\tau)^{-1} a_\sigma^{-1}} = \alpha_{\sigma\tau}\sigma(\alpha_\tau)^{-1}\alpha_\sigma^{-1},$$

which is the negative of the 2-cocycle (13.3). $\square$

Corollary 175. *Let W/k be a torsor under the connected algebraic group G/k. Then the extension $\pi_1(W/k)$ splits if and only if the torsor W is compatibly divisible by any isogeny $h : G_h \to G$, i.e., there are G_h-torsors W_h with $W_{\mathrm{id}} = W$ and for all $\varphi_{h',h} : G_{h'} \to G_h$ with $h' = h \circ \varphi_{h',h}$ we have $\varphi_{h',h,*} W_{h'} = W_h$.*

Proof. By Proposition 174 the extension $\pi_1(W/k)$ splits if and only if $[W]$ lies in the image of $\mathrm{H}^1(k, (G_h)) \to \mathrm{H}^1(k, G)$, which means that there is a (G_h)-torsor (W_h) that has $W_{\mathrm{id}} = W$, and thus translates immediately into the property of W being compatibly divisible. $\square$

Remark 176. (1) Proposition 174 differs by a minus sign from [HaSz09] Proposition 2.2, which deals with the special case of an abelian variety. It is very likely that the sign difference is hidden in a different definition of the boundary map δ_2. The sign does not affect Corollary 175.

(2) In case G is a commutative group such that multiplication by n is finite étale for all $n > 0$, the compatibly divisible elements of $\mathrm{H}^1(k, G)$ form the maximal divisible subgroup

$$\mathrm{Div}(\mathrm{H}^1(k, G)) \tag{13.4}$$

which might be strictly smaller than the subgroup of divisible elements

$$\mathrm{div}(\mathrm{H}^1(k,G)) = \bigcap_{n \geq 1} n\,\mathrm{H}^1(k,G). \tag{13.5}$$

(3) There is a direct way to prove Corollary 175. A section of $\pi_1(W/k)$ defines and is uniquely determined by its tower of neighbourhoods, which is nothing but $\varprojlim_h W_h$. This alternative argument proves Corollary 175 independently of the computation of δ_2 in Proposition 174.

Corollary 177. *Let W/k be a torsor under the connected commutative algebraic group G/k such that multiplication by n is finite étale for all $n > 0$. Then*

$$0 \to \mathrm{Div}(\mathrm{H}^1(k,G)) \to \mathrm{H}^1(k,G) \xrightarrow{\delta_2} \mathrm{H}^2(k,\pi_1(\overline{G},\bar{e}))$$

is exact. In particular, the extension $\pi_1(W/k)$ splits if and only if W belongs to the maximal divisible subgroup $\mathrm{Div}(\mathrm{H}^1(k,G))$ of $\mathrm{H}^1(k,G)$. □

13.3   Examples

We analyse special cases of algebraic groups G/k with respect to claims of the section conjecture for torsors under G.

Torsors under tori. Let $\mathbb{T}/k$ be an algebraic torus. In light of Corollary 175 we need to compute the maximal divisible subgroup $\mathrm{Div}(\mathrm{H}^1(k,\mathbb{T}))$. Let k'/k be a finite Galois extension which splits the torus $\mathbb{T}$. Then Hilbert's Theorem 90 shows in the exact sequence

$$0 \to \mathrm{H}^1(k'/k,\mathbb{T}(k')) \to \mathrm{H}^1(k,\mathbb{T}) \to \mathrm{H}^1(k',\mathbb{T})^{\mathrm{Gal}(k'/k)} = 0$$

that $\mathrm{H}^1(k,\mathbb{T})$ is torsion and killed by the degree $[k':k]$, so that $\mathrm{Div}(\mathrm{H}^1(k,\mathbb{T})) = 0$. Thus we conclude the following result.

Theorem 178. *Let W/k be a torsor under a torus $\mathbb{T}/k$. The fundamental group extension $\pi_1(W/k)$ admits a section if and only if W has a k-rational point, i.e., the analogue of the weak section conjecture holds for torsors under tori even over an arbitrary field of characteristic 0.*

Proof. Immediately from $\mathrm{Div}(\mathrm{H}^1(k,\mathbb{T})) = 0$ and Corollary 177. An alternative proof of this result can be obtained from the content of [Sk01] Theorem 2.4.1. □

An example for torsors under tori. Examples for torsors under tori are given by norm varieties by which we mean here the following. Let K/k be a finite field extension and $\mathrm{R}_{K|k}\,\mathbb{G}_\mathrm{m}$ be the Weil restriction of scalars with the norm map

$$N : \mathrm{R}_{K|k}\,\mathbb{G}_\mathrm{m} \to \mathbb{G}_\mathrm{m}.$$

For $a \in k^*$ and $n \in \mathbb{N}$ we define the norm variety $W_{a,n}$ by the fibre product

$$
\begin{array}{ccc}
W_{a,n} & \longrightarrow & R_{K|k}\,\mathbb{G}_m \\
\downarrow & & \downarrow N \\
\mathbb{G}_m & \xrightarrow{\;a\cdot[n]\;} & \mathbb{G}_m
\end{array}
$$

where $a \cdot [n]$ is the endomorphism of

$$\mathbb{G}_m = \mathrm{Spec}(k[T, T^{-1}])$$

defined by $T \mapsto aT^n$. The special case

$$\mathbb{T}_n = W_{1,n}$$

has canonically the structure of an algebraic torus and $W_{a,n}$ is a $\mathbb{T}_n$-torsor. By Theorem 178 we know that $W_{a,n}$ has a k-rational point if and only if $\pi_1(W_{a,n}/k)$ admits splittings.

The notion of evaluating units allows us even to find concrete rational points starting with a section s of $\pi_1(W_{a,n}/k)$. Let $b_1, \ldots, b_r$ be a k-basis of K. Then $W_{a,n}$ has a description as solutions $(t, x_1, \ldots, x_r)$ of the equation

$$aT^n = N_{K|k}\Big(\sum_i b_i X_i\Big)$$

with $t \neq 0$. That $\sum_i b_i x_i$ does not vanish follows automatically from the equation. The function

$$F = \sum_i b_i X_i$$

is a unit on $W_{a,n} \times_k K$. By Chap. 5, the evaluation in the section s yields for each fixed m values

$$t = T(s) \in k^*/(k^*)^m,$$

$$f = F(s) \in K^*/(K^*)^m$$

such that

$$at^n = N_{K|k}(f) \tag{13.6}$$

holds in $k^*/(k^*)^m$. If m is a multiple of n, then we can find representatives t, f such that (13.6) holds even in k^*. If we expand $f = \sum_i b_i x_i$ with $x_i \in k$ then

$$(t, x_1, \ldots, x_r) \in W_{a,n}(k).$$

Torsors under semisimple algebraic groups over number fields. A semisimple
algebraic group G/k has a finite étale universal isogeny

$$h \; : \; G^{sc} \to G \tag{13.7}$$

by a simply connected semisimple algebraic group G^{sc}/k. The sequence (13.1) now
reads

$$G^{sc}(k) \xrightarrow{h} G(k) \xrightarrow{\delta_1} H^1\left(k, \pi_1(\overline{G}, \bar{e})\right) \to H^1\left(k, G^{sc}\right) \xrightarrow{h_*} H^1(k, G) \xrightarrow{\delta_2} H^2(k, \pi_1(\overline{G}, \bar{e}))$$

and the following conjecture comes into play, see [Gi10] for a survey.

Conjecture 179 (Serre's conjecture II, see [Se97] III §3). *Let k be a perfect field
with* $\mathrm{cd}(\mathrm{Gal}_k) \le 2$. *Then* $H^1(k, G^{sc}) = 0$ *for every simply connected semisimple
algebraic group* G^{sc}.

Theorem 180. *Let G/k be a semisimple algebraic group over a field of characteristic 0. We assume that one of the following is true:*

(a) *The field k is a finite extension of $\mathbb{Q}_p$ for some prime number p.*
(b) *The field k is a totally imaginary algebraic number field*
(c) *The universal cover G^{sc} of the group G is a product of absolutely almost
    simple and simply connected classical linear groups and the fields k has
    cohomological dimension* $\mathrm{cd}(\mathrm{Gal}_k) \le 2$.

Then the following holds.

(1) *The weak section conjecture for torsors under G holds, i.e., a G-torsor W
    admits a k-rational point if and only if $\pi_1(W/k)$ splits.*
(2) *The profinite Kummer map $\kappa \; : \; G(k) \to \mathscr{S}_{\pi_1(G/k)}$ is surjective, i.e., every
    section of $\pi_1(G/k)$ is Diophantine.*

Proof. The conclusion holds thanks to (13.1) if and only if $H^1(k, G^{sc}) = 0$. When
(a) holds, this is a theorem of Kneser [Kn65], while in case of (b) this is a famous
theorem by Kneser [Kn69] (classical groups), Harder [Ha65-75] (trialitarian, G_2,
F_4, E_6, and E_7) and completed by Chernousov [Ch89] (for E_8), see [PlRa94] §6 or
[Gi10] §6.3 for references. When (c) holds, this is a result of Bayer and Parimala
[BaPa95], see [Gi10] Theorem 5.1.   □

Torsors under linear algebraic groups over number fields. A connected algebraic group G/k has a characteristic filtration

$$G \supseteq G_{\mathrm{lin}} \supseteq R(G) \supseteq R_{\mathrm{u}}(G) \supseteq 1 \tag{13.8}$$

by the maximal connected linear algebraic subgroup G_{lin}, the radical $R(G)$ of G_{lin}
and the unipotent radical $R_{\mathrm{u}}(G)$ of G_{lin}. The respective filtration quotients are an
abelian variety G/G_{lin}, a linear semisimple group $G_{\mathrm{lin}}/R(G)$ a torus $R(G)/R_{\mathrm{u}}(G)$
and a unipotent linear group $R_{\mathrm{u}}(G)$.

Lemma 181. *Let $1 \to G' \to G \to G'' \to 1$ be a short exact sequence of connected algebraic groups over a field k of characteristic 0. Then*

$$0 \to \pi_1(\overline{G'}, \bar{e}) \to \pi_1(\overline{G}, \bar{e}) \to \pi_1(\overline{G''}, \bar{e}) \to 0$$

is an exact sequence of abelian profinite groups.

Proof. We may assume $k \subset \mathbb{C}$ and replace k by $\mathbb{C}$. Since

$$\pi_2^{\mathrm{top}}(G''(\mathbb{C})) = 0$$

vanishes, the homotopy sequence of the fibration $G \to G''$ for topological homotopy groups reads

$$0 \to \pi_1^{\mathrm{top}}(G'(\mathbb{C})) \to \pi_1^{\mathrm{top}}(G(\mathbb{C})) \to \pi_1^{\mathrm{top}}(G''(\mathbb{C})) \to 0$$

from which the lemma follows by profinite completion since $- \otimes_{\mathbb{Z}} \hat{\mathbb{Z}}$ is exact. □

Theorem 182. *Let G be a connected linear algebraic group over a field k which is either*

(a) a finite extension of $\mathbb{Q}_p$ for some prime number p,
(b) or a totally imaginary algebraic number field.

Then the analogue of the weak section conjecture holds for torsors under G, i.e., a G-torsor W admits a k-rational point if and only if $\pi_1(W/k)$ splits.

Proof. We have to show that a G-torsor W with a section s of $\pi_1(W/k)$ admits a rational point. We argue by dévissage with respect to the filtration (13.8). The section s induces a section of $\pi_1(V/k)$ where $V = W/R(G)$ is a torsor under the semisimple algebraic group $S = G/R(G)$. By Theorem 180, first V is a trivial S-torsor, and moreover $s = s_a$ for a rational point $a \in S(k)$.

As in Proposition 31, the section s induces a section also called s of $\pi_1(W_a/k)$ where W_a is the fibre of $W \to V$ above a. The variety W_a is naturally a torsor under $R(G)$. Next, the section s provides a section of $\pi_1(U/k)$ where $U = W_a/R_u(G)$ is a torsor under the torus $\mathbb{T} = R(G)/R_u(G)$. By Theorem 178, the torsor is trivial, hence we find a rational point $b \in U(k)$. The fibre $(W_a)_b$ above b in $W_a \to U$ is a torsor under the unipotent group $R_u(G)$ and thus admits itself a rational point

$$c \in (W_a)_b(k) \subseteq W(k).$$

This provides a rational point of $W(k)$ and proves the theorem. □

Torsors under abelian varieties. For the purpose of the section conjecture the case of torsors under abelian varieties is crucial. Unfortunately here the results are less convincing.

Proposition 183. *Let A be an abelian variety over a finite extension $k/\mathbb{Q}_p$. Then*

$$\mathrm{Div}(\mathrm{H}^1(k, A)) \cong (\mathbb{Q}_p/\mathbb{Z}_p)^{[k:\mathbb{Q}_p] \cdot \dim(A)}$$

and, in particular, there are always nontrivial A-torsors W with split $\pi_1(W/k)$, so that the analogue of the weak section conjecture never holds for torsors under abelian varieties over k.

Proof. Let A^t/k be the dual abelian variety. By local Tate duality we have

$$\mathrm{Div}\left(\mathrm{H}^1(k, A)\right) = \mathrm{Div}\left(\mathrm{Hom}(A^t(k), \mathbb{Q}/\mathbb{Z})\right),$$

which by the Theorem of Mattuck–Tate [Ma55] is isomorphic to

$$\mathrm{Div}\left(\mathrm{Hom}(o_k^{\dim(A)}, \mathbb{Q}_p/\mathbb{Z}_p)\right) \cong (\mathbb{Q}_p/\mathbb{Z}_p)^{[k:\mathbb{Q}_p]\cdot\dim(A)}.$$

The rest follows from Corollary 177.   □

Remark 184. (1) This was obtained independently with the same proof by Harari and Szamuely in [HaSz09] §3.
(2) At least we can deduce again that the period of a $[W] \in \mathrm{H}^1(k, A)$ with split $\pi_1(W/k)$ must be a power of p. This reproves part of Theorem 114.

Let now A/k be an abelian variety over an algebraic number field. In case the A-torsor W is locally trivial, namely an element $[W] \in \mathrm{III}(A/k)$ of the Tate–Shafarevich group, then the weak section conjecture for such torsors leads to Bashmakov's problem: are there nontrivial elements

$$[W] \in \mathrm{III}(A/k) \cap \mathrm{Div}\left(\mathrm{H}^1(k, A)\right),$$

or equivalently, are there A-torsors W that are counter-examples to the Hasse principle and such that $\pi_1(W/k)$ splits? This question was asked by Bashmakov in 1964 in an attempt to solve a divisibility question of another nature posed by Cassels. Only recently Bashmakov's problem has been decided as follows for elliptic curves over $\mathbb{Q}$.

Theorem 185. *Let $E/\mathbb{Q}$ be an elliptic curve, and let p be a prime number. Then the following holds.*

(1) (Harari–Szamuely) If $E/\mathbb{Q}$ has analytic rank 1, then $\mathrm{Div}(\mathrm{H}^1(\mathbb{Q}, E)) = 0$.
(2) If $E/\mathbb{Q}$ has analytic rank 0, then $\mathrm{Div}(\mathrm{H}^1(\mathbb{Q}, E)) = \mathbb{Q}/\mathbb{Z}$.
(3) If $E/\mathbb{Q}$ has analytic rank 0 and good reduction at the odd prime $p \geq 3$ and the $\mathrm{Gal}_\mathbb{Q}$-representation on the p-torsion of E is irreducible, then the p-primary part of

$$\mathrm{III}(E/\mathbb{Q}) \cap \mathrm{Div}\left(\mathrm{H}^1(\mathbb{Q}, E)\right)$$

is trivial. This applies if $p > 7$, and E is semistable of analytic rank 0 with good reduction in p.
(4) If $E/\mathbb{Q}$ is the Jacobian of the Selmer curve, see [Se51],

$$S = \{3X^3 + 4Y^3 + 5Z^3 = 0\}$$

then

$$\text{III}(E/\mathbb{Q}) \cap \text{Div}\left(\text{H}^1(\mathbb{Q}, E)\right) \cong \mathbb{Z}/3\mathbb{Z}$$

generated by the class of S. In particular the extension $\pi_1(S/\mathbb{Q})$ splits despite S being a counter-example to the Hasse principle.

Proof. All of this (and more) was proven in [CiSx12a] §7+8.   □

All the results in this chapter so far concern the weak form of the section conjecture. Concerning the profinite Kummer map in the case of an abelian variety over an algebraic number field we have the following.

Theorem 186. *There is no abelian variety A/k over an algebraic number field k such that every section of*

$$1 \to \pi_1(\overline{A}) \to \pi_1(A) \to \text{Gal}_k \to 1$$

comes from a rational point of A.

Proof. This is [CiSx12b] Theorem 1: it follows from the continuous Kummer sequence that κ is bijective if and only if $A(k)$ is finite and $\text{Div}(\text{H}^1(k, A)) = 0$. But, by [CiSx11] Proposition 42, a finite Mordell–Weil group implies that $\text{II}^1(k, A)$ contains a subgroup $(\mathbb{Q}/\mathbb{Z})^d$ with $d \geq \dim A \cdot [k : \mathbb{Q}]$, a contradiction.

13.4 The Weak Section Conjecture for Curves of Genus 0

Let U be an affine smooth curve over a field k of characteristic 0 with smooth projective completion X of genus 0. Then X is a form of $\mathbb{P}^1$ over k and, as such, a Brauer–Severi variety for a quaternion algebra. The presence of k-rational points on X is thus decided by the vanishing of a class A_X in the 2-torsion $\text{Br}(k)[2]$ of the Brauer group $\text{Br}(k)$ of k.

The weak section conjecture predicts that $\pi_1(U/k)$ splits if and only if A_X is trivial, because then $X \cong \mathbb{P}^1_k$ forces X and thus U to have infinitely many rational points. More precisely we may ask, whether the extension $\pi_1(U/k)$ encodes the class A_X by a direct group theoretic construction.

The universal torus torsor. There is a torus $\mathbb{T}$ and a principal homogeneous $\mathbb{T}$-space W together with a map $\iota : U \to W$ universal with respect to such maps. The construction of W and $\mathbb{T}$ relies on group algebras as follows. The Gal_k-module of characters is $X^*(\mathbb{T}) = \mathcal{O}^*(\overline{U})/\bar{k}^*$ and

$$\mathbb{T} = \text{Spec}\left(\bar{k}[X^*(\mathbb{T})]\right)/\text{Gal}_k .$$

For the universal torsor W we can take

$$W = \text{Spec}\left(\bar{k}[\mathcal{O}^*(\overline{U})]/(a - \text{pr}^*(a); \text{ all } a \in \bar{k}^*)\right)/\text{Gal}_k$$

where pr $: U \to \operatorname{Spec}(k)$ is the projection map. The map ι is induced by the Gal_k-equivariant map $\mathcal{O}^*(\overline{W}) \to \mathcal{O}^*(\overline{U})$ on units, and induces an isomorphism

$$\iota^* : \mathcal{O}^*(\overline{W}) \xrightarrow{\sim} \mathcal{O}^*(\overline{U}).$$

By Kummer theory the natural map of extensions $\pi_1(U/k) \to \pi_1(W/k)$ can thus be identified with the map to the maximal geometrically abelian quotient $\pi_1^{\mathrm{ab}}(U/k)$,

Comparison of relative Brauer groups. The relative Brauer group

$$\operatorname{Br}(U/k) = \ker(\operatorname{pr}^* : \operatorname{Br}(k) \to \operatorname{Br}(U))$$

is generated by the Azumaya algebra A_X by purity and a theorem of Amitsur and Lichtenbaum, see [GiSz06] Theorem 5.4.10. In particular, the group $\operatorname{Br}(U/k)$ is cyclic of order ≤ 2.

The map $\iota : U \to W$ leads to a map of Leray spectral sequences for the projection to $\operatorname{Spec}(k)$. Taking into account that $\operatorname{H}^1(\overline{U}, \mathbb{G}_\mathrm{m})$ and $\operatorname{H}^1(\overline{W}, \mathbb{G}_\mathrm{m})$ vanish, we get a diagram with exact rows

$$
\begin{array}{ccccc}
0 & \xrightarrow{d_2^{0,1}} & \operatorname{H}^2(k, \mathcal{O}^*(\overline{W})) & \longrightarrow & \operatorname{Br}(W) \\
 & & \downarrow{\iota^*} & & \downarrow{\iota^*} \\
0 & \xrightarrow{d_2^{0,1}} & \operatorname{H}^2(k, \mathcal{O}^*(\overline{U})) & \longrightarrow & \operatorname{Br}(U)
\end{array}
$$

in which the first vertical map ι^* is an isomorphism by $\mathcal{O}^*(\overline{W}) = \mathcal{O}^*(\overline{U})$. It follows that the relative Brauer groups fit in the following diagram with exact rows.

$$
\begin{array}{ccccccc}
0 & \longrightarrow & \operatorname{Br}(W/k) & \longrightarrow & \operatorname{H}^2(k, \mathbb{G}_\mathrm{m}) & \longrightarrow & \operatorname{H}^2(k, \mathcal{O}^*(\overline{W})) \\
 & & \downarrow & & \| & & \downarrow{\iota^*} \\
0 & \longrightarrow & \operatorname{Br}(U/k) & \longrightarrow & \operatorname{H}^2(k, \mathbb{G}_\mathrm{m}) & \longrightarrow & \operatorname{H}^2(k, \mathcal{O}^*(\overline{U})).
\end{array}
$$

Hence the map $\operatorname{Br}(W/k) \to \operatorname{Br}(U/k)$ is an isomorphism. This allows to conclude the following result.

Proposition 187 (Weak section conjecture for genus 0 curves). *Let k be a field of characteristic 0. Let U/k be an affine smooth curve with smooth projective completion X of genus 0. Then the following are equivalent.*

(a) $\pi_1(U/k)$ *admits a section.*
(b) $\pi_1^{\mathrm{ab}}(U/k)$ *admits a section.*
(c) U *contains k-rational points.*

Proof. We only need to show (b) implies (c). Let $\iota : U \to W$ be the universal map to a torsor under a torus. By Theorem 178, a section of $\pi_1^{ab}(U/k) = \pi_1(W/k)$ trivialises the class of torsors represented by W and so forces W to have a rational point. Consequently then $\mathrm{Br}(W/k) = \mathrm{Br}(U/k) = \langle A_X \rangle$ vanishes, and $A_X = 0$ shows that $X \cong \mathbb{P}^1_k$. The proof is achieved because the infinitely many k-rational points of X cannot avoid U altogether.   □

Direct construction of the Brauer class. We will describe a direct construction of the Brauer class A_X. Let $Y = X \setminus U$ be the reduced boundary. The maximal unramified abelian extension of $\overline{U}$ has Galois group

$$\pi_1^{ab}(\overline{U}) = \hat{\mathbb{Z}}(1)[Y(\bar{k})]/\hat{\mathbb{Z}}(1)$$

with the quotient being by the diagonally embedded copy of $\hat{\mathbb{Z}}(1)$. With a Gal_k-equivariant map $\varphi : \pi_1^{ab}(\overline{U}) \to \mu_{2^m}$ we may push the extension $\pi_1^{ab}(U/k)$ and obtain an extension $\varphi_*([\pi_1^{ab}(U/k)])$ of Gal_k by μ_{2^m}, so an element

$$\alpha_\varphi \in \mathrm{Br}(k)[2^m] = \mathrm{H}^2(k, \mu_{2^m}).$$

Proposition 188. *(1) For $m \gg 1$ the image of the boundary map*

$$d_2^{0,1} : \mathrm{H}^0\left(k, \mathrm{H}^1(\overline{U}, \mu_{2^m})\right) \to \mathrm{H}^2(k, \mu_{2^m})$$

of the Hochschild–Serre spectral sequence for $\pi_1(U) \to \mathrm{Gal}_k$ with μ_{2^m} coefficients agrees with $\mathrm{Br}(U/k)$ and A_X is the generator of the image.
(2) Let $y \in Y$ be a closed point and $m \in \mathbb{N}$ such that $2^{m+1} \nmid \deg(y)$. Then $A_X = \alpha_\varphi$ for the map

$$\varphi : \hat{\mathbb{Z}}(1)[Y(\bar{k})]/\hat{\mathbb{Z}}(1) \to \mu_{2^m}$$

that sums up the $\hat{\mathbb{Z}}(1)$-coefficients of the geometric points in y and reduces modulo 2^m.

Proof. The Hochschild–Serre spectral sequence agrees with the Leray spectral sequence for $U \to \mathrm{Spec}(k)$ and yields the exact sequence

$$\mathrm{H}^0\left(k, \mathrm{H}^1(\overline{U}, \mu_{2^m})\right) \xrightarrow{d_2^{0,1}} \mathrm{H}^2(k, \mu_{2^m}) \xrightarrow{\mathrm{pr}^*} \mathrm{H}^2(U, \mu_{2^m}) \to \mathrm{H}^1\left(k, \mathrm{H}^1(\overline{U}, \mu_{2^m})\right).$$

The map $d_2^{0,1}$ is simply the map $\varphi \mapsto \alpha_\varphi$ described above. The Kummer sequence on U yields the exactness of

$$0 \to \mathrm{Pic}(U) \otimes \mathbb{Z}/2^m\mathbb{Z} \to \mathrm{H}^2(U, \mu_{2^m}) \to \mathrm{Br}(U)[2^m] \to 0.$$

Now (1) is a consequence of (2) because certainly $\mathrm{im}(d_2^{0,1}) \subset \mathrm{Br}(U/k)$.

For (2) we may by naturality in U assume that $Y = \{y\}$ consists of just one closed point. Then $\mathrm{H}^0\left(k, \mathrm{H}^1(\overline{U}, \mu_{2^m})\right)$ is cyclic of order 2^m and generated by the given map φ. It remains to show (1) in this case and this is equivalent to the composition

$$\mathrm{Pic}(U) \otimes \mathbb{Z}/2^m\mathbb{Z} \to \mathrm{H}^2(U, \mu_{2^m}) \to \mathrm{H}^1\left(k, \mathrm{H}^1(\overline{U}, \mu_{2^m})\right)$$

being injective. As the Leray spectral sequence is compatible with connecting homomorphisms, we can compute this composition as

$$\mathrm{Pic}(U) \otimes \mathbb{Z}/2^m\mathbb{Z} \to \mathrm{H}^1\left(k, \mathcal{O}^*(\overline{U})\right) \otimes \mathbb{Z}/2^m\mathbb{Z} \to \mathrm{H}^1\left(k, \mathrm{H}^1(\overline{U}, \mu_{2^m})\right).$$

The first map is an isomorphism since $\mathrm{Pic}(\overline{U}) = 0$. We compute the second map with the help of the following short exact sequences

$$0 \to \bar{k}^* \to \mathcal{O}^*(\overline{U}) \to \mathrm{Div}^0_{\overline{X}, \overline{Y}} \to 0,$$

$$0 \to \mathrm{Div}^0_{\overline{X}, \overline{Y}} \xrightarrow{2^m} \mathrm{Div}^0_{\overline{X}, \overline{Y}} \to \mathrm{H}^1(\overline{U}, \mu_{2^m}) \to 0.$$

Here $\mathrm{Div}^0_{\overline{X}, \overline{Y}}$ is the Gal_k-module of divisors of degree 0 on $\overline{X}$ with support in $\overline{Y}$. Hilbert's Theorem 90 yields an injection

$$\mathrm{H}^1(k, \mathcal{O}^*(\overline{U})) \otimes \mathbb{Z}_2 \hookrightarrow \mathrm{H}^1(k, \mathrm{Div}^0_{\overline{X}, \overline{Y}}) \otimes \mathbb{Z}_2$$

and the cohomology sequences lead to an injective map

$$\mathrm{H}^1(k, \mathrm{Div}^0_{\overline{X}, \overline{Y}}) \otimes \mathbb{Z}/2^m\mathbb{Z} \hookrightarrow \mathrm{H}^1\left(k, \mathrm{H}^1(\overline{U}, \mu_{2^m})\right)$$

and it remains to justify that the natural map

$$\mathrm{H}^1(k, \mathrm{Div}^0_{\overline{X}, \overline{Y}}) \otimes \mathbb{Z}_2 \to \mathrm{H}^1(k, \mathrm{Div}^0_{\overline{X}, \overline{Y}}) \otimes \mathbb{Z}/2^m\mathbb{Z}$$

is an isomorphism. The defining degree sequence

$$0 \to \mathrm{Div}^0_{\overline{X}, \overline{Y}} \to \mathrm{Div}_{\overline{X}, \overline{Y}} \xrightarrow{\deg} \mathbb{Z} \to 0$$

computes $\mathrm{H}^1(k, \mathrm{Div}^0_{\overline{X}, \overline{Y}}) \otimes \mathbb{Z}_2$ as $\mathbb{Z}_2 / \deg(y)\mathbb{Z}_2$ which is killed by 2^m by assumption and completes the proof. $\qquad\square$

13.5 Zero Cycles and Abelian Sections

The theory of zero cycles on a smooth projective variety X/k is controlled to a large extend by the arithmetic of its universal Albanese torsor Alb_X^1, see [Wi08]. For a variety X with nontrivial torsion in its Néron–Severi group $\mathrm{NS}_{\overline{X}}$, the abelianized fundamental group extension $\pi_1^{\mathrm{ab}}(X/k)$ goes slightly beyond $\pi_1(\mathrm{Alb}_X^1/k)$, see Proposition 69. We show that there is even a section of $\pi_1^{\mathrm{ab}}(X/k)$ associated to each k-rational zero-cycle of degree 1 on X and that the difference between $\pi_1^{\mathrm{ab}}(X/k)$ and $\pi_1(\mathrm{Alb}_X^1/k)$ defines a constraint for a k-rational class of zero-cycles on $\overline{X}$ to come from an actual k-rational cycle. We keep working over a field of characteristic 0.

Abelian sections and the albanese torsor. Let X/k be a smooth projective geometrically connected variety. We recall from Proposition 69 the exact sequence of Gal_k-modules

$$0 \to \left(\mathrm{NS}_{X,\mathrm{tors}}\right)^D(\bar{k}) \to \pi_1^{\mathrm{ab}}(\overline{X}) \to \pi_1(\overline{\mathrm{Alb}_X}) \to 0, \tag{13.9}$$

where Alb_X is the Albanese variety of X. Let $\alpha : X \to \mathrm{Alb}_X^1$ be the universal map into a torsor under an abelian variety, where Alb_X^1 is a torsor under Alb_X, see [Wi08].

Corollary 189. *If $\pi_1^{\mathrm{ab}}(X/k)$ splits, then $[\mathrm{Alb}_X^1] \in \mathrm{H}^1(k, \mathrm{Alb}_X)$ belongs to the maximal divisible subgroup* $\mathrm{Div}\left(\mathrm{H}^1(k, \mathrm{Alb}_X)\right)$. *The converse holds if the torsion subgroup* $\mathrm{NS}_{X,\mathrm{tors}}$ *of the Néron–Severi group of X vanishes.*

Proof. This follows immediately from (13.9) together with Corollary 177. □

Remark 190. Based on this criterion, in [HaSz09] §6 Harari and Szamuely, with the help of Flynn for the numerical examples, constructed smooth projective curves X of genus 2 over $\mathbb{Q}$ such that X has local points everywhere, $\pi_1^{\mathrm{ab}}(X/k)$ does not split and hence X does not admit a zero-cycle of degree 1.

Sections associated to zero cycles. The Chow group of zero cycles $\mathrm{CH}_0(X)$ up to rational equivalence of a proper geometrically connected variety X/k can be thought of as the abelian analogue of the set of rational points $X(k)$. On the other hand, the Albanese variety is related to the group of k-rational classes of zero cycles $\mathrm{H}^0(k, \mathrm{CH}_0(\overline{X}))$. We are going to improve the natural map

$$\mathrm{H}^0\left(k, \mathrm{CH}_0(\overline{X})^{\deg=1}\right) \to \mathrm{Alb}_X^1(k) \xrightarrow{\kappa} \mathscr{S}_{\pi_1(\mathrm{Alb}_X^1/k)}$$

to a map

$$\mathrm{CH}_0(X) \to \mathscr{S}_{\pi_1^{\mathrm{ab}}(X/k)}.$$

Let X/k be a geometrically connected projective variety. For $d \geq 0$, the dth symmetric product is the quotient $X^{(d)} = X^d/S_d$ by the symmetric group on d

elements which acts by permuting the factors. The quotient is a projective variety, and we understand $X^{(0)} = \mathrm{Spec}(k)$. We fix a geometric point $\bar{x} \in X$ and endow $X^{(d)}$ with the image

$$\bar{x} = (\bar{x}, \ldots, \bar{x}) \in X^d \to X^{(d)}.$$

The disjoint union $\bigsqcup_{d \geq 0} X^{(d)}$ forms a commutative monoid with $\mathbf{1}$ under concatenation

$$X^{(d_1)} \times_k X^{(d_2)} \to X^{(d_1 + d_2)}.$$

The Grothendieck group of $\bigsqcup_{d \geq 0} X^{(d)}(k)$ is the group $Z_0(X)$ of zero cycles on X.

The monoid structure preserves the base points $\bar{x}$. Due to the Künneth formula, concatenation also induces a map

$$\pi_1(X^{(d_1)}, \bar{x}) \times_{\mathrm{Gal}_k} \pi_1(X^{(d_2)}, \bar{x}) = \pi_1\left(X^{(d_1)} \times_k X^{(d_2)}, (\bar{x}, \bar{x})\right) \to \pi_1(X^{(d_1+d_2)}, \bar{x})$$

which induces the structure of a commutative monoid with $\mathbf{1}$ on $\bigsqcup_{d \geq 0} \mathscr{S}_{\pi_1(X^{(d)}/k)}$. Moreover, the profinite Kummer map provides a map of monoids

$$\kappa = \bigsqcup_{d \geq 0} \kappa_{X^{(d)}} : \bigsqcup_{d \geq 0} X^{(d)}(k) \to \bigsqcup_{d \geq 0} \mathscr{S}_{\pi_1(X^{(d)}/k)}.$$

Proposition 191. *The Grothendieck group $K\left(\bigsqcup_{d \geq 0} \mathscr{S}_{\pi_1(X^{(d)}/k)}\right)$ of the commutative monoid $\bigsqcup_{d \geq 0} \mathscr{S}_{\pi_1(X^{(d)}/k)}$ sits in a short exact sequence*

$$0 \to \mathrm{H}^1(k, \pi_1^{\mathrm{ab}}(\overline{X}, \bar{x})) \to K(\bigsqcup_{d \geq 0} \mathscr{S}_{\pi_1(X^{(d)}/k)}) \xrightarrow{\deg} D \cdot \mathbb{Z} \to 0$$

with

$$D = \gcd\{d \geq 0 \; ; \; \mathscr{S}_{\pi_1(X^{(d)}/k)} \neq \emptyset\}$$

and $\deg(s) = d$ for $s \in \mathscr{S}_{\pi_1(X^{(d)}/k)}$. Furthermore, for $d \geq 1$, we have

$$\deg^{-1}(d) = \mathscr{S}_{\pi_1^{\mathrm{ab}}(X^{(d)}/k)}.$$

Proof. It is enough to recall from descent theory of finite étale covers, see [SGA1] IX §5 page 252, that for $d \geq 2$ there is an isomorphism

$$\pi_1^{\mathrm{ab}}(\overline{X}, \bar{x}) \xrightarrow{\sim} \pi_1(\overline{X^{(d)}}, \bar{x})$$

induced by the natural map $\overline{X} \to \overline{X^{(d)}}$ given by $a \mapsto (a, \bar{x}, \ldots, \bar{x})$.   □

In the next theorem we associate to a zero cycle up to rational equivalence on a smooth projective variety a section of the abelianized fundamental group.

Theorem 192. *Let X/k be a geometrically connected smooth projective variety.*

(1) The profinite Kummer map induces a group homomorphism

$$\mathrm{CH}_0(X) \to K(\bigsqcup_{d \geq 0} \mathscr{S}_{\pi_1(X^{(d)}/k)})$$

compatible with the degree maps.

(2) With the universal map $\alpha : X \to \mathrm{Alb}^1_X$ into the Albanese torsor we have a commutative diagram

$$
\begin{array}{ccc}
\mathrm{CH}_0(X)^{\deg=1} & \xrightarrow{\hspace{4cm}} & \mathscr{S}_{\pi_1^{\mathrm{ab}}(X/k)} \\
\downarrow & & \downarrow{\scriptstyle \alpha_*} \\
H^0\left(k, \mathrm{CH}_0(\overline{X})^{\deg=1}\right) & \longrightarrow \mathrm{Alb}^1_X(k) \longrightarrow & \mathscr{S}_{\pi_1(\mathrm{Alb}^1_X/k)}.
\end{array}
$$

$$(13.10)$$

Proof. (1) The profinite Kummer map clearly induces a group homomorphism

$$Z_0(X) \to K(\bigsqcup_{d \geq 0} \mathscr{S}_{\pi_1(X^{(d)}/k)})$$

compatible with the degree maps. We need to show that it factors through rational equivalence. Since rational equivalence is generated by relations on smooth, projective curves C with a finite map $f : C \to X$, push forward functoriality

$$
\begin{array}{ccc}
Z_0(C) & \longrightarrow & K\left(\bigsqcup_{d \geq 0} \mathscr{S}_{\pi_1(C^{(d)}/k)}\right) \\
\downarrow{\scriptstyle f_*} & & \downarrow{\scriptstyle f_*} \\
Z_0(X) & \longrightarrow & K\left(\bigsqcup_{d \geq 0} \mathscr{S}_{\pi_1(X^{(d)}/k)}\right)
\end{array}
$$

reduces to the case of a smooth projective curve $C = X$. For $d \geq 2$ we have a commutative diagram

$$
\begin{array}{ccc}
Z_0(C)^{\deg=d} & =\!=\!= \quad C^{(d)}(k) \longrightarrow & \mathscr{S}_{\pi_1(C^{(d)}/k)} \\
\downarrow & \downarrow & \| \\
\mathrm{CH}_0(C)^{\deg=d} & \longrightarrow \mathrm{Pic}^d_C(k) \longrightarrow & \mathscr{S}_{\pi_1(\mathrm{Pic}^d_C/k)}
\end{array}
$$

which settles the case of a smooth projective curve. Assertion (2) follows at once from (1) and Proposition 191.   □

Corollary 193. *If the smooth projective variety X/k has a rational zero cycle of degree 1, then $\pi_1^{ab}(X/k)$ splits.*

Proof. This follows immediately from the construction of an abelian section associated to a zero cycle, or alternatively by the usual corestriction argument.   □

We may say that a class of smooth projective varieties satisfies a *weak section conjecture for zero cycles* if for every variety X/k in this class a splitting of $\pi_1^{ab}(X/k)$ implies the existence of a zero cycle of degree 1 on X.

Let s be a section of $\pi_1^{ab}(X/k)$. In view of diagram (13.10) an intermediate step for the construction of a zero cycle of degree 1 consists of finding a k-rational cycle on $\overline{X}$ of degree 1 thanks to the existence of the section $\alpha_*(s)$ of $\pi_1(\mathrm{Alb}_X^1/k)$. Let $\mathrm{Alb}_X^\bullet = \bigsqcup_{d\in\mathbb{Z}} \mathrm{Alb}_X^d$ be the extension

$$0 \to \mathrm{Alb}_X \to \mathrm{Alb}_X^\bullet \to \mathbb{Z} \to 0$$

with Alb_X^1 the universal Albanese torsor of X. Since by a theorem of Rojtman, see [Bl79], the kernel of $\mathrm{CH}_0(\overline{X}) \to \mathrm{Alb}_X^\bullet(\bar{k})$ is uniquely divisible, the map

$$H^0(k, \mathrm{CH}_0(\overline{X})^{\deg=1}) \twoheadrightarrow \mathrm{Alb}_X^1(k)$$

is surjective and this first step asks to show that Alb_X^1 is a trivial torsor under Alb_X. This was discussed in Sect. 13.3 but in an unsatisfactory manner.

The next step asks for a criterion that decides, whether a k-rational zero cycle on $\overline{X}$ can be lifted to a zero cycle on X itself. The lifting along

$$a_* : \mathscr{S}_{\pi_1^{ab}(X/k)} \to \mathscr{S}_{\pi_1(\mathrm{Alb}_X/k)}$$

is obstructed by an obstruction with values in $H^2(k, \mathrm{NS}_{X,\mathrm{tors}}^D)$ due to (13.9). We find a map

$$v : H^0(k, \mathrm{CH}_0(\overline{X})^{\deg=1}) \to \mathscr{S}_{\pi_1(\mathrm{Alb}_X/k)} \to H^2(k, \mathrm{NS}_{X,\mathrm{tors}}^D)$$

which annihilates exactly those k-rational zero cycles of degree 1 whose associated section of $\pi_1(\mathrm{Alb}_X^1/k)$ lifts to a section of $\pi_1^{ab}(X/k)$.

Chapter 14
Nilpotent Sections

The next difficult characteristic quotient of a profinite group beyond the maximal abelian quotient might be the maximal pro-nilpotent quotient or its truncated versions of bounded nilpotency. These quotients have been studied in the realm of the section conjecture by Ellenberg around 2000, unpublished, and later by Wickelgren in her thesis [Wg09], and in [Wg10, Wg12a, Wg12b] with special emphasis on the interesting case $\mathbb{P}^1 - \{0, 1, \infty\}$.

The (relative) pro-algebraic version has played an important role in at least two strands of mathematics: (1) on the Hodge theoretic side in the study conducted by Hain of the Teichmüller group and the section conjecture for the generic curve [Ha11b], and (2) on the arithmetic side in the non-abelian Chabauty method of Kim [Ki05] for Diophantine finiteness problems.

We will examine in detail the Lie algebra associated to the maximal pro-ℓ quotient of the geometric fundamental group, see Sect. 14.3, and in particular prove Proposition 207 about the sub Lie algebra of invariants under a finite abelian group action. This will be crucial for counting pro-ℓ sections over a finite field in Sect. 15.3.

The nilpotent section conjecture is known to fail by work of Hoshi [Ho10]. We try to explain that examples for this failure should be seen as *accidents* due to an accidental coincidence of very special properties. In Sect. 14.7, we extend the range of examples, show that in most of these examples the spaces of pro-p sections are in fact uncountable, and suggest a way of reviving the pro-p version of the section conjecture by asking a virtually pro-p section conjecture.

14.1 Primary Decomposition

Let X/k be a geometrically connected variety. We may push the extension $\pi_1(X/k)$ by the maximal pro-nilpotent quotient $\pi_1(\overline{X}) \twoheadrightarrow \pi_1^{\mathrm{nilp}}(\overline{X})$ to obtain the maximal nilpotent extension $\pi_1^{\mathrm{nilp}}(X/k)$. As any finite nilpotent group is canonically the

J. Stix, *Rational Points and Arithmetic of Fundamental Groups*, Lecture Notes in Mathematics 2054, DOI 10.1007/978-3-642-30674-7_14,
© Springer-Verlag Berlin Heidelberg 2013

direct product of its unique p-Sylow groups we obtain in the limit a canonical isomorphism

$$\pi_1^{\mathrm{nilp}}(\overline{X}) = \prod_p \pi_1^{\mathrm{pro}\text{-}p}(\overline{X}) \tag{14.1}$$

and also a primary decomposition

$$\pi_1^{\mathrm{nilp}}(X/k) = \prod_p \pi_1^{\mathrm{pro}\text{-}p}(X/k) \tag{14.2}$$

that has to be read as a fibre product over Gal_k. For the corresponding section spaces and Kummer maps this leads to

$$\kappa_{\mathrm{nilp}} = (\kappa_p)_p : X(k) \to \mathscr{S}_{\pi_1^{\mathrm{nilp}}(X/k)} = \prod_p \mathscr{S}_{\pi_1^{\mathrm{pro}\text{-}p}(X/k)}. \tag{14.3}$$

If X/k moreover is abelian injective, then because $\pi^{\mathrm{ab}}(X/k)$ is a quotient extension of $\pi_1^{\mathrm{nilp}}(X/k)$ we have

$$X(k) = \kappa_{\mathrm{nilp}}(X(k)) \subset \prod_p \kappa_p(X(k)).$$

Let for the moment k be an algebraic number field and X/k a smooth hyperbolic curve. Then, by Theorem 76, we have $X(k) = \kappa_p(X(k))$, and the section conjecture raises the question whether only diagonal tuples of pro-p sections lift to actual sections along

$$\mathscr{S}_{\pi_1(X/k)} \to \mathscr{S}_{\pi_1^{\mathrm{nilp}}(X/k)},$$

or better: the section conjecture could be modified to ask for a definition of diagonal tuples as the image and find a Diophantine description of this set. For the section conjecture to hold in its original form it would be desirable if the diagonal tuples would feature a certain independence of p.

14.2   Obstructions from the Descending Central Series

The obstructions against lifting of an abelian section s^{ab} to a nilpotent section form a hierarchy of obstruction classes $\delta_n(s^{\mathrm{ab}})$ with δ_n only being defined if all the previous obstructions δ_i vanish for $i < n$ and also depending on the chosen partial lifts. This study was initiated by Ellenberg and in the thesis of Wickelgren [Wg09].

Definition 194. The *descending central filtration* $C_\bullet \Gamma$ on a profinite group Γ is defined inductively by

$$C_{-1}\Gamma = \Gamma \quad \text{and} \quad C_{-(n+1)}\Gamma = [\Gamma, C_{-n}\Gamma]$$

for $n \geq 2$ where $[A, B]$ is the profinite subgroup generated by the corresponding commutators $[a, b]$ with $a \in A$ and $b \in B$.

The strange numbering takes into account the weight of the associated graded when $\Gamma = \pi_1(\overline{X})$, at least when X is smooth and projective.

Definition 195. Let X/k be a geometrically connected variety and let $\ell \neq \text{char}(k)$ be a prime number. For every $n \in \mathbb{N}$ we have

(1) The *geometrically n-step nilpotent quotient extension* of $\pi_1(X/k)$

$$C_{\geq -n}\big(\pi_1(X/k)\big)$$

as the pushout by the characteristic quotient

$$\pi_1(\overline{X}) \twoheadrightarrow \pi_1(\overline{X})/C_{-(n+1)}\pi_1(\overline{X}).$$

(2) The *geometrically n-step pro-ℓ nilpotent quotient extension* of $\pi_1(X/k)$

$$C_{\geq -n}\big(\pi_1^{\text{pro-}\ell}(X/k)\big)$$

as the pushout by the characteristic quotient

$$\pi_1^{\text{pro-}\ell}(\overline{X}) \twoheadrightarrow \pi_1^{\text{pro-}\ell}(\overline{X})/C_{-(n+1)}\pi_1^{\text{pro-}\ell}(\overline{X}).$$

Definition 196. The following commutative diagram defines truncated nilpotent Kummer maps

$$\kappa_{\text{ab}}, \quad \kappa_{\text{nilp}}, \quad \kappa_n, \quad \kappa_\ell, \quad \text{and} \quad \kappa_{\ell,n},$$

the abelian, nilpotent, n-step nilpotent, pro-ℓ, n-step nilpotent pro-ℓ Kummer map respectively. The diagram moreover shows how these Kummer maps factorize each other:

Dévissage for truncated nilpotent sections. We abbreviate $\overline{\pi} = \pi_1(\overline{X})$ and for the maximal pro-ℓ quotient $\overline{\pi}^\ell = \pi_1^{\text{pro-}\ell}(\overline{X})$. The central extension

$$1 \to \text{gr}_{-n}^C \overline{\pi} \to \overline{\pi}/C_{-(n+1)}\overline{\pi} \xrightarrow{\text{pr}} \overline{\pi}/C_{-n}\overline{\pi} \to 1 \quad . \tag{14.4}$$

yields as an application of Sect. 1.3 the following exact sequence of sets in the sense of Proposition 31:

$$1 \; \longrightarrow \; H^0\left(k, \mathrm{gr}^C_{-n}\,\overline{\pi}\right) \; \longrightarrow \; H^0\left(k, \overline{\pi}/C_{-(n+1)}\overline{\pi}\right) \; \longrightarrow \; H^0\left(k, \overline{\pi}/C_{-n}\overline{\pi}\right)$$

$$\longrightarrow \; H^1\left(k, \mathrm{gr}^C_{-n}\,\overline{\pi}\right) \; \longrightarrow \; \mathscr{S}_{C_{\geq -n}\pi_1(X/k)} \; \xrightarrow{\mathrm{pr}_*} \; \mathscr{S}_{C_{\geq -(n-1)}\pi_1(X/k)}$$

$$\longrightarrow \; H^2\left(k, \mathrm{gr}^C_{-n}\,\overline{\pi}\right).$$

Because the extension (14.4) is central, twisting has no effect on its coefficients and so all fibres of pr_* are homogeneous $H^1\left(k, \mathrm{gr}^C_{-n}\,\overline{\pi}\right)$-sets.

Proposition 197. *Let X/k be a smooth geometrically connected curve over a finitely generated field. Then we have an exact sequence*

$$1 \to H^1\left(k, \mathrm{gr}^C_{-n}\,\overline{\pi}\right) \to \mathscr{S}_{C_{\geq -n}\pi_1(X/k)} \xrightarrow{\mathrm{pr}_*} \mathscr{S}_{C_{\geq -(n-1)}\pi_1(X/k)} \xrightarrow{\delta_n} H^2\left(k, \mathrm{gr}^C_{-n}\,\overline{\pi}\right).$$

The obstruction to lifting a section s_{n-1} of $C_{\geq -(n-1)}\pi_1(X/k)$ to $C_{\geq -n}\pi_1(X/k)$ is given by the class $\delta_n(s_{n-1})$.

Proof. In the arithmetic case of a finitely generated field k the theory of weights applies saying that $\pi_1^{\mathrm{ab},\ell}(\overline{X}) = \mathrm{gr}^C_{-1}\,\overline{\pi}$ has weights in the interval $[-2, -1]$ because X is smooth. Consequently $\mathrm{gr}^C_{-n}\,\overline{\pi}$ has weights in the interval $[-2n, -n]$ for $n \geq 1$, and this implies that

$$H^0\left(k, \mathrm{gr}^C_{-n}\,\overline{\pi}\right) = 0,$$

because for a curve $\mathrm{gr}^C_{-n}\overline{\pi}$ is torsion free. By dévissage we also have $H^0\left(k, \overline{\pi}/C_{-(n+1)}\overline{\pi}\right) = 0$. $\square$

Remark 198. As usual, the obstruction to lift along central extensions can be related to Massey products which are higher order versions of the cup-product. The obstructions δ_n are related to Massey products in [Wg09] and partially computed for

$$X = \mathbb{P}^1 - \{0, 1, \infty\}$$

in [Wg12a], [Wg12b].

Dévissage for truncated pro-ℓ sections We project to the ℓ-Sylow part. Let X/k be a smooth projective curve with good reduction $\mathscr{X} \to B$ for some irreducible regular scheme of finite type B with function field k and such that ℓ is invertible on B. The diagram

$$\begin{array}{ccccccccc}
1 & \longrightarrow & \pi_1^{\mathrm{pro}\text{-}\ell} & \longrightarrow & \pi_1(X)/\ker\left(\pi_1(\overline{X}) \to \pi_1^{\mathrm{pro}\text{-}\ell}(\overline{X})\right) & \longrightarrow & \mathrm{Gal}_k & \longrightarrow & 1 \\
& & \| & & \downarrow & & \downarrow{\scriptstyle j_*} & & \\
1 & \longrightarrow & \pi_1^{\mathrm{pro}\text{-}\ell} & \longrightarrow & \pi_1(\mathscr{X})/\ker\left(\pi_1(\overline{X}) \to \pi_1^{\mathrm{pro}\text{-}\ell}(\overline{X})\right) & \longrightarrow & \pi_1(B) & \longrightarrow & 1
\end{array}$$

describes a pullback of extensions, see [Sx05] Proposition 2.6. By Proposition 91 any section s of $\pi_1^{\text{pro-}\ell}(X/k)$ is unramified at points $b \in B$ of codimension 1, and thus by Zariski–Nagata purity, see [SGA1] X Theorem 3.1, every pro-ℓ section is unramified on B and descends uniquely to a section

$$s_B : \pi_1(B) \to \pi_1(\mathcal{X})/\ker\left(\pi_1(\overline{X}) \to \pi_1^{\text{pro-}\ell}(\overline{X})\right).$$

The map $s \mapsto s_B$ yields an inverse to the pullback map

$$j^* : \mathscr{S}_{\pi_1^{\text{pro-}\ell}(\mathcal{X}/B)} \xrightarrow{\;\sim\;} \mathscr{S}_{\pi_1^{\text{pro-}\ell}(X/k)}.$$

The central extension

$$1 \to \mathrm{gr}^C_{-n}\,\overline{\pi}^\ell \to \overline{\pi}^\ell/C_{-(n+1)}\overline{\pi}^\ell \xrightarrow{\;\mathrm{pr}\;} \overline{\pi}^\ell/C_{-n}\overline{\pi}^\ell \to 1 \tag{14.5}$$

yields as an application of Sect. 1.3 the following pro-ℓ version.

Proposition 199. *Let X/k have good reduction over the base B with function field k. Then we have an exact sequence*

$$1 \to \mathrm{H}^1(B_{\text{ét}}, \mathrm{gr}^C_{-n}\,\overline{\pi}^\ell) \to \mathscr{S}_{C_{\geq -n}\pi_1^{\text{pro-}\ell}(X/k)} \xrightarrow{\;\mathrm{pr}_*\;} \mathscr{S}_{C_{\geq -(n-1)}\pi_1^{\text{pro-}\ell}(X/k)} \xrightarrow{\;\delta_n\;} \mathrm{H}^2(B_{\text{ét}}, \mathrm{gr}^C_{-n}\,\overline{\pi}^\ell).$$

In particular, the obstruction to lifting a section s_{n-1} of $C_{\geq -(n-1)}\pi_1^{\text{pro-}\ell}(X/k)$ to a section of $C_{\geq -n}\pi_1^{\text{pro-}\ell}(X/k)$ is given by the class $\delta_n(s_{n-1})$.

Proof. Étale cohomology of $B_{\text{ét}}$ and group cohomology of $\pi_1(B)$ compare as follows. We have

$$\mathrm{H}^1(\pi_1(B), \mathrm{gr}^C_{-n}\,\overline{\pi}^\ell) = \mathrm{H}^1(B_{\text{ét}}, \mathrm{gr}^C_{-n}\,\overline{\pi}^\ell)$$

and an inclusion

$$\mathrm{H}^2(\pi_1(B), \mathrm{gr}^C_{-n}\,\overline{\pi}^\ell) \subseteq \mathrm{H}^2(B_{\text{ét}}, \mathrm{gr}^C_{-n}\,\overline{\pi}^\ell).$$

Now the proof is essentially the same as for Proposition 197.   □

Remark 200. Hain considers in [Ha11a] a fully pro-algebraic analogue of non-abelian cohomology as discussed in Sect. 14.2. He replaces the base group Gal_k by an ℓ-adic representation

$$\rho : \mathrm{Gal}_k \to R(\mathbb{Q}_\ell)$$

in a reductive algebraic group R and considers the relative algebraic unipotent completion

$$\tilde{\rho} : \pi_1(X) \to G(\mathbb{Q}_\ell)$$

of $\pi_1(X)$ relative ρ that takes values in the universal (in some sense) unipotent extension G of R. This makes non-abelian cohomology somewhat computable since we can now work with Lie algebras, more precisely Lie algebras in the Tannaka category of R-modules. Hain's result on the section conjecture for the generic curve in [Ha11b] builds on this pro-algebraic version of non-abelian cohomology.

14.3 The Lie Algebra

Following Magnus and Lazard, see [La94], we associate to the descending central series $C_\bullet \Gamma$ of a pro-ℓ group Γ the graded $\mathbb{Z}_\ell$-Lie-algebra

$$\text{Lie}(\Gamma) = \bigoplus_{n \geq 1} \text{Lie}_n(\Gamma) = \bigoplus_{n \geq 1} \text{gr}_{-n}^C \left(C_{-n}\Gamma / C_{-(n+1)}\Gamma \right) \qquad (14.6)$$

with the Lie bracket being induced by the commutator in the group Γ.

Form now on let X/k be a smooth, projective curve of genus at least 2. For a prime number ℓ different form the characteristic of k, the group $\overline{\pi}^\ell = \pi_1(\overline{X})^{\text{pro-}\ell}$ is a Poincaré duality group and coincides with the pro-ℓ completion of the surface group

$$\Pi = \Pi_g = \langle x_1, \ldots, x_{2g} | [x_1, x_2] \ldots [x_{2g-1}, x_{2g}] \rangle.$$

In this case we set

$$\mathfrak{p} = \text{Lie}(\overline{\pi}^\ell). \qquad (14.7)$$

The graded piece $\mathfrak{p}_n = \text{Lie}_n(\overline{\pi}^\ell)$ in degree $-n$ is a free $\mathbb{Z}_\ell$-module of finite rank, see [La66] Proposition 1. We set

$$\mathfrak{p}_K = \mathfrak{p} \otimes_{\mathbb{Z}_\ell} K \qquad (14.8)$$

for the change of coefficients to a field extension $K/\mathbb{Q}_\ell$. The conjugation action by $\pi_1(X)$ descends to a Gal_k action on the graded Lie algebra $\mathfrak{p}_K$ over K.

The Poincaré series. We are interested to compute the Poincaré series of $\mathfrak{p}$ as a power series

$$[\mathfrak{p}] = \sum_{n \geq 1} [\mathfrak{p}_n] T^n \qquad (14.9)$$

with coefficients in the Grothendieck ring of $\mathbb{Z}_\ell[\text{Gal}_k]$-modules which are free of finite rank as $\mathbb{Z}_\ell$-modules. For the free Lie algebra this was achieved in characteristic 0 by Brandt [Br44] Theorem III, and rediscovered more conceptually by many, e.g. [By03] Theorem 5.4. We follow the same path, especially we argue in strong analogy with the computation of the generating function

$$\sum_{n \geq 1} \dim_K \mathfrak{p}_{K,n} T^n$$

by Labute [La67] Theorem 2.

Let L_H be the free $\mathbb{Z}_\ell$-Lie algebra generated by the $\mathbb{Z}_\ell[\mathrm{Gal}_k]$-module

$$H = \overline{\pi}^{\mathrm{ab},\ell} = \mathrm{gr}^C_{-1} \overline{\pi}^\ell = \bigoplus_{i=1}^{2g} \mathbb{Z}_\ell x_i.$$

There is a natural short exact sequence of graded $\mathbb{Z}_\ell$-Lie algebras

$$0 \to \mathfrak{r} \to L_H \to \mathfrak{p} \to 0.$$

Let $\rho \in \mathfrak{r}_2$ be the image of the relation $[x_1, x_2] \ldots [x_{2g-1}, x_{2g}]$. We have the following special case of Labute's results on one-relator groups.

Theorem 201 (Labute [La67]).

(1)   As a module under the universal enveloping algebra $U(\mathfrak{p})$ the module $\mathfrak{r}/[\mathfrak{r},\mathfrak{r}]$ is free of rank 1 and generated by ρ in degree 2.
(2)   We have a short exact sequence

$$0 \to \mathfrak{r}/[\mathfrak{r},\mathfrak{r}] \to U(\mathfrak{p})^{2g} \to U(\mathfrak{p}) \to \mathbb{Z}_\ell \to 0$$

which is a free resolution of $\mathbb{Z}_\ell$ with trivial action by free $U(\mathfrak{p})$-modules of finite rank.   □

By Theorem 201 and Poincaré–Birkhoff–Witt we have the following identifications of graded Gal_k-modules.

$$\mathfrak{r}/[\mathfrak{r},\mathfrak{r}] \cong U(\mathfrak{p}) \otimes \mathbb{Z}_\ell(-1) \tag{14.10}$$

$$U(\mathfrak{r}) \cong \mathrm{Ass}_{\mathfrak{r}/[\mathfrak{r},\mathfrak{r}]} \tag{14.11}$$

$$U(\mathfrak{r}) \otimes U(\mathfrak{p}) \cong U(L_H) \cong \mathrm{Ass}_H \tag{14.12}$$

Here Ass_V is the free associative algebra on the $\mathbb{Z}_\ell$-module V, and we have made use of the fact, that $\mathfrak{r}$ as a subalgebra of a free Lie algebra is again free on $\mathfrak{r}/[\mathfrak{r},\mathfrak{r}]$ after [La67] Proposition 2.

The Poincaré series of $\mathfrak{p}$ will be computed through the Poincaré series of $U(\mathfrak{p})$ which is as follows.

$$\frac{1}{1 - [H] \cdot T} = [\mathrm{Ass}_H] = [U(L_H)] = [U(\mathfrak{r})] \cdot [U(\mathfrak{p})] = [\mathrm{Ass}_{\mathfrak{r}/[\mathfrak{r},\mathfrak{r}]}] \cdot [U(\mathfrak{p})]$$

$$= \frac{1}{1 - [\mathfrak{r}/[\mathfrak{r},\mathfrak{r}]]} \cdot [U(\mathfrak{p})] = \frac{[U(\mathfrak{p})]}{1 - [U(\mathfrak{p})] \cdot [\mathbb{Z}_\ell(-1)] \cdot T^2}$$

This solves for $[U(\mathfrak{p})]$ as

$$[U(\mathfrak{p})] = \frac{1}{1 - [H] \cdot T + [\mathbb{Z}_\ell(-1)] \cdot T^2} \tag{14.13}$$

Adams operations. The formula for $[\mathfrak{p}_n]$ which can be extracted from (14.13) requires Adams operations Ψ^d on the corresponding Grothendieck ring of $\mathbb{Z}_\ell[\mathrm{Gal}_k]$-modules which are free of finite rank as $\mathbb{Z}_\ell$-modules, see [Be84, Se77]. In particular, for any V in the Grothendieck ring which we put in degree d we have

$$[\mathrm{Sym}^\bullet(V)] = \exp\left(\sum_{m\geq 1} \Psi^m([V])\frac{T^{dm}}{m}\right)$$

and

$$T\frac{d\log}{dT}[\mathrm{Sym}^\bullet(V)] = \sum_{m\geq 1} \Psi^m(d[V])\cdot T^{dm}$$

The Poincaré–Birkhoff–Witt Theorem shows

$$[U(\mathfrak{p})] = \prod_d [\mathrm{Sym}^\bullet \mathfrak{p}_d]$$

and thus for the logarithmic derivative

$$\sum_{n\geq 1}\sum_{d\mid n} \Psi^{n/d}(d[\mathfrak{p}_d])\cdot T^n = \frac{[H]\cdot T - 2[\mathbb{Z}_\ell(-1)]\cdot T^2}{1-[H]\cdot T + [\mathbb{Z}_\ell(-1)]\cdot T^2} =: \sum_{n\geq 1} S_n T^n.$$

Working with formal roots of the denominator

$$1 - [H]\cdot T + [\mathbb{Z}_\ell(-1)]\cdot T^2 = (1-\alpha T)(1-\beta T)$$

we easily deduce a linear recursion formula for the $S_n = \alpha^n + \beta^n$ as follows:

$$S_{n+2} = [H]\cdot S_{n+1} - [\mathbb{Z}_\ell]\cdot S_n$$

and $S_1 = [H]$ while $S_0 = 2[\mathbf{1}]$ is twice the trivial 1-dimensional representation.
The Möbius inversion formula in this case reads

$$\sum_{d\mid n}\mu(d)\Psi^d(S_{n/d}) = \sum_{d\mid n}\mu(d)\Psi^d\left(\sum_{e\mid \frac{n}{d}} \Psi^{n/ed}(e\cdot[\mathfrak{p}_e])\right) = \sum_{ed\mid n}\mu(d)\Psi^{n/e}(e\cdot[\mathfrak{p}_e])$$

$$(14.14)$$

$$= \sum_{e\mid n}\Psi^{n/e}(e\cdot[\mathfrak{p}_e])\sum_{d\mid \frac{n}{e}}\mu(d) = n[\mathfrak{p}_n]. \qquad (14.15)$$

The analogue of [La70] Theorem (1), requires an explicit formula for the S_n which can be proven by induction via an informed Ansatz mimicking the formula [La70] Theorem (1). We obtain

$$S_m = \sum_{i=0}^{\lfloor m/2\rfloor} (-1)^i \frac{m}{m-i}\binom{m-i}{i}[H]^{2m-i}\cdot[\mathbb{Z}_\ell(-i)]$$

and thus

$$[\mathfrak{p}_n] = \frac{1}{n} \sum_{d|n} \mu(n/d) \Psi^{n/d} \Big(\sum_{i=0}^{\lfloor d/2 \rfloor} (-1)^i \frac{d}{d-i} \binom{d-i}{i} [H]^{2d-i} \cdot [\mathbb{Z}_\ell(-i)] \Big).$$

(14.16)

Remark 202. (1) It is unclear to me whether this description of $[\mathfrak{p}_n]$ is of any use. For example, it seems impossible to decide by means of (14.16) whether $\mathrm{H}^0(G, \mathfrak{p}) \neq 0$ if G is a finite cyclic group acting on $\mathfrak{p}$ via graded Lie algebra automorphisms.

(2) A slight generalization of (14.16) and the use of Adams operations occurs in §2.3 and in particular formula (2.3.3) of [AN95].

14.4   Finite Dimensional Subalgebras and Invariants

The key property of $\mathfrak{p}_K$ that forces $\dim_K \mathrm{H}^0(G, \mathfrak{p}_K)$ to be infinite for a finite group G acting on $\mathfrak{p}_K$ is the following bound on cohomological dimension.

Lemma 203. *Let $K/\mathbb{Q}_\ell$ be a field. The cohomological dimension of a sub-Lie algebra $\mathfrak{g} \subset \mathfrak{p}_K$ is at most 2.*

Proof. By the Poincaré–Birkhoff–Witt Theorem $U(\mathfrak{p}_K)$ is a free $U(\mathfrak{g})$-module. Thus the resolution

$$0 \to \mathfrak{r}/[\mathfrak{r}, \mathfrak{r}] \otimes K \to U(\mathfrak{p}_K)^{2g} \to U(\mathfrak{p}_K) \to K \to 0$$

derived from Theorem 201 (2) shows that K with trivial $\mathfrak{g}$-action has projective dimension at most 2. Hence $\mathrm{H}^q(\mathfrak{g}, M) = \mathrm{Ext}^q_{U(\mathfrak{g})}(K, M)$ vanishes for $q \geq 3$.   □

Lemma 204. *Let $\mathfrak{g} = \bigoplus_{n \geq 1} \mathfrak{g}_n$ be a graded Lie algebra over the field K of dimension $4 \leq \dim_K \mathfrak{g} < \infty$. Then there is a graded abelian Lie algebra $\mathfrak{a} \subset \mathfrak{g}$ with $\dim_K \mathfrak{a} \geq 3$, i.e., we have $[\mathfrak{a}, \mathfrak{a}] = 0$.*

Proof. Let N be maximal with $\mathfrak{g}_N \neq 0$. It follows that the piece $\mathfrak{g}_N$ is central in $\mathfrak{g}$ and we are done if $\dim_K \mathfrak{g}_N \geq 2$. So we assume now $\dim_K \mathfrak{g}_N = 1$.

Let m be maximal with $m < N$ and $\mathfrak{g}_m \neq 0$. We set $\mathfrak{g}_{<m} = \bigoplus_{i < m} \mathfrak{g}_i$ and note that

$$[,] : \mathfrak{g}_{<m} \otimes \mathfrak{g}_m \to \mathfrak{g}$$

describes a bilinear pairing with values in $\mathfrak{g}_N$. If $\mathfrak{g}_{<m} \neq 0$, then because by assumption either $\dim_K \mathfrak{g}_{<m} \geq 2$ or $\dim_K \mathfrak{g}_m \geq 2$ we find homogeneous lines $\mathfrak{a}_{<m} \subseteq \mathfrak{g}_{<m}$, and $\mathfrak{a}_m \subseteq \mathfrak{g}_m$ with $[\mathfrak{a}_{<m}, \mathfrak{a}_m] = 0$. Hence

$$\mathfrak{a} = \mathfrak{a}_{<m} \oplus \mathfrak{a}_m \oplus \mathfrak{g}_N$$

is abelian of dimension 3.

It remains to discuss the case $\mathfrak{g}_{<m} = 0$. We then argue with the alternating pairing

$$[,] : \mathfrak{g}_m \otimes \mathfrak{g}_m \to \mathfrak{g}_N.$$

As by assumption $\dim_K \mathfrak{g}_m \geq 3$ we can find an isotropic subspace $\mathfrak{a}_m \subset \mathfrak{g}_m$ of dimension 2. Then $\mathfrak{a} = \mathfrak{a}_m \oplus \mathfrak{g}_N$ is abelian of dimension 3. This settles the claim in all cases.   □

Proposition 205. *Let* $\mathfrak{g} \subset \mathfrak{p}_K$ *be a graded sub-Lie algebra of finite K-dimension. Then* $\dim_K \mathfrak{g} \leq 3$ *and* $\dim_K \mathfrak{g}_n \leq 2$ *for all* $n \geq 1$.

Proof. We argue by contradiction. If $\dim_K \mathfrak{g} \geq 4$, then by Lemma 204 we find an abelian Lie algebra $\mathfrak{a} \subseteq \mathfrak{g} \subset \mathfrak{p}$ with $\dim_K \mathfrak{a} = 3$. But as $\mathrm{H}^3(\mathfrak{a}, K) = K$ does not vanish, this contradicts Lemma 203.

The assertion $\dim_K \mathfrak{g}_n \leq 2$ follows because the only potential exception would be a graded sub-Lie algebra $\mathfrak{g} \subseteq \mathfrak{p}_K$ with $\mathfrak{g} = \mathfrak{g}_n$ of dimension 3 for some n. But such a $\mathfrak{g}$ were abelian and thus leads to the same contradiction.   □

Corollary 206. *The Lie algebra $\mathfrak{g}$ generated by a subspace $\mathfrak{g}_n \subseteq \mathfrak{p}_{K,n}$ with dimension* $\dim_K \mathfrak{g}_n \geq 3$ *is infinite dimensional and* $\mathfrak{g}_{dn} \neq 0$ *for all* $d \geq 1$.   □

The Lie algebra of invariants. Instead of Gal_k we now discuss the case of a finite abelian group G acting on $\mathfrak{p}$ by graded Lie algebra automorphisms.

Proposition 207. *Let G be a finite abelian group of order invertible in K which acts on $\mathfrak{p}_K$ by graded Lie algebra automorphisms. Then the G-invariants $\mathrm{H}^0(G, \mathfrak{p}_K)$ form a graded sub-Lie algebra of infinite K-dimension.*

Proof. It is clear that $\mathrm{H}^0(G, \mathfrak{p}_K)$ forms a graded sub-Lie algebra. Let N be the exponent of G. In order to determine the dimension of $\mathrm{H}^0(G, \mathfrak{p}_K)$ we can assume that K contains all Nth roots of unity.

Let V_χ be the χ-isotypical component of a G-representation V with respect to the character χ. If for some n and some character χ we have $\dim_K \mathfrak{p}_{n,\chi} \geq 3$, then by Corollary 206 the Lie algebra $\langle \mathfrak{p}_{n,\chi} \rangle \subseteq \mathfrak{p}_K$ generated by $\mathfrak{p}_{n,\chi}$ is infinite dimensional and nontrivial in every degree which is a multiple of n. But G acts on $\langle \mathfrak{p}_{n,\chi} \rangle_{dn}$ by χ^d so that for every $r \in \mathbb{N}$ we have

$$0 \neq \langle \mathfrak{p}_{n,\chi} \rangle_{rNn} \subset \mathrm{H}^0(G, \mathfrak{p}_K).$$

That there is a character χ and $n \in \mathbb{N}$ with $\dim_K \mathfrak{p}_{n,\chi} \geq 3$ follows from the pigeon hole principle and the estimate for $\dim_K \mathfrak{p}_n$ in the following lemma.   □

Lemma 208. *We have the following estimate*

$$\left| \mathrm{rk}_{\mathbb{Z}_\ell} \mathfrak{p}_n - \frac{\alpha^n}{n} \right| \leq \frac{\alpha}{n(\alpha - 1)} \alpha^{n/p} + 1$$

where p is the smallest prime factor of n and $\alpha = g + \sqrt{g^2 - 1} \approx 2g$. In particular, we have $\mathrm{rk}_{\mathbb{Z}_\ell}\, \mathfrak{p}_n \to \infty$ *for* $n \to \infty$.

Proof. We remind that we are working under the hypothesis that $g \geq 2$. We define $\beta = g - \sqrt{g^2 - 1}$ so that $(1 - \alpha T)(1 - \beta T) = 1 - 2gT + T^2$ and $\mathrm{rk}_{\mathbb{Z}_\ell}\, S_n = \alpha^n + \beta^n$. It follows from (14.15) that

$$\mathrm{rk}_{\mathbb{Z}_\ell}\, \mathfrak{p}_n = \frac{1}{n} \cdot \sum_{d \mid n} \mu(d) \left(\alpha^{n/d} + \beta^{n/d} \right).$$

By the triangle inequality and because $|\beta| < 1$ we have

$$\left| \mathrm{rk}_{\mathbb{Z}_\ell}\, \mathfrak{p}_n - \frac{\alpha^n}{n} \right| \leq \frac{1}{n} \cdot \left(\sum_{d \mid n, d < n} \alpha^d + \sum_{d \mid n} \beta^d \right)$$

$$< \frac{1}{n} \left(\sum_{i=1}^{n/p} \alpha^i + \sum_{i=1}^{n} 1 \right) < \frac{\alpha}{n(\alpha - 1)} \alpha^{n/p} + 1. \qquad \square$$

For an application of Proposition 207 see Theorem 226 in Sect. 15.3. This application was the main stimulus behind our discussion of the Lie algebra $\mathfrak{p}$ in Sect. 14.3.

14.5   Nilpotent Sections in the Arithmetic Case

In the remaining sections of this chapter we examine the space of nilpotent sections for a smooth projective curve X/k over an algebraic number field k, or over a finite extension $k/\mathbb{Q}_p$. The behaviour is fundamentally different for the maximal geometrically pro-ℓ quotient for $\ell \neq p$ and for the maximal geometrically pro-p quotient.

The maximal pro-ℓ quotient with good reduction. Let $k/\mathbb{Q}_p$ be a finite extension with ring of integers $\mathfrak{o}_k$ and residue field $\mathbb{F}$. Let X/k be a smooth, projective curve with good reduction $\mathscr{X}/\mathrm{Spec}(\mathfrak{o}_k)$ and special fibre $Y = \mathscr{X}_{\mathbb{F}}$ over $\mathbb{F}$.

For $\ell \neq p$, every section of $\pi_1^{\mathrm{pro}\text{-}\ell}(X/k)$ is unramified over $\mathfrak{o}_k$ by a pro-ℓ version of Proposition 91. For sections associated to rational points this was noted in [KiTa08] Theorem 0.1, see Sect. 8.5. Moreover, the specialisation map

$$\mathscr{S}_{\pi_1^{\mathrm{pro}\text{-}\ell}(X/k)} \to \mathscr{S}_{\pi_1^{\mathrm{pro}\text{-}\ell}(Y/\mathbb{F})}$$

is bijective. The pro-ℓ Kummer map sits in a diagram

$$
\begin{array}{ccc}
X(k) & \xrightarrow{\;\kappa_\ell\;} & \mathscr{S}_{\pi_1^{\mathrm{pro}\text{-}\ell}}(X/k) \\[2mm]
\downarrow & & \| \\[2mm]
Y(\mathbb{F}) & \xrightarrow{\;\kappa_\ell\;} & \mathscr{S}_{\pi_1^{\mathrm{pro}\text{-}\ell}}(Y/\mathbb{F})
\end{array}
$$

and thus factors over the finite set. On the other hand, by Theorem 226 below, we know that $\mathscr{S}_{\pi_1^{\mathrm{pro}\text{-}\ell}}(Y/\mathbb{F})$ is uncountable. We conclude that the pro-ℓ section conjecture fails badly for proper smooth p-adic curves of good reduction with $\ell \neq p$.

Remark 209. It has been observed[1] by Tamagawa, see [Ho09] Remark 10 (i), that the pro-ℓ Kummer map

$$
\kappa_\ell : Y(\mathbb{F}_q) \to \mathscr{S}_{\pi_1^{\mathrm{pro}\text{-}\ell}}(Y/\mathbb{F}_q)
$$

may fail to be injective for hyperbolic curves Y over a finite field $\mathbb{F}_q$.

14.6   Pro-p Counter-Examples After Hoshi

We are going to explain the counter-examples to a pro-p version of the section conjecture over algebraic number fields found by Hoshi, see [Ho10].

Theorem 210 (Hoshi, [Ho10] Theorem A). *Let $p \geq 3$ be a regular prime and let $k/\mathbb{Q}(\zeta_p)$ be a Galois extension unramified outside p with Galois group a finite p-group.*
    Let $\beta : X \to \mathbb{P}_k^1$ be a finite map of connected proper smooth curves such that

  (i) the genus of X is ≥ 2,
  (ii) $X(k)$ is nonempty,
  (iii) $\overline{\beta} : \overline{X} \to \mathbb{P}_k^1$ is Galois of p-power degree and unramified outside 0, 1, and ∞,
  (iv) and the hyperbolic curve $X \setminus \beta^{-1}(\{0, 1, \infty\})$ has good reduction outside p.

Then there exists a finite extension k'/k unramified outside p with pro-p Galois hull, such that the pro-p Kummer map

$$
\kappa_p : X(k') \to \mathscr{S}_{\pi_1^{\mathrm{pro}\text{-}p}}(X/k)(k')
$$

is not surjective, i.e., there are non-Diophantine pro-p sections after a finite pro-p extension unramified outside p. Moreover, if $v|p$ is a place of k' with completion k'_v, then also the local pro-p Kummer map

---

[1] I thank Yuichiro Hoshi for bringing Tamagawa's observation to my attention.

$$\kappa_p \; : \; X(k'_v) \to \mathscr{S}_{\pi_1^{\text{pro-}p}(X/k)}(k'_v)$$

is not surjective.

Remark 211. Hoshi also constructs an explicit series of examples. Let $p \geq 11$ be a regular prime and let $k/Q(\zeta_p)$ be as in Theorem 210. Let $X_{\text{Fermat},p}/k$ be the Fermat curve

$$\{A^p + B^p = C^p\} \subset \mathbb{P}^2_k.$$

Then as a consequence of [Ho10] Theorem B, we obtain even that $\mathscr{S}_{\pi_1^{\text{pro-}p}}(X_{\text{Fermat},p}/k)$ is at least countable infinite.

We develop, complement and generalize the ideas of [Ho10] in the sequel.

Lemma 212. *Let k be an algebraic number field, and let $S \subseteq \text{Spec}(o_k[\frac{1}{p}])$ be a dense open arithmetic curve with a geometric point $\bar{s} \in S$.*

Let X/k be a smooth, projective geometrically connected curve of genus ≥ 2 with

(i) good reduction over S,
(ii) and Gal_k acts on $\pi_1^{\text{ab}}(\overline{X}, \bar{x}) \otimes \mathbb{F}_p = H_1(\overline{X}, \mathbb{F}_p)$ through a p-group.

Then the pro-p outer Galois action

$$\rho_{X/k} : \text{Gal}_k \to \text{Out}\left(\pi_1^{\text{pro-}p}(\overline{X})\right)$$

factors over $\pi_1^{\text{pro-}p}(S, \bar{s})$.

Proof. That $\rho_{X/k}$ is unramified above S, i.e., factors over $\pi_1(S, \bar{s})$ follows from [Sx05] Propositions 2.6 and 2.7. By a profinite version of a theorem of Hall, see [Ha59] Theorem 12.2.2., the kernel of

$$\text{Out}\left(\pi_1^{\text{pro-}p}(\overline{X})\right) \to \text{Aut}\left(H_1(\overline{X}, \mathbb{F}_p)\right)$$

is a pro-p group. Hence (ii) implies that the image of $\rho_{X/k}$ is a pro-p group. □

In the situation of Lemma 212 the extension $\pi_1^{\text{pro-}p}(X/k)$ is the pullback of the extension

$$1 \to \pi_1^{\text{pro-}p}(\overline{X}) \to \text{Aut}\left(\pi_1^{\text{pro-}p}(\overline{X})\right) \times_{\text{Out}\left(\pi_1^{\text{pro-}p}(\overline{X})\right)} \pi_1^{\text{pro-}p}(S, \bar{s}) \to \pi_1^{\text{pro-}p}(S, \bar{s}) \to 1.$$
(14.17)

Let $\mathscr{S}$ denote the $\pi_1^{\text{pro-}p}(\overline{X})$-conjugacy classes of sections of (14.17). As in the context of base change, we obtain a natural map

$$\mathscr{S} \to \mathscr{S}_{\pi_1^{\text{pro-}p}(X/k)}.$$

Lemma 213. *With X/k as in Lemma 212, the map $\mathscr{S} \to \mathscr{S}_{\pi_1^{\text{pro-}p}(X/k)}$ is bijective.*

Proof. Since p is invertible in S, it follows that all ramification at places $v \in S$ is at most tame. We conclude by Proposition 91 that every pro-p section of $\pi_1^{\text{pro-}p}(X/k)$ is unramified above S. As (14.17) is a sequence of pro-p groups, every section further descends to $\mathscr{S}$. $\square$

Lemma 214. *Let p be an odd prime number, and let $S \subseteq \text{Spec}(\mathfrak{o}_k[\frac{1}{p}])$ be a dense open arithmetic curve with function field $k/\mathbb{Q}(\zeta_p)$ and geometric point $\bar{s} \in S$. Then $\pi_1^{\text{pro-}p}(S, \bar{s})$ is a free pro-p group if and only if the following conditions all hold:*

(i) $S = \text{Spec}(\mathfrak{o}_k[\frac{1}{p}])$,
(ii) p is inert in $k/\mathbb{Q}$,
(iii) $\text{Pic}(\mathfrak{o}_k) \otimes \mathbb{Z}_p$ is generated by the only prime $\mathfrak{p}|p$.

Proof. This is well known. We give a proof for the convenience of the reader. The group $\pi_1^{\text{pro-}p}(S, \bar{s})$ is free pro-p if and only if

$$\text{H}^2(\pi_1^{\text{pro-}p}(S, \bar{s}), \mathbb{Z}/p\mathbb{Z}) = \text{H}^2(S, \mathbb{Z}/p\mathbb{Z}) \cong \text{H}^2(S, \mu_p)$$

vanishes. From the Kummer sequence we obtain an exact sequence

$$0 \to \text{Pic}(S) \otimes \mathbb{F}_p \to \text{H}^2(S, \mu_p) \to \text{Br}(S)[p] \to 0.$$

As $\text{Br}(S)[p]$ is isomorphic to the elements in $\bigoplus_{v \notin S} \mathbb{Q}/\mathbb{Z}$ of sum 0 where v ranges over all finite places of k outside S, the Brauer term vanishes if and only if (i) and (ii) hold. Clearly then (iii) is equivalent to the vanishing of the Picard term. $\square$

Theorem 215. *Let p be an odd prime number, and let $k/\mathbb{Q}(\zeta_p)$ be an algebraic number field, such that*

(i) p is inert in $k/\mathbb{Q}$,
(ii) $\text{Pic}(\mathfrak{o}_k) \otimes \mathbb{Z}_p$ is generated by the only prime $\mathfrak{p}|p$.

Let X/k be a smooth, projective geometrically connected curve of genus ≥ 2 with

(iii) good reduction over $S = \text{Spec}(\mathfrak{o}_k[\frac{1}{p}])$,
(iv) Gal_k acts on $\pi_1^{\text{ab}}(\overline{X}, \bar{x}) \otimes \mathbb{F}_p = \text{H}_1(\overline{X}, \mathbb{F}_p)$ through a p-group,
(v) and positive Mordell-Weil rank, i.e., $\text{Pic}_X^0(k)$ is infinite.

Then the set $\mathscr{S}_{\pi_1^{\text{pro-}p}(X/k)}$ is uncountably infinite and the pro-p Kummer map

$$\kappa_p : X(k) \to \mathscr{S}_{\pi_1^{\text{pro-}p}(X/k)}$$

is not surjective, i.e., there are non-Diophantine pro-p sections.

Proof. The extension (14.17) can be pushed to the geometrically maximal abelian quotient and gives an extension

$$1 \to \pi_1^{\text{ab,pro-}p}(\overline{X}) \to \frac{\text{Aut}\left(\pi_1^{\text{pro-}p}(\overline{X})\right) \times_{\text{Out}\left(\pi_1^{\text{pro-}p}(\overline{X})\right)} \pi_1^{\text{pro-}p}(S, \bar{s})}{\ker\left(\pi_1^{\text{pro-}p}(\overline{X}) \twoheadrightarrow \pi_1^{\text{ab,pro-}p}(\overline{X})\right)} \to \pi_1^{\text{pro-}p}(S, \bar{s}) \to 1$$

$$(14.18)$$

the $\pi_1^{\mathrm{ab},\mathrm{pro}\text{-}p}(\overline{X})$-conjugacy classes of which we denote by $\mathscr{S}^{\mathrm{ab}}$. The analogue of Lemma 213 holds and yields a natural bijective map

$$\mathscr{S}^{\mathrm{ab}} \to \mathscr{S}_{\pi_1^{\mathrm{ab},\mathrm{pro}\text{-}p}}(X/k) = \mathscr{S}_{\pi_1^{\mathrm{pro}\text{-}p}(\mathrm{Pic}_X^1/k)}.$$

The Albanese torsor map $\alpha : X \hookrightarrow \mathrm{Pic}_X^1$ provides a commutative diagram

$$
\begin{array}{ccccc}
X(k) & \xrightarrow{\ \kappa_{X,p}\ } & \mathscr{S}_{\pi_1^{\mathrm{pro}\text{-}p}}(X/k) & \xleftarrow{\ \sim\ } & \mathscr{S} \\
\downarrow{\scriptstyle \alpha} & & \downarrow{\scriptstyle \alpha_*} & & \downarrow{\scriptstyle \alpha_*} \\
\mathrm{Pic}_X^1(k) & \longrightarrow & \mathscr{S}_{\pi_1^{\mathrm{pro}\text{-}p}(\mathrm{Pic}_X^1/k)} & \xleftarrow{\ \sim\ } & \mathscr{S}^{\mathrm{ab}}
\end{array}
\qquad (14.19)
$$

With (i) and (ii) we deduce form Lemma 214 that $\pi_1^{\mathrm{pro}\text{-}p}(S,\bar{s})$ is a free pro-p group, hence the map

$$\alpha_* : \mathscr{S} \twoheadrightarrow \mathscr{S}^{\mathrm{ab}}$$

is surjective and a forteriori, by Lemma 213 and its abelianized analogue, the map

$$\alpha_* : \mathscr{S}_{\pi_1^{\mathrm{pro}\text{-}p}}(X/k) \twoheadrightarrow \mathscr{S}_{\pi_1^{\mathrm{pro}\text{-}p}(\mathrm{Pic}_X^1/k)}$$

is surjective. Moreover, the spaces of sections in (14.19) are in fact non-empty. The cohomological description in the abelian case now shows a bijection

$$\mathscr{S}_{\pi_1^{\mathrm{pro}\text{-}p}(\mathrm{Pic}_X^1/k)} \cong \mathrm{H}^1\left(k, \pi_1^{\mathrm{ab},\mathrm{pro}\text{-}p}(\overline{X})\right) = \mathrm{H}^1\left(k, \mathrm{T}_p(\mathrm{Pic}_X^0)\right)$$

which via the Kummer sequence contains $\mathrm{Pic}_X^0(k) \otimes \mathbb{Z}_p$ and thus by assumption (v) is uncountably infinite. Consequently, also the space of pro-p sections $\mathscr{S}_{\pi_1^{\mathrm{pro}\text{-}p}}(X/k)$ is uncountably infinite, which shows in particular the presence of non-Diophantine pro-p sections. $\qquad\square$

In [Ho10] Hoshi finds an ingenious anabelian way to ensure property (iv) of Theorem 215 that we are going to explain now.

Proposition 216 (Hoshi, [Ho10] Lemma 2.1). *Let $\beta : X \to Y$ be a finite map of geometrically connected proper smooth curves over an algebraic number field k such that*

(i) there is a hyperbolic dense open $V \subset Y$ such that $\beta|_U : U = \beta^{-1}(V) \to V$ is finite étale,

(ii) Gal_k acts on $\mathrm{H}_1(\overline{V}, \mathbb{F}_p)$ through a p-group,

(iii) and, geometrically, the map $\overline{\beta} : \overline{U} \to \overline{V}$ is Galois of p-power degree.

Then Gal_k acts on $\mathrm{H}_1(\overline{X}, \mathbb{F}_p)$ via a finite p-group.

Proof. Since $H_1(\overline{X}, \mathbb{F}_p)$ is a quotient module of $H_1(\overline{U}, \mathbb{F}_p)$ it suffices to show that the action of Gal_k on the latter is via a p-group. As above, this is equivalent to the outer pro-p Galois action

$$\rho_{U/k} \;:\; \mathrm{Gal}_k \to \mathrm{Out}\left(\pi_1^{\mathrm{pro}\text{-}p}(\overline{U})\right)$$

factoring over a pro-p group.

By (i) and (iii) there is a finite p-group G and the following diagram with exact rows and column.

$$
\begin{array}{ccccccccc}
 & & & & 1 & & & & \\
 & & & & \uparrow & & & & \\
 & & & & G & & & & \\
 & & & & \uparrow & & & & \\
1 & \longrightarrow & \pi_1^{\mathrm{pro}\text{-}p}(\overline{V}) & \longrightarrow & \pi_1^{(\mathrm{pro}\text{-}p)}(V) & \longrightarrow & \mathrm{Gal}_k & \longrightarrow & 1 \\
 & & \uparrow & & \uparrow \beta_* & & \| & & \\
1 & \longrightarrow & \pi_1^{\mathrm{pro}\text{-}p}(\overline{U}) & \longrightarrow & \pi_1^{(\mathrm{pro}\text{-}p)}(U) & \longrightarrow & \mathrm{Gal}_k & \longrightarrow & 1 \\
 & & \uparrow & & & & & & \\
 & & 1 & & & & & &
\end{array}
$$

$$(14.20)$$

Let Z_U (resp. Z_V) be the centraliser of $\pi_1^{\mathrm{pro}\text{-}p}(\overline{U})$ in $\pi_1^{(\mathrm{pro}\text{-}p)}(U)$ (resp. of $\pi_1^{\mathrm{pro}\text{-}p}(\overline{V})$ in $\pi_1^{(\mathrm{pro}\text{-}p)}(V)$). Since U and V are hyperbolic, $\pi_1^{\mathrm{pro}\text{-}p}(\overline{U})$ and $\pi_1^{\mathrm{pro}\text{-}p}(\overline{V})$ have trivial center, Z_U and Z_V inject as normal closed subgroups in Gal_k, and we have

$$\mathrm{im}(\rho_{U/k}) = \mathrm{Gal}_k /Z_U \quad \text{and} \quad \mathrm{im}(\rho_{V/k}) = \mathrm{Gal}_k /Z_V.$$

To the exact column of (14.20) belongs an outer action

$$\rho : G \to \mathrm{Out}\left(\pi_1^{\mathrm{pro}\text{-}p}(\overline{U})\right)$$

and a natural isomorphism

$$\pi_1^{\mathrm{pro}\text{-}p}(\overline{V}) = \mathrm{Aut}\left(\pi_1^{\mathrm{pro}\text{-}p}(\overline{U})\right) \times_{\mathrm{Out}\left(\pi_1^{\mathrm{pro}\text{-}p}(\overline{U})\right),\rho} G.$$

It follows that an automorphism of $\pi_1^{\mathrm{pro}\text{-}p}(\overline{U})$ extends in at most one way to an automorphism of $\pi_1^{\mathrm{pro}\text{-}p}(\overline{V})$, see [Sx02] Lemma 4.2.9 for another case of this argument. Consequently, we have an inclusion $Z_U \subseteq Z_V$ and an exact sequence

$$1 \to Z_V/Z_U \to \mathrm{im}(\rho_{U/k}) \to \mathrm{im}(\rho_{V/k}) \to 1.$$

By assumption (ii) and the profinite version of [Ha59] Theorem 12.2.2, the image $\mathrm{im}(\rho_{V/k})$ is pro-p, so that it suffices to analyse Z_V/Z_U. We define a map

$$\psi : Z_V \to G$$

which sends $a \in Z_V$ of the form $a = \gamma u$ with $\gamma \in \pi_1^{\mathrm{pro}\text{-}p}(\overline{V})$ and $u \in \pi_1^{(\mathrm{pro}\text{-}p)}(U)$ to

$$\psi(a) = \gamma \cdot \pi_1^{\mathrm{pro}\text{-}p}(\overline{U}).$$

The map ψ is well defined, as with another decomposition $a = \gamma' u'$ we have

$$\gamma^{-1}\gamma' = u(u')^{-1} \in \pi_1^{\mathrm{pro}\text{-}p}(\overline{V}) \cap \pi_1^{(\mathrm{pro}\text{-}p)}(U) = \pi_1^{\mathrm{pro}\text{-}p}(\overline{U})$$

and so

$$\gamma \cdot \pi_1^{\mathrm{pro}\text{-}p}(\overline{U}) = \gamma' \cdot \pi_1^{\mathrm{pro}\text{-}p}(\overline{U}).$$

If $b \in Z_V$ is another element with decomposition $b = \delta v$ with $\delta \in \pi_1^{\mathrm{pro}\text{-}p}(\overline{V})$ and $v \in \pi_1^{(\mathrm{pro}\text{-}p)}(U)$, then

$$ab = a(\delta v) = \delta a v = \delta(\gamma u)v = (\delta\gamma)(uv),$$

so that

$$\psi(ab) = \psi(b)\psi(a).$$

Hence, the map ψ is a homomorphism $Z_V \to G^{\mathrm{opp}}$ to G with the opposite group law, which is still a p-group by assumption (iii). Since

$$\ker(\psi) = Z_V \cap \left(\pi_1^{(\mathrm{pro}\text{-}p)}(U)\right) = Z_U,$$

the quotient Z_V/Z_U is a finite p-group.   □

Remark 217. (1) Theorem 210 now is deduced as follows. The assumptions on k implies that

$$\pi_1^{\mathrm{pro}\text{-}p}(\mathrm{Spec}(\mathrm{o}_k[\tfrac{1}{p}])) \subseteq \pi_1^{\mathrm{pro}\text{-}p}(\mathrm{Spec}(\mathbb{Z}[\zeta_p, \tfrac{1}{p}]))$$

is an open subgroup, which is a free pro-p group due to $p \geq 3$ being a regular prime, see Lemma 214. This guarantees properties (i) and (ii) of Theorem 215, while property (iii) is also assumed from the start in Theorem 210. In order to assure (iv) we apply Proposition 216 with

$$V = \mathbb{P}_k^1 - \{0, 1, \infty\} \subset Y = \mathbb{P}_k^1,$$

and the map $\beta : X \to \mathbb{P}^1_k$ of Theorem 210. This allows to conclude as in the proof of Theorem 215 that for the Albanese torsor map $\alpha : X \hookrightarrow \mathrm{Pic}^1_X$ the induced abelianization map on sections

$$\alpha_* : \mathscr{S}_{\pi_1^{\mathrm{pro}\text{-}p}(X/k)} \twoheadrightarrow \mathscr{S}_{\pi_1^{\mathrm{pro}\text{-}p}(\mathrm{Pic}^1_X/k)}$$

is surjective with both sets non-empty. Here Hoshi resorts to another argument to ensure that there are non-Diophantine sections in case property (v) of Theorem 215, the positive Mordell–Weil rank, fails. It is this step that requires to replace k by a finite pro-p extension k'/k which is unramified outside p. For details on the latter argument see [Ho10] §4.

(2) For a regular prime $p \geq 11$ and $k = \mathbb{Q}(\zeta_p)$, the pth Fermat curve

$$X_{\mathrm{Fermat},p} = \{A^p + B^p = C^p\} \subset \mathbb{P}^2_k$$

actually satisfies all assumptions of Theorem 215, with (v) following by [GrRo78] and with Belyi map given by

$$(A, B, C) \mapsto (A^p, B^p, C^p) \in \{(u,v,w) \, ; \, u + v = w\} \cong \mathbb{P}^1_k.$$

Hence we have also explained part of [Ho10] Theorem B.

(3) We cannot help but think that the above counter-examples to the pro-p version of the section conjecture come to life due to a coincidence of a number of *accidents*. For example, the freeness of the pro-p fundamental group of the arithmetic base curve $S \subset \mathrm{Spec}(\mathfrak{o}_k[\frac{1}{p}])$ forces S to be almost all of $\mathrm{Spec}(\mathfrak{o}_k)$. On the other hand, having good reduction almost everywhere is a rare commodity among smooth projective curves. We could artificially force the Galois action on $\mathrm{H}_1(\overline{X}, \mathbb{F}_p)$ to be unipotent, or even trivial, but only at the expense of enlarging k in an uncontrolled manner with respect to the pro-p freeness condition.

Nevertheless, the hope to prove the section conjecture immediately via a pro-p approach is destroyed. In light of the above described accidents, we might ask, whether still pro-p methods can prove the section conjecture, if we do not apply them directly to a given curve X/k but to an auxiliary finite, maybe even cyclic, étale cover $X' \to X$ such that we leave the realm of the accidental failure of the pro-p section conjecture.

14.7 Variations on Pro-p Counter-Examples After Hoshi

We aim at a generalization of Hoshi's approach to counter-examples for the pro-p section conjecture which makes use of more precise knowledge of pro-p arithmetic fundamental groups.

Theorem 218. *Let p be a prime number, and let k be an algebraic number field. Let B be a dense open in $\mathrm{Spec}(\mathfrak{o}_k)$ with complement S, such that p is invertible on B. We moreover assume that there is a subset $S_0 \subseteq S$ of the places of k above p with*

(i) $\sum_{v \in S_0} \frac{1 - \#\mu_p(k_v)}{1 - p} = \frac{1 - \#\mu_p(k)}{1 - p}$,

(ii) *and the map $\mathrm{H}^1(B, \mu_p) \to \prod_{v \in S_0} \mathrm{H}^1(k_v, \mu_p)$ is injective.*

Let X/k be a smooth, projective geometrically connected curve of genus ≥ 2 with

(iii) *good reduction over B,*

(iv) *and Gal_k acts on $\pi_1^{\mathrm{ab}}(\overline{X}, \bar{x}) \otimes \mathbb{F}_p = \mathrm{H}_1(\overline{X}, \mathbb{F}_p)$ through a p-group.*

Let us furthermore assume that

(v) $X(k_v) \neq \emptyset$ *for all $v \in S \setminus S_0$,*

(vi) *and S_0 misses at least one place $\mathfrak{p} \mid p$ of k or the auxiliary set T below in the proof is bigger than S.*

Then the set $\mathscr{S}_{\pi_1^{\mathrm{pro}\text{-}p}}(X/k)$ is uncountably infinite and the pro-p Kummer map

$$\kappa_p : X(k) \to \mathscr{S}_{\pi_1^{\mathrm{pro}\text{-}p}}(X/k)$$

is not surjective, i.e., there are non-Diophantine pro-p sections.

Proof. According to [NSW08] Theorem 10.9.1, properties (i) and (ii) are the precise criterion to put us in the *degenerate case* as defined in [NSW08] Definition 10.9.3, which means that there is a finite set of places T containing S and a natural isomorphism

$$\underset{v \in S \setminus S_0}{\text{\Large∗}} \mathrm{Gal}_{k_v}^{\mathrm{pro}\text{-}p} * \underset{T \setminus S}{\text{\Large∗}} \mathbb{Z}_p \xrightarrow{\sim} \pi_1^{\mathrm{pro}\text{-}p}(B), \tag{14.21}$$

which sends $\mathrm{Gal}_{k_v}^{\mathrm{pro}\text{-}p}$ (resp. $1 \in \mathbb{Z}_p$) for $v \in S \setminus S_0$ (resp. for $v \in T \setminus S$) to the decomposition group in $\pi_1^{\mathrm{pro}\text{-}p}(B)$ (resp. to the Frobenius) of a place above v.

Properties (iii) and (iv) imply by Lemma 212 that the outer pro-p Galois representation

$$\rho_{X/k} : \mathrm{Gal}_k \to \mathrm{Out}\left(\pi_1^{\mathrm{pro}\text{-}p}(\overline{X})\right)$$

factors through $\pi_1^{\mathrm{pro}\text{-}p}(B)$. We denote again by $\mathscr{S}$ the $\pi_1^{\mathrm{pro}\text{-}p}(\overline{X})$-conjugacy classes of sections of the extension

$$1 \to \pi_1^{\mathrm{pro}\text{-}p}(\overline{X}) \to \mathrm{Aut}\left(\pi_1^{\mathrm{pro}\text{-}p}(\overline{X})\right) \times_{\mathrm{Out}\left(\pi_1^{\mathrm{pro}\text{-}p}(\overline{X})\right)} \pi_1^{\mathrm{pro}\text{-}p}(B) \to \pi_1^{\mathrm{pro}\text{-}p}(B) \to 1 \tag{14.22}$$

which again pulls back to the extension $\pi_1^{\mathrm{pro}\text{-}p}(X/k)$ to yield a base change map

$$\mathscr{S} \to \mathscr{S}_{\pi_1^{\mathrm{pro}\text{-}p}}(X/k)$$

whose bijectivity is assured by property (iii) and Lemma 213.

The same conclusion holds for the following local analogues. First, for a place v of k, the local outer pro-p Galois representation of $X \otimes k_v / k_v$ still has a pro-p group as its image, so that the extension $\pi_1^{\text{pro-}p}(X \otimes k_v / k_v)$ is the pullback of the extension

$$1 \to \pi_1^{\text{pro-}p}(\overline{X}) \to \text{Aut}\left(\pi_1^{\text{pro-}p}(\overline{X})\right) \times_{\text{Out}\left(\pi_1^{\text{pro-}p}(\overline{X})\right)} \text{Gal}_{k_v}^{\text{pro-}p} \to \text{Gal}_{k_v}^{\text{pro-}p} \to 1.$$
$$(14.23)$$

Let $\mathscr{S}_v$ denote the $\pi_1^{\text{pro-}p}(\overline{X})$-conjugacy classes of sections of (14.23). Then there is again a base change map

$$\mathscr{S}_v \to \mathscr{S}_{\pi_1^{\text{pro-}p}(X/k)}(k_v)$$

which is clearly bijective. Secondly, for a place $v \in B$ where X has good reduction and $v \nmid p$, the local outer pro-p Galois representation is even unramified and thus factors over

$$\text{Gal}_{k_v}^{\text{pro-}p,\text{nr}} = \mathbb{Z}_p.$$

The extension $\pi_1^{\text{pro-}p}(X \otimes k_v / k_v)$ then is even a pullback of the extension

$$1 \to \pi_1^{\text{pro-}p}(\overline{X}) \to \text{Aut}\left(\pi_1^{\text{pro-}p}(\overline{X})\right) \times_{\text{Out}\left(\pi_1^{\text{pro-}p}(\overline{X})\right)} \mathbb{Z}_p \to \mathbb{Z}_p \to 1 \qquad (14.24)$$

describing the situation for the special fibre of the good reduction at v. Let $\mathscr{S}_v^{\text{nr}}$ denote the $\pi_1^{\text{pro-}p}(\overline{X})$-conjugacy classes of sections of (14.24). There is again a base change map

$$\mathscr{S}_v^{\text{nr}} \to \mathscr{S}_{\pi_1^{\text{pro-}p}(X/k)}(k_v)$$

which is bijective by Proposition 91. We obtain the following commutative diagram

$$
\begin{array}{ccccc}
X(k) & \xhookrightarrow{\;\; \kappa_{X,p} \;\;} & \mathscr{S}_{\pi_1^{\text{pro-}p}(X/k)} & \xleftarrow{\;\;\sim\;\;} & \mathscr{S} \\
\downarrow & & \downarrow & & \downarrow \\
\displaystyle\prod_{v \in T \setminus S_0} X(k_v) & \xrightarrow{\;\; \kappa_{X \otimes k_v,p} \;\;} & \displaystyle\prod_{v \in T \setminus S_0} \mathscr{S}_{\pi_1^{\text{pro-}p}(X/k)}(k_v) & \xleftarrow{\;\;\sim\;\;} & \displaystyle\prod_{v \in S \setminus S_0} \mathscr{S}_v \times \prod_{v \in T \setminus S} \mathscr{S}_v^{\text{nr}}
\end{array}
$$

where by the degenerate structure (14.21) of $\pi_1^{\text{pro-}p}(B)$ the localisation map

$$\mathscr{S} \to \prod_{v \in S \setminus S_0} \mathscr{S}_v \times \prod_{v \in T \setminus S} \mathscr{S}_v^{\text{nr}}$$

is surjective. Property (v) prevents $\prod_{v \in S \setminus S_0} \mathscr{S}_v$ from being empty, while

$$\prod_{v \in T \setminus S} \mathscr{S}_v^{\text{nr}}$$

is always nonempty. If there is a place $\mathfrak{p}|p$ in $S \setminus S_0$, then by Theorem 76 the map

$$\kappa_p \;:\; X(k_\mathfrak{p}) \hookrightarrow \mathscr{S}_{\pi_1^{\text{pro-}p}(X/k)}(k_\mathfrak{p})$$

is injective, and a consequently $\mathscr{S}_\mathfrak{p}$ is uncountable. If on the other hand we have $v \in T \setminus S$, then $\mathscr{S}_v^{\text{nr}}$ is uncountable by Theorem 226. By property (vi) at least one of these places exists and in any case $\mathscr{S}$ is uncountable. We deduce that again $\mathscr{S}_{\pi_1^{\text{pro-}p}(X/k)}$ is uncountable, so that there must be in particular non-Diophantine pro-p sections for X/k.   □

Remark 219. (1) If k in Theorem 218 contains ζ_p, then by [NSW08] Theorem 10.9.1 we have necessarily $S_0 = \{\mathfrak{p}\}$ with $\mathfrak{p}|p$.

(2) In Theorem 210, the conditions imposed on the number field k imply by Lemma 214 that there is a unique place $\mathfrak{p}$ of k with $\mathfrak{p}|p$ and that for

$$B = \text{Spec}(\mathfrak{o}_k[\tfrac{1}{p}])$$

property (i) and (ii) of Theorem 218 holds with respect to $S_0 = \{\mathfrak{p}\} = S$. The auxiliary set T contains S properly. Hence also property (vi) holds. Theorem 210 is in fact a special case of Theorem 218.

Nevertheless, although Theorem 218 provides more flexibility in the construction of counter-examples with regard to the number field and the locus of good reduction, however, I see no other method than Hoshi's to establish the key property (iv), see Proposition 216.

Example 220. Here is a concrete example for the failure of the pro-3 section conjecture that lies beyond Theorem 210. In the notation of Theorem 218, we set $k = \mathbb{Q}(\zeta_3)$ and

$$B = \text{Spec}(\mathbb{Z}[\zeta_3, \tfrac{1}{6}])$$

with $S_0 = \{3\}$ and $S = \{2, 3, \infty\}$. Then (i) holds and (ii) is equivalent to the restriction

$$\text{res}_3 \;:\; \mathcal{O}^*(B)/(\mathcal{O}^*(B))^3 \to \mathbb{Q}_3(\zeta_3)^*/(\mathbb{Q}_3(\zeta_3)^*)^3$$

being injective. Since 2 is inert in $\mathbb{Q}(\zeta_3)/\mathbb{Q}$ and

$$3 = -\zeta_3^2 \cdot (1 - \zeta_3)^2$$

we find that the classes of $\zeta_3, 2, 3$ form a basis of the left hand side. If we look at the filtration of the right hand side given by the subspaces

$$\ker\left(N : \mathbb{Z}_3[\zeta_3]^*/(\mathbb{Z}_3[\zeta_3]^*)^3 \to \mathbb{Z}_3^*/(\mathbb{Z}_3^*)^3\right) \subset \mathbb{Z}_3[\zeta_3]^*/(\mathbb{Z}_3[\zeta_3]^*)^3$$

then res_3 becomes upper triangular with $\zeta_3, 2, 3$ being nontrivial in the respective filtration quotients. It follows as in the proof of Theorem 218 from [NSW08]

Theorem 10.9.1 that we are in the degenerate case. Moreover we have $\#(T \setminus S) = 1$, and formula (14.21) in this particular case reads

$$\left(\mathbb{Z}_3(1) \rtimes_4 \mathbb{Z}_3\right) \ast \mathbb{Z}_3 = \mathrm{Gal}^{\mathrm{pro}\text{-}3}_{\mathbb{Q}_2(\zeta_3)} \ast \mathbb{Z}_3 \xrightarrow{\sim} \pi_1^{\mathrm{pro}\text{-}3}\left(\mathrm{Spec}(\mathbb{Z}[\zeta_3, \tfrac{1}{6}])\right).$$

Here $\mathbb{Z}_3(1) \rtimes_4 \mathbb{Z}_3$ is the semidirect product, where the generator of $\mathbb{Z}_3$ acts via the 3-adic automorphism of multiplication by $4 = 1 + 3$ on $\mathbb{Z}_3(1) = \mathbb{Z}_3$.

The example is now provided by the smooth projective curve $C = C_0 \times_{\mathbb{Q}} \mathbb{Q}(\zeta_3)$ of genus 3 given by

$$C_0 = \{Y^3 Z = X(X - Z)(X - 3Z)(X - 9Z)\} \subset \mathbb{P}^2_{\mathbb{Q}}.$$

Indeed, the curve C_0 has a $\mathbb{Q}$-rational point, namely $[0 : 1 : 0]$, and good reduction outside $2 \cdot 3$ as can be seen easily from the jacobian criterion applied to the integral curve $\mathscr{C}_0 \subset \mathbb{P}^2_{\mathbb{Z}}$ given by the same equation. The example at this point relies on the *accident* that the only primes which divide differences of the numbers $0, 1, 3, 9$ are 2 and 3. Moreover, the curve C_0 is a μ_3 torsor over $\mathbb{P}^1_{\mathbb{Q}}$ described by taking a cube root of

$$T(T - 1)(T - 3)(T - 9)$$

hence finite étale over $U = \mathbb{P}^1_{\mathbb{Q}} - \{0, 1, 3, 9\}$ with geometric monodromy a 3-group. Because the ramification points are rational, we find

$$\mathrm{H}_1\left(\overline{U}, \mathbb{F}_3\right) = \mathbb{Z}/3\mathbb{Z}(1) \oplus \mathbb{Z}/3\mathbb{Z}(1) \oplus \mathbb{Z}/3\mathbb{Z}(1).$$

In particular, Gal_k acts through a 3-group and property (iv) of Theorem 218 holds. We may conclude that the pro-3 Kummer map

$$\kappa_3 \ : \ C(\mathbb{Q}(\zeta_3)) \to \mathscr{S}_{\pi_1^{\mathrm{pro}\text{-}3}(C/\mathbb{Q}(\zeta_3))}$$

is injective with finite image in an uncountable space of pro-3 sections.

We end this chapter by posing a question which might revitalize work on the pro-p analogue of the section conjecture.

Question 221. Does every smooth projective geometrically connected curve X/k over an algebraic number field k admit a finite étale cover $h : X' \to X$ with X'/k geometrically connected, such that the pro-p Kummer map

$$\kappa_p : X'(k) \to \mathscr{S}_{\pi_1^{\mathrm{pro}\text{-}p}(X'/k)}$$

is bijective for X'/k?

Chapter 15
Sections over Finite Fields

Let $\mathbb{F}_q$ be a finite field with q elements of characteristic p. The absolute Galois group $\mathrm{Gal}_{\mathbb{F}_q}$ is profinite free and generated by the qth-power Frobenius Frob_q. Thus any extension of $\mathrm{Gal}_{\mathbb{F}_q}$ splits, and does so most likely in an abundant number of inequivalent ways. However, surprisingly, the analogue of the section conjecture holds true for abelian varieties over $\mathbb{F}_q$, see Theorem 222.

On the other hand, for hyperbolic curves our intuition is correct that the section conjecture fails, see Theorem 224, and by quite a big margin, because by Theorem 226 the space of sections is always uncountable. The first method of proof requires to determine all projective hyperbolic curves over finite fields that are space filling inside their Jacobian: by Theorem 223 these are only certain curves of genus 2 or 3 over $\mathbb{F}_2$. The claim on the cardinality of the space of sections benefits from the results on the graded Lie algebra of the descending central series in Sects. 14.3 and 14.4.

We also recall the characterisation of Diophantine sections among all sections due to Tamagawa.

15.1 Abelian Varieties over Finite Fields

Let $A/\mathbb{F}_q$ be an abelian variety. By a theorem of Lang and Tate the group $\mathrm{H}^1(\mathbb{F}_q, A)$ vanishes. Equivalently, every A-torsor admits an $\mathbb{F}_q$-rational point, and in particular genus 1 curves over $\mathbb{F}_q$ which are torsors under their respective Jacobian are elliptic curves. In terms of the section conjecture this means that torsors under abelian varieties over finite fields satisfy the weak section conjecture.

Moreover, the Kummer sequence of an étale isogeny $\varphi : B \to A$ computes

$$A(\mathbb{F}_q)/\varphi(B(\mathbb{F}_q)) \xrightarrow{\sim} \mathrm{H}^1\left(\mathbb{F}_q, \ker(\varphi)\right). \tag{15.1}$$

In the limit over all étale isogenies we obtain that the Kummer map of A

$$\kappa : A(\mathbb{F}_q) \xrightarrow{\sim} \mathrm{H}^1(\mathbb{F}_q, \pi_1(\overline{A}))$$

J. Stix, *Rational Points and Arithmetic of Fundamental Groups*, Lecture Notes in Mathematics 2054, DOI 10.1007/978-3-642-30674-7_15,
© Springer-Verlag Berlin Heidelberg 2013

is an isomorphism, because for the Lang isogeny

$$\wp = F - \mathrm{id} : A \to A$$

the image $\wp(A(\mathbb{F}_q))$ is trivial. Here F is the geometric qth-power Frobenius.

Theorem 222. *The section conjecture is true for abelian varieties over a finite field.*

Proof. By Corollary 71, the Kummer map for an abelian variety A coincides with the profinite Kummer map normalized by the section associated to $0 \in A(k)$. $\square$

15.2 Space Filling Curves in Their Jacobian

In compliance with intuition, the section conjecture for curves of genus ≥ 2 over $\mathbb{F}_q$ is wrong in all cases. But the case of elliptic curves should serve as a warning that we need to justify why a specific section has no Diophantine origin.

Let $X/\mathbb{F}_q$ be a smooth, projective curve of genus $g \geq 2$. Since $\mathrm{Gal}_{\mathbb{F}_q} \cong \hat{\mathbb{Z}}$ the extension $\pi_1(X/\mathbb{F}_q)$ always splits. Hence, if $X(\mathbb{F}_q) = \emptyset$ then the section conjecture fails for obvious reasons. Otherwise we choose $P \in X(\mathbb{F}_q)$ and define the Albanese map $X \hookrightarrow \mathrm{Pic}_X^0$ with respect to this point P. The natural map

$$\mathscr{S}_{\pi_1(X/\mathbb{F}_q)} \twoheadrightarrow \mathscr{S}_{\pi_1^{\mathrm{ab}}(X/\mathbb{F}_q)} = \mathscr{S}_{\pi_1(\mathrm{Pic}_X^0/\mathbb{F}_q)} \cong \mathrm{Pic}_X^0(\mathbb{F}_q)$$

is surjective, because $\mathrm{Gal}_{\mathbb{F}_q}$ is a free profinite group. Therefore, if

$$X(\mathbb{F}_q) \subseteq \mathrm{Pic}_X^0(\mathbb{F}_q)$$

is a proper subset, then we can lift an abelian section associated to a point from $\mathrm{Pic}_X^0(\mathbb{F}_q) \setminus X(\mathbb{F}_q)$ to a non-Diophantine section. Again, the section conjecture fails for such curves. The argument works for most curves as we will see below by (an elementary improvement of) a known estimate for the size

$$h = \# \mathrm{Pic}_X^0(\mathbb{F}_q). \tag{15.2}$$

Consequently, the section conjecture fails for these curves already on the abelian level.

An improved estimate for the Picard group. We venture to determine all curves X over $\mathbb{F}_q$ of genus at least 2 such that

$$X(\mathbb{F}_q) = \mathrm{Pic}_X^0(\mathbb{F}_q).$$

We improve the estimate from [LMD90] Theorem 2 by combining the better halfs of intermediate estimates. Let

$$D_n = \#\{D \geq 0 \,;\, \deg(D) = n\}$$

be the number of effective $\mathbb{F}_q$-rational divisors of degree n on X. We fix a factorisation of the characteristic polynomial of the geometric Frobenius F

$$\det\left(1 - F^*T \,|\, \mathrm{H}^1(\overline{X}, \mathbb{Q}_\ell)\right) = \prod_{i=1}^{2g}(1 - \alpha_i T)$$

such that $\alpha_{i+g} = \overline{\alpha}_i = q/\alpha_i$ for $i = 1, \ldots, g$. Then [LMD90] Theorem 1 reads

$$\sum_{n=0}^{g-2} D_n + \sum_{n=0}^{g-1} q^{g-1-n} D_n = h \cdot \sum_{i=1}^{g} \frac{1}{|1 - \alpha_i|^2}. \tag{15.3}$$

We set $N = \#X(\mathbb{F}_q)$ and observe that $D_0 = 1$ and $D_n \geq \#X(\mathbb{F}_q)$ for $n \geq 1$. Combining [LMD90] §4 (5)

$$\sum_{i=1}^{g} \frac{1}{|1 - \alpha_i|^2} \leq \frac{(g+1)(q+1) - N}{(q-1)^2} \tag{15.4}$$

with (15.3) we obtain the estimate

$$h \geq (q-1)^2 \cdot \frac{1 + q^{g-1} + N(g - 2 + \frac{q^{g-1}-1}{q-1})}{(g+1)(q+1) - N} = (*) \tag{15.5}$$

that is slightly better than the estimates in [LMD90] Theorem 2. The denominator of $(*)$ is positive by the Hasse–Weil bound, because

$$(g+1)(q+1) = 1 + 2g\sqrt{q} + q + g(\sqrt{q}-1)^2 > 1 + 2g\sqrt{q} + q \geq N.$$

Reduction to small cases. We set $n(q-1) = N$ and analyse $(*) > N$ to be equivalent to

$$1 + q^{g-1} + n\left((g-2)(q-1) + q^{g-1} - 1\right) > n\left((g+1)\cdot\frac{q+1}{q-1} - n\right)$$

$$\iff n^2 + n\left(q^{g-1} + (g-2)(q-1) + 1 - (g+1)\cdot\frac{q+1}{q-1}\right) + 1 + q^{g-1} > 0$$

$$\iff n^2 + n\left(q^{g-1} + (g+q)(q-1-\frac{q+1}{q-1}) - q^2 + 4\right) + 1 + q^{g-1} > 0.$$

$$\iff n^2 + n\left(q^2(q^{g-3} - 1) + (g+q)\frac{q(q-3)}{q-1} + 4\right) + 1 + q^{g-1} > 0. \tag{15.6}$$

The coefficient of the linear term in n is monotone increasing as a function in g for $q \geq 3$, and is monotone increasing as a function in $q > 1$ for $g \geq 3$. The value of this coefficient for $g = q = 3$ is 4, hence the entire inequality holds true except possibly if $g = 2$ or $q = 2$. If $g = 2$, then (15.6) reads

$$n^2 + n\frac{(q+1)(q-4)}{q-1} + 1 + q = (n-1)^2 + n\frac{(q+2)(q-3)}{q-1} + q > 0$$

which is true for all $n \geq 0$ if $q \geq 3$. Therefore, if $h = N$ we necessarily have $q = 2$. If $q = 2$ the estimate (15.6) reduces to

$$n^2 + n(2^{g-1} - 2g - 4) + 1 + 2^{g-1} > 0$$

which is true for all $n \geq 0$ if $g \geq 4$. The remaining case for (g, q) with potentially curves of genus g over $\mathbb{F}_q$ with $h = N$ therefore are $(2,2)$ and $(3,2)$.

The case of genus 2. Here we can compute exactly. Let $\alpha, \overline{\alpha} = q/\alpha, \beta, \overline{\beta} = q/\beta$ be the eigenvalues of Frobenius of the smooth projective curve $X/\mathbb{F}_q$ of genus 2. We set $\gamma = \alpha + q/\alpha$ and $\delta = \beta + q/\beta$, such that γ, δ are real and the following list of formulae hold:

$$\gamma^2 = \alpha^2 + \overline{\alpha}^2 + 2q,$$

$$\delta^2 = \beta^2 + \overline{\beta}^2 + 2q,$$

$$N = 1 + q - (\gamma + \delta),$$

$$N_2 := \#X(\mathbb{F}_{q^2}) = 1 + q^2 - (\gamma^2 + \delta^2 - 4q),$$

$$L(T) = \det\left(\mathbf{1} - \text{Frob}^* \, T \,|\, \text{H}^1(\overline{X}, \mathbb{Q}_\ell)\right) = (1 - \gamma T + qT^2)(1 - \delta T + qT^2),$$

$$h = L(1) = (1 - \gamma + q)(1 - \delta + q) = (1 + q)N + \gamma\delta.$$

We can eliminate γ, δ in the linear combination

$$(1+q-N)^2 + N_2 = (\gamma+\delta)^2 + 1 + q^2 - (\gamma^2 + \delta^2 - 4q) = 2h + 1 + 4q + q^2 - 2(1+q)N$$

to obtain the equation

$$N_2 = 2h + 2q - N^2.$$

Now in case $h = N$ the condition $N_2 \geq N$ reads as follows.

$$N_2 = 2N + 2q - N^2 \geq N$$

$$\Longleftrightarrow \quad 2q + 1/4 \geq (N - 1/2)^2$$

which for $q = 2$ yields two cases: $N = 1$ and $N = 2$. A computation using SAGE, see [S$^+$08], gives us a complete list of isomorphism classes of examples as follows.

type	q	g	$N = h$	N_2	$L(T)$	equation
I	2	2	1	5	$1 - 2T + 2T^2 - 4T^3 + 4T^4$	$Y^2 + Y = X^5 + X^3 + 1$
II	2	2	2	4	$1 - T - 2T^3 + 4T^4$	$Y^2 + XY = X^5 + X^2 + X$
III	2	2	2	4	$1 - T - 2T^3 + 4T^4$	$Y^2 + Y = \frac{1}{X^3+X+1}$

All genus 2 curves are hyperelliptic. The examples of type II and III differ in their ramification behaviour of the hyperelliptic degree 2 cover of $\mathbb{P}^1$, namely, type II has 2 while type III has 3 branch points. The curve of type I was already described in [McR71] Theorem 4.4 in the attempt to describe the function field analogue of imaginary quadratic fields of class number 1.

The case of genus 3. A search via SAGE, see [S$^+$08], among smooth quartic curves in $\mathbb{P}^2$ over $\mathbb{F}_2$, which covers all canonically embedded non-hyperelliptic curves of genus 3 adds three more examples to our list. We do not give equations of the 40 incarnations of our examples that by symmetry possibly reduce to 5 different examples. Instead, we content ourselves with the L-Polynomial, which at least distinguishes three non-isomorphic cases.

type	q	g	$N = h$	N_2	$N_3 = \#X(\mathbb{F}_8)$	$L(T)$
IV	2	3	3	5	15	$1 - 6T^3 + 8T^6$
V	2	3	3	7	18	$1 + T^2 - 9T^3 + 2T^4 + 8T^6$
VI	2	3	3	3	12	$1 - T^2 - 3T^3 - 2T^4 + 8T^6$

We summarize our calculations as follows.

Theorem 223. *There are smooth projective curves $X/\mathbb{F}_q$ of genus $g \geq 2$ such that*

$$\#X(\mathbb{F}_q) = \# \operatorname{Pic}^0_X(\mathbb{F}_q)$$

if and only if $q = 2$ and $g = 2$ or 3.

The section conjecture fails. At this point we can answer the original question on whether the section conjecture is bound to fail for curves over $\mathbb{F}_q$ of genus ≥ 2.

Theorem 224. *Let $X/\mathbb{F}_q$ be a smooth, projective curve of genus $g \geq 2$. Then there are sections of $\pi_1(X/\mathbb{F}_q)$ which are not Diophantine, i.e., which do not come from a $\mathbb{F}_q$-rational point of X.*

Proof. If (g, q) does not belong to the list $(2, 2)$, $(3, 2)$, then there is a point in $\operatorname{Pic}^0_X(\mathbb{F}_q)$ which is not in the image of the Albanese map $X \hookrightarrow \operatorname{Pic}^0_X$. A lift of the associated abelian section to a section of $\pi_1(X/\mathbb{F}_q)$ cannot be Diophantine.

For the (g, q) from the above list we have to use an auxiliary section that of course always exists (besides, we are done if $X(\mathbb{F}_q)$ is empty) and construct a neighbourhood $f : X' \to X$ of this section with genus ≥ 4. Then $\pi_1(X'/\mathbb{F}_q)$ admits a section s' which is not Diophantine. If $f(s') \in \mathscr{S}_{\pi_1(X/\mathbb{F}_q)}$ were Diophantine, then by the discussion of fibres above sections also s' must be Diophantine. This contradiction proves the theorem. $\qquad\square$

Corollary 225. *(1) Every smooth, projective curve $X/\mathbb{F}_q$ of genus ≥ 2 admits a finite étale cover $X' \to X$ by a geometrically connected $X'/\mathbb{F}_q$ without $\mathbb{F}_q$-rational point.*

(2) More precisely, we may choose $X' \to X$ such that $\overline{X}' \to \overline{X}$ is Galois with abelian Galois group G, except possibly for the cases where $q = 2$ and $g = 2$ or 3 when G might have to be metabelian.

Proof. (1) Let s be a section of $\pi_1(X/\mathbb{F}_q)$ that does not belong to a rational point. Because only finitely many sections belong to rational points, and because a section is uniquely determined by the neighbourhood to which it lifts, we may choose a neighbourhood $X' \to X$ of s such that no s_a for $a \in X(\mathbb{F}_q)$ lifts to X' and thus $X'(\mathbb{F}_q)$ is empty. Being a neighbourhood makes $X'/\mathbb{F}_q$ geometrically connected.

To prove (2) we make the procedure of (1) explicit. If $\#X(\mathbb{F}_q) < \#\mathrm{Pic}^0_X(\mathbb{F}_q)$ then we may embed $X \hookrightarrow \mathrm{Pic}^0_X$ such that X misses $0 \in \mathrm{Pic}^0_X$. The pullback of the Lang-isogeny

$$\wp = F - \mathrm{id} : \mathrm{Pic}^0_X \to \mathrm{Pic}^0_X$$

yields a geometrically connected finite étale cover $X' \to X$ such that $X'(\mathbb{F}_q) = \emptyset$ because for $P \in \mathrm{Pic}^0_X(\mathbb{F}_q)$ we have $\wp(P) = 0$.

If $\#X(\mathbb{F}_q) = \#\mathrm{Pic}^0_X(\mathbb{F}_q)$ then $q = 2$ and $g \leq 3$ by Theorem 223. We embed $X \hookrightarrow \mathrm{Pic}^0_X$ and exploit the pullback $X' \to X$ of the multiplication by 3 map on Pic^0_X such that the genus of X' exceeds 3. By Theorem 224 we choose a non-Diophantine section s' of $\pi_1(X'/\mathbb{F}_q)$. Let s be the image of s' in $\mathscr{S}_{\pi_1(X/\mathbb{F}_q)}$ such that $X' \to X$ is a neighbourhood of s by construction. Then we proceed as before with respect to the section s' to get a finite étale cover $X'' \to X'$ that is geometrically abelian and $X''(\mathbb{F}_q) = \emptyset$. Again by construction $X'' \to X$ is a neighbourhood of s. In the last step we replace X'' by the neighbourhood $X''' \to X$ of s defined by the Galois closure of $\overline{X}'' \to \overline{X}$ and its $\mathrm{Gal}_{\mathbb{F}_q}$-conjugates. The neighbourhood $X''' \to X$ has geometrically metabelian Galois group and proves part (2). $\square$

15.3 Counting Sections

Let $X/\mathbb{F}_q$ be a smooth, projective curve of genus $g \geq 2$. Because $\mathrm{Gal}_{\mathbb{F}_q}$ is a free profinite group and because every finite nilpotent group is canonically the product of its Sylow subgroups we have surjections

$$\mathscr{S}_{\pi_1(X/\mathbb{F}_q)} \twoheadrightarrow \mathscr{S}_{\pi_1^{\mathrm{nilp}}(X/\mathbb{F}_q)} \twoheadrightarrow \prod_{\ell \neq p} \mathscr{S}_{\pi_1^{\mathrm{pro}\text{-}\ell}(X/\mathbb{F}_q)}. \tag{15.7}$$

Theorem 226. *Let $X/\mathbb{F}_q$ be a smooth, projective curve of genus $g \geq 2$. For every $\ell \neq p$ the cardinality of (pro-ℓ) sections of $\pi_1(X/\mathbb{F}_q)$ is uncountable*

$$\#\mathscr{S}_{\pi_1(X/\mathbb{F}_q)} = \#\mathscr{S}_{\pi_1^{\mathrm{pro}\text{-}\ell}(X/\mathbb{F}_q)} = 2^{\aleph_0},$$

while for every $n \geq 1$ the set $\mathscr{S}_{C_{\geq -n}\left(\pi_1(X/\mathbb{F}_q)\right)}$ is finite.

Proof. Since $\mathscr{S}_{\pi_1(X/\mathbb{F}_q)}$ is a projective limit of finite sets along a countable index system, and using (15.7) we have

$$2^{\aleph_0} \geq \#\mathscr{S}_{\pi_1(X/\mathbb{F}_q)} \geq \#\mathscr{S}_{\pi_1^{\mathrm{nilp}}(X/\mathbb{F}_q)} \geq \#\mathscr{S}_{\pi_1^{\mathrm{pro}\text{-}\ell}(X/\mathbb{F}_q)}.$$

Because $\mathrm{Gal}_{\mathbb{F}_q}$ has cohomological dimension 1, there is no H^2 and Proposition 197 specialises to a short exact sequence

$$1 \to \mathrm{H}^1\left(\mathbb{F}_q, \mathrm{gr}_{-n}^C \overline{\pi}\right) \to \mathscr{S}_{C_{\geq -n}\pi_1(X/\mathbb{F}_q)} \xrightarrow{\mathrm{pr}_*} \mathscr{S}_{C_{\geq -(n-1)}\pi_1(X/\mathbb{F}_q)} \to 1$$

and in its pro-ℓ version to the short exact sequence

$$1 \to \mathrm{H}^1\left(\mathbb{F}_q, \mathbb{Z}_\ell \otimes \mathrm{gr}_{-n}^C \overline{\pi}\right) \to \mathscr{S}_{C_{\geq -n}\pi_1^{\mathrm{pro}\text{-}\ell}(X/\mathbb{F}_q)} \xrightarrow{\mathrm{pr}_*} \mathscr{S}_{C_{\geq -(n-1)}\pi_1^{\mathrm{pro}\text{-}\ell}(X/\mathbb{F}_q)} \to 1.$$

It remains to show, that $\mathrm{H}^1\left(\mathbb{F}_q, \mathrm{gr}_{-n}^C \overline{\pi}\right)$ is finite for every n with its ℓ-part being nontrivial for infinitely many n.

We first discuss the prime to p part of $\mathrm{H}^1\left(\mathbb{F}_q, \mathrm{gr}_{-n}^C \overline{\pi}\right)$. Let $F \in \mathrm{Gal}_{\mathbb{F}_q}$ be the arithmetic Frobenius. The cohomology is computed by the exact sequence

$$0 \to \mathrm{H}^0\left(\mathbb{F}_q, \mathrm{gr}_{-n}^C \overline{\pi}\right) \to \mathrm{gr}_{-n}^C \overline{\pi} \xrightarrow{1-F} \mathrm{gr}_{-n}^C \overline{\pi} \to \mathrm{H}^1\left(\mathbb{F}_q, \mathrm{gr}_{-n}^C \overline{\pi}\right) \to 0.$$

As $\mathbb{Z}_\ell \otimes \mathrm{gr}_{-n}^C \overline{\pi}$ has weight $-n$ we find that $(1 - F) \otimes \mathbb{Q}_\ell$ is bijective and thus

$$\#\mathrm{H}^1\left(\mathbb{F}_q, \mathbb{Z}_\ell \otimes \mathrm{gr}_{-n}^C \overline{\pi}\right) = \ell\text{-part of } \det\left(1 - F | \mathbb{Z}_\ell \otimes \mathrm{gr}_{-n}^C \overline{\pi}\right)$$

is finite. The formula (14.16) for $[\mathbb{Z}_\ell \otimes \mathrm{gr}_{-n}^C \overline{\pi}] = [\mathfrak{p}_n]$ shows that the characteristic polynomial of F on $\mathbb{Z}_\ell \otimes \mathrm{gr}_{-n}^C \overline{\pi}$ has integral coefficients that are independent of $\ell \neq p$. In particular, $c_n = \det\left(1 - F | \mathbb{Z}_\ell \otimes \mathrm{gr}_{-n}^C \overline{\pi}\right)$ is an integer independent of n which gives a finite upper bound

$$\#\mathrm{H}^1\left(\mathbb{F}_q, \hat{\mathbb{Z}}' \otimes \mathrm{gr}_{-n}^C \overline{\pi}\right) \leq |c_n|$$

for the prime to p part of $\mathrm{H}^1\left(\mathbb{F}_q, \mathrm{gr}_{-n}^C \overline{\pi}\right)$. For the p-part of $\mathrm{H}^1\left(\mathbb{F}_q, \mathrm{gr}_{-n}^C \overline{\pi}\right)$ we may argue similarly because étale p-adic cohomology is the slope zero part of crystalline cohomology and thus the eigenvalues of F on $\pi_1^{\mathrm{ab}}(\overline{X}) \otimes \mathbb{Q}_p$ are still Weil-numbers of weight -1. At least the finiteness of

$$\#\mathrm{H}^1\left(\mathbb{F}_q, \mathbb{Z}_p \otimes \mathrm{gr}_{-n}^C \overline{\pi}\right) < \infty,$$

for $n \geq 1$, follows as for the ℓ-part.

It remains to show the occasional nonvanishing of $\mathrm{H}^1(\mathbb{F}_q, \mathbb{Z}_\ell \otimes \mathrm{gr}^C_{-n} \overline{\pi})$, which occurs if and only if the action of F on $\mathbb{Q}_\ell \otimes \mathrm{gr}^C_{-n} \overline{\pi}$ has an eigenvalue

$$\alpha \equiv 1 \mod \mathfrak{l}$$

for some place $\mathfrak{l} \mid \ell$ in an algebraic extension of $\mathbb{Q}$. The Galois action on

$$\mathfrak{p}_{\mathbb{Q}_\ell} = \bigoplus_{n \geq 1} \mathbb{Q}_\ell \otimes \mathrm{gr}^C_{-n} \overline{\pi}$$

is through the image of $\mathrm{Gal}_{\mathbb{F}_q}$ acting on $H = \pi_1^{\mathrm{ab}}(\overline{X}) \otimes \mathbb{Z}_\ell$. Since $\mathrm{GL}(H)$ is a virtually pro-ℓ group, the Galois image is isomorphic to $\mathbb{Z}_\ell \times \mathbb{Z}/N\mathbb{Z}$ for a finite N prime to ℓ.

The subspace of $\mathfrak{p}_{\mathbb{Q}_\ell}$ on which F acts with eigenvalues $\alpha \equiv 1 \mod \mathfrak{l}$ for some $\mathfrak{l} \mid \ell$ coincides with the invariants $\mathrm{H}^0(G, \mathfrak{p}_{\mathbb{Q}_\ell})$ under the factor $G = \mathbb{Z}/N\mathbb{Z}$ of the Galois image in $\mathrm{GL}(H)$. The claim on the cardinality of $\mathscr{S}_{\pi_1^{\mathrm{pro\text{-}}\ell}(X/\mathbb{F}_q)}$ follows because $\mathrm{H}^0(G, \mathfrak{p}_{\mathbb{Q}_\ell})$ is infinite dimensional by Proposition 207.   □

We can rephrase Theorem 226 as follows: for hyperbolic curves over finite fields the section conjecture goes wrong even by a very big margin.

15.4 Diophantine Sections After Tamagawa

The fundamental result of [Ta97] that forms the starting point for anabelian geometry over finite fields is the following group theoretic characterisation of the set of Diophantine sections among all sections. Let $\mathrm{Frob} \in \mathrm{Gal}_{\mathbb{F}_q}$ be the arithmetic Frobenius $a \mapsto a^q$ and let $F = \mathrm{Frob}^{-1}$ be the geometric Frobenius.

Theorem 227 (Tamagawa). *Let $X/\mathbb{F}_q$ be a smooth, projective curve of genus ≥ 1. A section s of $\pi_1(X/\mathbb{F}_q)$ is Diophantine if and only if every neighbourhood of s admits an $\mathbb{F}_q$-rational point. Moreover, if $X' \to X$ is a neighbourhood of s given by the open subgroup*

$$s(\mathrm{Gal}_{\mathbb{F}_q}) \subset \pi_1(X') \subset \pi_1(X),$$

then with $\pi_1(\overline{X}') = \pi_1(X') \cap \pi_1(\overline{X})$ and therefore $\pi_1^{\mathrm{ab}}(\overline{X}')$ as a Galois-module via conjugation with $s(\mathrm{Gal}_{\mathbb{F}_q})$ we have

$$\#X'(\mathbb{F}_q) = 1 + q - \mathrm{tr}(\mathrm{Frob} \mid \pi_1^{\mathrm{ab}}(\overline{X}') \otimes \mathbb{Q}_\ell)$$

for any $\ell \neq p$. Consequently, the condition that X' admits an $\mathbb{F}_q$-rational point is encoded in the group theory of the section.

Proof. This is [Ta97] Proposition 0.7, which deals with the case $g \geq 2$. But the case $g = 1$ is obvious as every section is Diophantine and every finite étale cover is of genus 1 itself and admits a rational point.

The formula for $\#X'(\mathbb{F}_q)$ is nothing but the Grothendieck–Lefschetz trace formula for Frobenius applied to $X'/\mathbb{F}_q$ when we remark that

$$
\begin{aligned}
\mathrm{tr}(F^* \mid \mathrm{H}^1_c(\overline{X'}, \mathbb{Q}_\ell)) &= \mathrm{tr}(\mathrm{Frob}^{-1} \mid \mathrm{H}^1(\overline{X'}, \mathbb{Q}_\ell)) \\
&= \mathrm{tr}(\pi_1(1 \times \mathrm{Frob}^{-1}) \mid \mathrm{Hom}(\pi_1^{\mathrm{ab}}(\overline{X'}), \mathbb{Q}_\ell)) \\
&= \mathrm{tr}(\pi_1(1 \times \mathrm{Frob}^{-1}) \mid \pi_1^{\mathrm{ab}}(\overline{X'}) \otimes \mathbb{Q}_\ell) \\
&= \mathrm{tr}(s(\mathrm{Frob})(-)s(\mathrm{Frob})^{-1} \mid \pi_1^{\mathrm{ab}}(\overline{X'}) \otimes \mathbb{Q}_\ell) \\
&= \mathrm{tr}(\mathrm{Frob} \mid \pi_1^{\mathrm{ab}}(\overline{X'}) \otimes \mathbb{Q}_\ell)
\end{aligned}
$$

where F is the geometric and Frob is the arithmetic Frobenius in $\mathrm{Gal}_{\mathbb{F}_q}$. □

Chapter 16
On the Section Conjecture over Local Fields

The analogue of the section conjecture can be explored over a local field k. In the archimedean case only $k = \mathbb{R}$ makes sense. In this case, the space of sections is in bijection with the set of connected components of real points, see Theorem 229. Several proofs of this fact are known and presented here.

For a finite extension $k/\mathbb{Q}_p$ the results are less complete. A section in this case gives rise to a valuation of the function field that extends the p-adic valuation on k, such that the image of the section lies in the decomposition subgroup of the valuation, see Theorem 235. However, up to now we cannot exclude that this valuation might be an exotic valuation not corresponding to a rational point.

16.1 The Real Section Conjecture

Let $X/\mathbb{R}$ be a geometrically connected real variety. Since finite étale covers induce finite topological covers on the underlying real analytic spaces of real points, two points $a_1, a_2 \in X(\mathbb{R})$ in the same connected component lift to the same finite étale covers as real points. It follows that the profinite Kummer map factors as a map

$$\kappa_{\mathbb{R}} : \pi_0(X(\mathbb{R})) \to \mathscr{S}_{\pi_1(X/\mathbb{R})}. \tag{16.1}$$

Since $\pi_0(X(\mathbb{R}))$ is a finite set, a suitable modification of Proposition 54 applies to describe when $\kappa_{\mathbb{R}}$ is surjective or injective.

Lemma 228. *Let s be a section of $\pi_1(X/\mathbb{R})$ and let $X_s = \varprojlim X'$ be the decomposition tower of s. Then the following holds.*

(1) We have $s = s_a$ for a component $s \in \pi_0(X(\mathbb{R}))$ if and only if $X_s(\mathbb{R}) \neq \emptyset$.
(2) The component $a \in \pi_0(X(\mathbb{R}))$ in (1) is unique if and only if

$$\mathrm{im}\left(X_s(\mathbb{R}) \to X(\mathbb{R})\right)$$

consists of only one connected component. □

J. Stix, *Rational Points and Arithmetic of Fundamental Groups*, Lecture Notes in Mathematics 2054, DOI 10.1007/978-3-642-30674-7_16,
© Springer-Verlag Berlin Heidelberg 2013

Theorem 229 (Real section conjecture, Mochizuki). *Let $X/\mathbb{R}$ be a smooth geometrically connected curve of genus ≥ 1. Then the map*

$$\kappa_{\mathbb{R}} \ : \ \pi_0(X(\mathbb{R})) \to \mathscr{S}_{\pi_1(X/\mathbb{R})}$$

*is a bijection of finite sets. More generally, the same holds for geometrically
connected algebraic $K(\pi, 1)$-spaces, in particular for abelian varieties.*

Several proofs for Theorem 229 are now available and will be described below. A
discrete version relating $\pi_0(X(\mathbb{R}))$ to sections of the discrete orbifold fundamental
group extension

$$1 \to \pi_1^{\text{top}}(X(\mathbb{C}), \bar{x}) \to \pi_1^{\text{orb}}(X(\mathbb{C})/\operatorname{Gal}_{\mathbb{R}}, \bar{x}) \to \operatorname{Gal}_{\mathbb{R}} \to 1$$

follows immediately from Sullivan's conjecture on homotopy fixed points of CW-
complexes with respect to an involution, see [Su71] p. 179 and Remark 25.

A theorem of Cox. The first proof due to Mochizuki [Mo03] Theorem 3.13 relies
on the following theorem of Cox.

Theorem 230 (Cox, [Cx79] Proposition 1.2, see [Sch94] Introduction). *Let $X/\mathbb{R}$
be a variety of dimension d. Then, for $q > 2d$, we have*

$$\operatorname{H}^q(X, \mathbb{Z}/2\mathbb{Z}) \cong \bigoplus_{i=0}^{d} \operatorname{H}^i(X(\mathbb{R}), \mathbb{Z}/2\mathbb{Z}).$$

The proof of the surjectivity part of Theorem 229 now proceeds as follows. By
Lemma 228 we only have to show that a section of $\pi_1(X/\mathbb{R})$ implies the existence
of a real point. If $\pi_1(X/\mathbb{R})$ splits then $\pi_1(X, \bar{x})$ contains 2-torsion and furthermore
the 2-cohomological dimension of $\pi_1(X, \bar{x})$ is infinite. It follows that there is a
connected finite étale cover $Y \to X$ such that for a fixed $q > 2 \dim X$ we have
$\operatorname{H}^q(\pi_1(Y, \bar{y}), \mathbb{Z}/2\mathbb{Z}) \neq 0$. Since X and thus also Y are algebraic $K(\pi, 1)$-spaces,
we furthermore have $\operatorname{H}^q(Y, \mathbb{Z}/2\mathbb{Z}) \neq 0$ so that Theorem 230 shows that $Y(\mathbb{R}) \neq \emptyset$
and a forteriori X admits a real point.

In the spirit of Sullivan: A profinite Smith fixed point theorem. A proof in the
spirit of Sullivan's conjecture, namely computing the pro-2 homotopy type of the
projective system $X_s = \varprojlim X'$ of all neighbourhoods of a section s of $\pi_1(X/\mathbb{R})$,
was given finally by Pál [Pa11]. The argument can be summarized as follows.
By Lemma 228 we only have to show that $X_s(\mathbb{R})$ consists of a unique connected
component. We fix a number $n > 2d$ where $d = \dim(X)$ and compute using
Theorem 230

$$\bigoplus_{i=0}^{d} \operatorname{H}^i(X_s(\mathbb{R}), \mathbb{Z}/2\mathbb{Z}) = \varinjlim_{X'} \bigoplus_{i=0}^{d} \operatorname{H}^i(X'(\mathbb{R})\mathbb{Z}/2\mathbb{Z})$$

$$= \varinjlim_{X'} \operatorname{H}^n(X'_{\text{ét}}, \mathbb{Z}/2\mathbb{Z}) = \operatorname{H}^n(X_{s,\text{ét}}, \mathbb{Z}/2\mathbb{Z}) = \operatorname{H}^n(\mathbb{R}, \mathbb{Z}/2\mathbb{Z}) = \mathbb{Z}/2\mathbb{Z}.$$

We conclude that $X_s(\mathbb{R})$ is non-empty and thus $\mathrm{H}^0(X_s(\mathbb{R}), \mathbb{Z}/2\mathbb{Z}) \neq 0$. Thus, for $i > 0$ we must have $\mathrm{H}^i\left(X_s(\mathbb{R}), \mathbb{Z}/2\mathbb{Z}\right) = 0$, a forteriori

$$\mathrm{H}^0(X_s(\mathbb{R}), \mathbb{Z}/2\mathbb{Z}) = \mathbb{Z}/2\mathbb{Z},$$

and $X_s(\mathbb{R})$ is connected and non-empty. This completes the proof of Theorem 229.

A theorem of Witt and Kummer theory. In [Sx10b] Appendix B, a proof of Theorem 229 was given in the case of a smooth projective curve $X/\mathbb{R}$ of genus ≥ 1, which parallels the approach to the period–index result in the p-adic case, see 10.1, and which relies mostly on Kummer theory on X. The argument that proves the surjectivity part is as follows. A section of $\pi_1(X/\mathbb{R})$ implies by [Sx10b] Proposition 12 that the Brauer obstruction map

$$b \;:\; \mathrm{Pic}_X(\mathbb{R}) \to \mathrm{Br}(\mathbb{R})$$

vanishes on torsion and thus vanishes. Consequently, the relative Brauer group

$$\mathrm{Br}(X/\mathbb{R}) = \ker\left(\mathrm{Br}(\mathbb{R}) \to \mathrm{Br}(X)\right)$$

vanishes and thus $X(\mathbb{R})$ is nonempty due to the following theorem of Witt.

Theorem 231 (Witt [Wi34]). *For a smooth, projective curve $X/\mathbb{R}$ the map*

$$\mathrm{Br}(X) \to \mathrm{H}^0\left(X(\mathbb{R}), \mathbb{Z}/2\mathbb{Z}\right)$$

given by pointwise evaluation in $\mathrm{Br}(\mathbb{R}) = \mathbb{Z}/2\mathbb{Z}$ *is an isomorphism.*

A cycle class of degree 1. The most elegant way to enforce the existence of a real point due to the presence of a section was found by Esnault and Wittenberg. Let $X/\mathbb{R}$ be a smooth, projective curve of genus ≥ 1 with a section s of $\pi_1(X/\mathbb{R})$. The mod 2 cycle class $\mathrm{cl}_{s,2} \in \mathrm{H}^2\left(X, \mathbb{Z}/2\mathbb{Z}(1)\right)$ lifts to the mod 4 cycle class $\mathrm{cl}_{s,4}$ and thus by diagram chase in

$$
\begin{array}{ccccccccc}
0 & \longrightarrow & \mathrm{Pic}(X)/4\,\mathrm{Pic}(X) & \xrightarrow{\;c_1\;} & \mathrm{H}^2\left(X, \mathbb{Z}/4\mathbb{Z}(1)\right) & \longrightarrow & \mathrm{Br}(X)[4] & \longrightarrow & 0 \\
& & \downarrow & & \downarrow & & \downarrow{\scriptstyle 2=0} & & \\
0 & \longrightarrow & \mathrm{Pic}(X)/2\,\mathrm{Pic}(X) & \xrightarrow{\;c_1\;} & \mathrm{H}^2\left(X, \mathbb{Z}/2\mathbb{Z}(1)\right) & \longrightarrow & \mathrm{Br}(X)[2] & \longrightarrow & 0
\end{array}
$$

and due to $\mathrm{Br}(X)$ being 2-torsion, shows the existence of a line bundle $\mathscr{L} \in \mathrm{Pic}(X)$ with

$$\mathrm{cl}_{s,2} = c_1(\mathscr{L}).$$

It follows that X contains a divisor of odd degree $\deg(\mathscr{L}) \equiv \deg(\mathrm{cl}_{s,2}) = 1$ modulo 2 the support of which must contain a real point of X.

A 2-step nilpotent version. The approach of Wickelgren yields the following more precise result, of which Theorem 229 is an immediate corollary.

Theorem 232 (Wickelgren [Wg10] Theorem 1.1). *Let $X/\mathbb{R}$ be a smooth projective geometrically connected curve of genus ≥ 1. Then the map*

$$\kappa_{\mathbb{R},2\text{-}nilp} : \pi_0(X(\mathbb{R})) \to \mathrm{im}\left(\mathscr{S}_{C_{\geq -2}\left(\pi_1^{\mathrm{pro}\text{-}2}(X/\mathbb{R})\right)} \to \mathscr{S}_{\pi_1^{\mathrm{ab,pro}\text{-}2}(X/\mathbb{R})}\right)$$

induced by the real profinite Kummer map $\kappa_{\mathbb{R}}$ is a bijection between the set of connected components of real points and the set of abelian sections of $\pi_1^{\mathrm{ab,pro}\text{-}2}(X/\mathbb{R})$ that lift to the next step in the descending pro-2-*central series.*

Remark 233. In [Wg10] Theorem 1.1 we find the unnecessary additional assumption that $X(\mathbb{R}) \neq \emptyset$.

Proof (Sketch of proof). The maximal divisible subgroup $\mathrm{Div}\left(\mathrm{H}^1(\mathbb{R}, \mathrm{Pic}_X^0)\right)$ vanishes, since the group $\mathrm{H}^1(\mathbb{R}, \mathrm{Pic}_X^0)$ is killed by 2, whence by Corollary 177 the Pic_X^0-torsor Pic_X^1 is trivial. It follows that the real curve X has a real divisor of odd degree and thus admits a real point.

It follows from [GrHa81], see [Wg10] Proposition 4.2, that the real profinite Kummer map for Pic_X^1 induces a short exact sequence

$$0 \to \mathrm{H}^1\left(\mathbb{R}, \pi_1^{\mathrm{ab,pro}\text{-}2}(\overline{X})\right) \to \bigoplus_{a \in \pi_0(X(\mathbb{R}))} \mathbb{F}_2 \cdot a \xrightarrow{\Sigma} \mathbb{F}_2 \to 0$$

where $\sum$ is the summation map. We deduce that $\mathscr{S}_{\pi_1^{\mathrm{ab,pro}\text{-}2}(X/\mathbb{R})}$ is an affine space over $\mathbb{F}_2$ freely spanned by the sections s_a for $a \in \pi_0(X(\mathbb{R}))$. In particular, the injectivity part of the real section conjecture holds.

In order to conclude the proof of Theorem 232 we need to compute the boundary map

$$\delta : \mathscr{S}_{\pi_1^{\mathrm{ab,pro}\text{-}2}(X/\mathbb{R})} \to \mathrm{H}^2\left(\mathbb{R}, \mathrm{gr}_{-2}^C \pi_1^{\mathrm{pro}\text{-}2}(\overline{X})\right)$$

explicitly, and then show that

$$\delta^{-1}(0) = \kappa_{\mathbb{R},2\text{-}nilp}\left(\pi_0(X(\mathbb{R}))\right).$$

This has been achieved by Kirsten Wickelgren in [Wg10] §5.   □

A finite nilpotent version? Let $W/\mathbb{R}$ be a torsor under an abelian variety $A/\mathbb{R}$. For $n \in \mathbb{Z}$ we denote by $\pi_1^{\mathrm{mod}\,n}(W/\mathbb{R})$ the mod n quotient extension of $\pi_1(W/\mathbb{R})$ obtained by pushing with the characteristic quotient

$$\pi_1(\overline{W}) \twoheadrightarrow \pi_1(\overline{W})/n\pi_1(\overline{W}).$$

It is not difficult to show that we have a bijection

$$\kappa_{W,2} : \pi_0(W(\mathbb{R})) \xrightarrow{\sim} \text{im}\left(\mathscr{S}_{\pi_1^{\text{mod }4}(W/\mathbb{R})} \to \mathscr{S}_{\pi_1^{\text{mod }2}(W/\mathbb{R})}\right)$$

of the set of connected components of $W(\mathbb{R})$ with the mod 2 sections that lift to mod 4 sections.

Question 234. Let $X/\mathbb{R}$ be a smooth projective geometrically connected curve. Is there a finite characteristic intermediate quotient

$$\pi_1^{\text{pro-}2}(\overline{X}) \xrightarrow{\varphi} G \to \pi_1^{\text{ab}}(\overline{X})/2\pi_1^{\text{ab}}(\overline{X})$$

of size $\#G$ bounded by a function of the genus of X (better: G verbal and independent of X in this verbal definition), such that the map

$$\pi_0(X(\mathbb{R})) \to \text{im}\left(\mathscr{S}_{\varphi_*\left(\pi_1^{\text{pro-}2}(X/\mathbb{R})\right)} \to \mathscr{S}_{\pi_1^{\text{ab,mod }2}(X/\mathbb{R})}\right)$$

induced by $\kappa_{\mathbb{R}}$ is a bijection? Natural candidates are filtration quotients of the 2-central series or of the 2-Zassenhaus filtration of $\pi_1^{\text{pro-}2}(\overline{X})$.

16.2 The *p*-adic Section Conjecture

Let k be a finite extension of $\mathbb{Q}_p$ with $\mathfrak{o}_k$ the ring of p-adic integers in k. For a smooth projective curve X/k of genus at least 2 the profinite Kummer map

$$\kappa : X(k) \to \mathscr{S}_{\pi_1(X/k)}$$

was shown to be injective in Chap. 7, and Proposition 96 shows that it is a continuous map of compact profinite spaces. Despite serious attempts we do not know whether every section in the p-adic case comes from a rational point.

The following result of a *valuative p-adic section conjecture*, which was obtained in joint work with Florian Pop, see [PoSx11], provides by far the most arithmetic description of a p-adic section. The idea is as follows. In the case of an affine curve over any field, not every section must belong to a rational point, because the arithmetic of the section might be localised in the boundary of a compactification. More concretely, the image of the section might be contained in the decomposition group of a rational point at infinity.

In many respects a smooth projective geometrically connected curve X/k over a finite extension $k/\mathbb{Q}_p$ is not complete. It is customary to consider a p-adic curve as the generic fibre of a model $\mathscr{X}$ over $\text{Spec}(\mathfrak{o}_k)$ which contains X as an open dense subset. But such a model is not unique due to for example blow-ups. Therefore the natural boundary of X/k consists in the special fibre of the projective limit

$$\varprojlim \mathscr{X}$$

of all proper flat models $\mathscr{X}/\operatorname{Spec}(\mathfrak{o}_k)$ of X/k. This is nothing but the Riemann–Zariski space of all valuations of the function field of X which extend the p-adic valuation on k.

The result is now the following, see [PoSx11] for details and a proof which in particular involves the period–index results as recalled in Theorem 114.

Theorem 235 (Valuative p-adic section conjecture, [PoSx11] Theorem 37). *Let X be a smooth, hyperbolic, geometrically connected curve over a finite extension k of $\mathbb{Q}_p$.*

Then for any section s of $\pi_1(X/k)$ there exists a valuation $\tilde{w}$ of the function field of the universal pro-étale cover of X which extends the p-adic valuation on k, such that the image $s(\operatorname{Gal}_k)$ is contained in the decomposition subgroup $D_{\tilde{w}|w} \subset \pi_1(X)$.

In view of Theorem 235, the p-adic analogue of the section conjecture now follows if only for a valuation associated to a rational point the associated projection map

$$D_{\tilde{w}|w} \twoheadrightarrow \operatorname{Gal}_k \tag{16.2}$$

splits. Moreover, since the decomposition group $D_{\tilde{w}|w}$ is pro-solvable, Theorem 235 significantly has reduced the group theoretic complexity of the task. Conversely, if this fails and there are other exotic valuations $\tilde{w}$ that through a splitting of (16.2) give rise to sections of $\pi_1(X/k)$, then the p-adic analogue of the section conjecture simply fails for a good arithmetic reason. In this sense Theorem 235 is optimal with respect to a p-adic analogue of the section conjecture.

Question 236. Here are two questions which might further enrich the picture of the space of sections in the p-adic case.

(1) Is the profinite space $\mathscr{S}_{\pi_1(X/K)}$ naturally a (smooth, analytic) p-adic space? This question relates to Minhyong Kim's Selmer varieties, a central notion in the non-abelian Chabauty method. At least for the pro-p part it is plausible to construct Selmer varieties over $\mathbb{Z}_p$ and therefore construct pro-p-adic analytic neighbourhoods of any section in $\mathscr{S}_{\pi_1^{\text{pro-}p}}(X/K)$.

(2) Is $\mathscr{S}_{\pi_1(X/K)}$ either empty or of the same cardinality as $\mathbb{Q}_p$? Note that there are countable profinite spaces.

Chapter 17
Fields of Cohomological Dimension 1

The fields of cohomological dimension 1 are exactly the fields with projective absolute Galois group. The analogue of the weak section conjecture holds for geometrically connected curves over such fields if and only if the field is PAC, see Theorem 238. At first glance, it seems strange to examine the assertions of the section conjecture for varieties over such fields. Among those fields are the finite fields, which were discussed at length in Chap. 15. In Theorem 243, we will construct an infinite algebraic extension of $\mathbb{F}_p$ such that the profinite Kummer map for every smooth projective subvariety of an abelian variety is injective with dense image with respect to the topology from Chap. 4, but never surjective.

Another source of fields with cohomological dimension 1 is given by the maximal cyclotomic extensions of algebraic number fields. A thorough understanding of the assertions of the section conjecture in this case might actually provide valuable insights, see Theorem 247, and may ultimately justify this chapter.

17.1 PAC Fields

A PAC field behaves like an algebraically closed field without necessarily being algebraically closed.

Definition 237. A *pseudo algebraically closed (PAC)* field is a field k such that every smooth geometrically connected curve over k has at least one and therefore at least countably many k-rational points.

The notion of PAC fields was introduced by Ax in [Ax68], who proves that PAC fields are of cohomological dimension ≤ 1, see [FrJa08] Theorem 11.6.2. The following arose in a discussion with Wittenberg.

Theorem 238. *Let k be a field of $\mathrm{cd}(k) \leq 1$. Then the following are equivalent.*

(a) *The weak section conjecture holds for smooth, geometrically connected curves over k.*
(b) *The field k is PAC.*
(c) *The weak section conjecture holds for any smooth, geometrically connected variety X/k.*

Proof. Obviously (b) implies (c) which implies (a). As the field k has $\mathrm{cd}(k) \leq 1$, its absolute Galois group Gal_k is projective and for any smooth geometrically connected curve U/k the extension $\pi_1(U/k)$ splits. If (a) holds this implies that $U(k)$ is non-empty and the field k reveals to be PAC, hence (b) holds.   □

The weak section conjecture being settled for fields of cohomological dimension at most 1, it is nevertheless quite unlikely that there should be a single PAC field over which the actual section conjecture holds. The section conjecture certainly fails for algebraically closed fields. In general we have the following proposition.

Proposition 239. *Let X/k be a geometrically connected variety over a PAC field k. The profinite Kummer map $\kappa : X(k) \to \mathscr{S}_{\pi_1(X/k)}$ has dense image with respect to the pro-discrete topology of Chap. 4 on $\mathscr{S}_{\pi_1(X/k)}$.*

Proof. This follows immediately from the criterion of Proposition 54 (5).   □

Countable large fields. PAC fields are large by intuition but also according to the following definition.

Definition 240. A *large field*[1] is a field k such that every geometrically irreducible variety with a smooth k-rational point has at least one (and thus at least countably many) other smooth k-rational points.

It is a result of Koenigsmann [Ko05] Proposition 3.1, that the section conjecture fails for countable large fields. We provide a proof of Koenigsmann's result in our language but exploit the same idea.

Proposition 241. *Let k be a large field and let X/k be a smooth geometrically connected variety of dimension > 0, such that $\kappa : X(k) \hookrightarrow \mathscr{S}_{\pi_1(X/k)}$ is injective. Then the closure $\overline{\kappa(X(k))}$ and thus $\mathscr{S}_{\pi_1(X/k)}$ is uncountable.*

Proof. Let $\mathscr{S}_n$ be the image of $X(k)$ in $\mathscr{S}_{Q_n(\pi_1(X/k))}$, so that

$$\overline{\kappa(X(k))} = \varprojlim_n \mathscr{S}_n.$$

We will recursively construct a sequence $n_1 < n_2 < \ldots$ and finite subsets $\Sigma_i \subset \mathscr{S}_{n_i}$ such that the projection $\Sigma_i \to \Sigma_{i-1}$ is surjective and all fibres have cardinality 2.

---

[1]The notion of a *large field* is unfortunately known under many synonyms by various authors. Large fields are also called *ample fields*, or *anti-mordellic fields*, or *fertile fields*, or ....

Then $\overline{\kappa(X(k))}$ contains $\Sigma = \varprojlim_i \Sigma_i$ which by a binary code is homeomorphic to $\mathbb{Z}_2$ and thus of uncountable cardinality.

Because $\mathscr{S}_n \to \mathscr{S}_{n-1}$ is surjective by definition, it suffices to solve the following problem. For $n \in \mathbb{N}$ and $a \in X(k)$ we have to find $n' > n$ and $b \in X(k)$ such that (i) the images of s_a and s_b agree in $\mathscr{S}_n$ but (ii) differ in $\mathscr{S}_{n'}$.

Let $X_n(s_a)$ be the neighbourhood of s_a associated to the open subgroup

$$\ker\left(\pi_1(\overline{X}) \twoheadrightarrow Q_n(\pi_1(\overline{X}))\right) \cdot s_a(\mathrm{Gal}_k) \subseteq \pi_1(X),$$

see Definition 46. Condition (i) means that b is in the image of $X_n(s_a)(k) \to X(k)$. As $X_n(s_a)(k)$ contains a lift of a and is smooth, by the field k being large, $X_n(s_a)$ contains at least countably many rational points. Only finitely many of these are mapped to a and we pick $b \in X(k)$ different from a in the image. Since we assume that $[s_a] \neq [s_b]$ we have to find for $n' \gg n$ that the images of a and b in $\mathscr{S}_{n'}$ are distinct by Lemma 44. □

As an immediate corollary we find Koenigsmann's [Ko05] Proposition 3.1.

Corollary 242 (Koenigsmann). *(1) Let k be a countable large field, and let X/k be a geometrically connected smooth variety of dimension > 0. If $X(k)$ is not empty, then the profinite Kummer map $\kappa : X(k) \to \mathscr{S}_{\pi_1(X/k)}$ is not bijective.*

(2) The section conjecture never holds for a geometrically connected variety over a countable PAC field.

Proof. (1) By Proposition 241 the map κ is not injective, or $\mathscr{S}_{\pi_1(X/k)}$ is at least uncountable and thus not in bijection with the countable set $X(k)$. Assertion (2) follows from (1). □

17.2 Infinite Algebraic Extensions of Finite Fields

In this section we will construct a countable PAC field k such that the profinite Kummer map κ is injective for all geometrically connected varieties which inject into an abelian variety. In particular, the map κ is injective for smooth projective curves of positive genus.

By naturality of the map κ it suffices to check injectivity for abelian varieties A/k alone. The extract of the Kummer sequence for an isogeny (15.1) yields a factorisation of κ as

$$A(k) \to \varprojlim_{\varphi : B \to A} A(k)/\varphi(B(k)) \hookrightarrow \mathrm{H}^1(k, \pi_1(\overline{A}))$$

so that

$$\ker(\kappa) = \bigcap_{\varphi:B\to A} \varphi(B(k)),$$

where φ ranges over all étale isogenies. By Theorem 222 for $k = \mathbb{F}_q$ a finite field, this intersection $\bigcap_{\varphi:B\to A} \varphi(B(\mathbb{F}_q))$ vanishes for abelian varieties over $\mathbb{F}_q$. We have to construct an infinite algebraic extension $\mathbb{F}/\mathbb{F}_q$ such that this property persists. We do this recursively as follows. Let

$$(\mathbb{F}_{q_i}, A_i, a_i)_{i\in\mathbb{N}}$$

be a list of all triples $(\mathbb{F}_q, A, a)$ up to isomorphy of a finite field $\mathbb{F}_q$ of characteristic p, an abelian variety A over $\mathbb{F}_q$, and a non-zero rational point $0 \neq a \in A(\mathbb{F}_q)$. We construct recursively for $n \in \mathbb{N}$ a finite field $k_n/\mathbb{F}_p$ and $i_n, d_n \in \mathbb{N}$ and an étale isogeny $\varphi_n : B_n \to A_{i_n}$ such that

  (i) k_n contains $\mathbb{F}_{q_i}$ for $i = i_1, \dots, i_n$
 (ii) Every residue extension above a_{i_n} in the scalar extension of φ_n to k_n is nontrivial and of degree dividing d_n
(iii) $d_{n-1} \mid d_n$
(iv) The degree of k_n/k_{n-1} is prime to d_n

We first choose i_{n+1} as the first unused index, i.e., as the minimum of

$$\{i \in \mathbb{N} \, ; i \neq i_r \text{ for all } r \leq n, \ \mathbb{F}_{q_i} \subseteq k_n\}.$$

Then we pick an isogeny $\varphi_{n+1} : B_{n+1} \to A_{i_{n+1}}$ with $\varphi_{n+1}(B_{n+1}(k_n)) = 0$ which is possible by (15.1) and Theorem 222. For example the corresponding Lang isogeny will do. We then set d_{n+1} as the least common multiple of d_n and the residue degrees above k_n of closed points above $a_{i_{n+1}}$ in the scalar extension of φ_{n+1} to k_n. For the field k_{n+1} we pick a prime number $\ell \nmid d_{n+1}$ and let k_{n+1}/k_n be the unique extension of degree ℓ. Properties (i), (iii) and (iv) are true by construction, while property (ii) for φ_{n+1} holds with k_{n+1} replaced by k_n. As the degree of k_{n+1}/k_n was chosen prime to the *forbidden degree* d_{n+1} the property (ii) for φ_{n+1} persists for the scalar extension to k_{n+1}.

In order to start the recursive construction, we start with $i_1 = 1$ and proceed as above with respect to the values $k_0 = \mathbb{F}_{q_1}$ and $d_0 = 1$.

We set $\mathbb{F} = \bigcup_n k_n$ which by (iv) is an infinite algebraic extension of $\mathbb{F}_p$. Any abelian variety A over $\mathbb{F}$ together with a nontrivial point $a \in A(\mathbb{F})$ is defined as $A_0/\mathbb{F}_0$ together with $a_0 \in A_0(\mathbb{F}_0)$ over some finite subfield $\mathbb{F}_0 \subset \mathbb{F}$. As soon as $\mathbb{F}_0 \subset k_n$ the triple $(\mathbb{F}_0, A_0, a_0)$ is eligible to be picked from the list of all triples in the above construction. The waiting list being finite, i.e., the number of potential picks from the list with smaller index, ultimately there will be an n such that

$$(\mathbb{F}_0, A_0, a_0) \cong (\mathbb{F}_{q_{i_n}}, A_{i_n}, a_{i_n}).$$

The isogeny $\varphi_n : B_n \to A_{i_n}$ has a fibre above a_{i_n} which does not allow an $\mathbb{F}$-rational point because $\mathbb{F}/k_n$ is a limit of finite extensions of degree prime to d_n, a multiple of the residue degrees in this fibre as a scheme over k_n.

Let $\varphi : B \to A$ be the base change of φ_n to $\mathbb{F}$. By the above, the point a represents a nontrivial class in $A(\mathbb{F})/\varphi(B(\mathbb{F}))$. As $a \in A(\mathbb{F})$ was arbitrary, we find that $\bigcap_\varphi \varphi(B(\mathbb{F})) = 0$ and the map

$$\kappa : A(\mathbb{F}) \hookrightarrow \mathrm{H}^1(\mathbb{F}, \pi_1(\overline{A}))$$

is injective, even for every abelian variety $A/\mathbb{F}$.

Theorem 243. *There is a PAC field $\mathbb{F}$ which is infinite algebraic over the finite field $\mathbb{F}_p$ such that for all geometrically connected varieties $X/\mathbb{F}$ of dimension > 0 which inject into an abelian variety the profinite Kummer map*

$$\kappa : X(\mathbb{F}) \hookrightarrow \mathscr{S}_{\pi_1(X/\mathbb{F})}$$

is injective with dense image but not surjective.

Proof. A field which is infinite algebraic over a finite field is PAC because rational points always exist by the Hasse–Weil bound. The field $\mathbb{F}$ as constructed above therefore is PAC and also makes κ injective. The rest follows from Proposition 239 and Corollary 242 (1). □

Remark 244. The field $\mathbb{F}$ constructed for Theorem 243 has absolute Galois group isomorphic to $\hat{\mathbb{Z}}$. If we arrange $q_1 = p$ then there is sequence of prime numbers (ℓ_n) and

$$\mathrm{Gal}(\mathbb{F}/\mathbb{F}_p) = \prod_n \mathbb{Z}/\ell_n\mathbb{Z}.$$

17.3   The Maximal Cyclotomic Extension of a Number Field

Another source of fields of cohomological dimension 1 is provided by certain infinite algebraic extensions of $\mathbb{Q}$.

Lemma 245. *Let $K/\mathbb{Q}$ be an algebraic extension. Then $\mathrm{cd}(K) \leq 1$ if and only if K is totally complex and for every prime p and place $v \mid p$ of K the degree of the completion $K_v/\mathbb{Q}_p$ is divisible by ℓ^∞ for every prime number ℓ (including p).*

Proof. By the local–global principle for the Brauer group of number fields, the condition on the local degrees shows that $\mathrm{Br}(K') = 0$ for every finite extension K'/K. Hence we may conclude by [Se97] II §3.1 Proposition 5. □

Lemma 245 applies to the maximal cyclotomic extensions $k^{\mathrm{cyc}} = k(\mu_\infty)$ of an algebraic number field k, which form a class of fields that behaves even better.

Proposition 246. *Let* $K = k^{\mathrm{cyc}}$ *be the maximal cyclotomic extension of an algebraic number field* k. *For any section* s *of* $\pi_1(X/K)$ *for a smooth, projective curve* X/K *the centraliser* $Z_{\pi_1(\overline{X})}(s)$ *is the trivial group.*

Proof. The proof proceeds as in Proposition 104 and reduces to the claim that for any abelian variety A/K we have $\mathrm{H}^0(K, \mathrm{T}A) = 0$. The restriction to abelian varieties is due to the restriction to projective curves in the statement of the proposition. We compute

$$\mathrm{H}^0(K, \mathrm{T}A) = \varprojlim_n A[n](K) = \varprojlim_n A(K)[n] = 0,$$

as torsion in $A(K)$ is finite by Ribet's theorem [KaLa81] Appendix Theorem 1. □

Theorem 247. *Let* X/k *be a proper, smooth curve of genus at least* 2 *over an algebraic number field* k *with maximal cyclotomic extension* $K = k^{\mathrm{cyc}}$.

(1) The restriction map

$$\mathscr{S}_{\pi_1(X/k)} \xrightarrow{\mathrm{res}} \left(\mathscr{S}_{\pi_1(X_K/K)} \right)^{\mathrm{Gal}(K/k)}$$

is a bijection.

(2) The profinite Kummer map $X(K) \to \mathscr{S}_{\pi_1(X_K/K)}$ *is injective.*

(3) The section conjecture holds for X/k *if and only if any Galois invariant section of* $\pi_1(X_K/K)$ *comes from point of* $X(K)$.

Proof. We abbreviate $\Delta = \mathrm{Gal}(K/k)$. By the non-abelian Hochschild–Serre spectral sequence of Sect. 1.4 we have an exact sequence

$$\mathrm{H}^1(K/k, Z_{\pi_1(\overline{X})}(s)) \to \mathscr{S}_{\pi_1(X/k)} \xrightarrow{\mathrm{res}} \left(\mathscr{S}_{\pi_1(X_K/K)} \right)^{\Delta} \xrightarrow{\delta} \bigsqcup_s \mathrm{Ext}\left(\Delta, Z_{\pi_1(\overline{X})}(s) \right).$$

The vanishing of the centraliser $Z_{\pi_1(\overline{X})}(s)$ for all sections s by Proposition 246 proves (1).

For a finite subextension k'/k of K/k we have $K = (k')^{\mathrm{cyc}}$. By (1) and the known injectivity for the section conjecture over algebraic number fields as in Proposition 75, we find an injective map

$$X(k') \hookrightarrow \mathscr{S}_{\pi_1(X_{k'}/k')} = \left(\mathscr{S}_{\pi_1(X_K/K)} \right)^{\mathrm{Gal}(K/k')} \hookrightarrow \mathscr{S}_{\pi_1(X_K/K)}.$$

In the limit over all k' we deduce (2). Assertion (3) follows immediately from (1) and (2).   □

Chapter 18
Cuspidal Sections and Birational Analogues

As Grothendieck noticed in his letter to Faltings, a k-rational point in the boundary of a good compactification contributes a packet of sections, see [Gr83] page 8, not necessarily accounted for by rational points of the variety. These sections will here be called cuspidal sections. In the general framework, cuspidal sections are best constructed by Deligne's theory of tangential base points, see [De89]. Cuspidal sections have been studied by Nakamura in the profinite setting [Na90a, Na90b, Na91], while Esnault and Hai [EsHa08] discussed the analogue in a tannakian setting.

We will define packets of cuspidal sections and show in Proposition 250 that for hyperbolic curves over arithmetic base fields these packets are either empty or uncountable. Then we describe and extend Nakamura's characterisation of cuspidal sections in terms of cyclotomically normalized pro-cyclic subgroups in order to cover the description foreseen by Grothendieck in his letter.

We will also discuss the birational analogue of the section conjecture when smooth geometrically connected varieties are replaced by their function fields: by passing to the limit over all open subschemes, all sections in this birational setup are predicted to become cuspidal. For curves over finite extensions of $\mathbb{Q}_p$ the birational section conjecture holds by Koenigsmann's Theorem, and we have even a 2-step nilpotent pro-p version due to Pop. Over an algebraic number field there are some partial results, see Theorem 265, and for curves over $\mathbb{Q}$ we even have a full group theoretic description of all birationally liftable sections in Theorem 269 with respect to certain GL_2-representations.

For ease of exposition we will work in characteristic 0 only.

18.1 Construction of Tangential Sections

Let X/k be a geometrically connected variety endowed with a normal crossing divisor $Y = \bigcup Y_\alpha$ with open complement $U = X \setminus Y$.

J. Stix, *Rational Points and Arithmetic of Fundamental Groups*, Lecture Notes in Mathematics 2054, DOI 10.1007/978-3-642-30674-7_18,
© Springer-Verlag Berlin Heidelberg 2013

Tangential data. We consider a point $y \in Y(k)$ and denote by A_y the set of branches of Y passing through y. By abuse of notation we denote an element of A_y by α if the corresponding branch belongs to Y_α. We set

$$(\mathrm{T}_y\, X)^0 = \mathrm{T}_y\, X \setminus \bigcup_{\alpha \in A_y} \mathrm{T}_y\, Y_\alpha,$$

and construct from a k-rational tangent vector $v \in (\mathrm{T}_y\, X)^0(k)$ a section s_v in the packet of sections at y to be defined below. Normal crossing provides the exact sequence of normal bundles

$$0 \to \bigcap_{\alpha \in A_y} \mathrm{T}_y\, Y_\alpha \to \mathrm{T}_y\, X \xrightarrow{\mathrm{pr}} \mathrm{N}_{Y|X,y} = \bigoplus_{\alpha \in A_y} \mathrm{N}_{Y_\alpha|X,y} \to 0$$

where $\mathrm{N}_{Y_\alpha|X,y}$ is the 1-dimensional normal space of the inclusion $Y_\alpha \hookrightarrow X$ in y, and more precisely to the branch corresponding to $\alpha \in A_y$. We set

$$\mathrm{N}^0_{Y_\alpha|X,y} = \mathrm{N}_{Y_\alpha|X,y} - \{0\} \cong \mathbb{G}_{\mathrm{m}}$$

so that the restriction of the projection map pr yields an affine space bundle

$$\mathrm{pr} : (\mathrm{T}_y\, X)^0 \to \mathrm{N}^0_{Y|X,y} = \bigoplus_{\alpha \in A_y} \mathrm{N}^0_{Y_\alpha|X,y}.$$

The section s_v will actually only depend on the normal data $\mathrm{pr}(v) \in \mathrm{N}^0_{Y|X,y}(k)$.

The packet of cuspidal sections. We fix a henselisation $\mathcal{O}^{\mathrm{h}}_{X,y}$ and set

$$X_y = \mathrm{Spec}(\mathcal{O}^{\mathrm{h}}_{X,y}) \tag{18.1}$$

for the étale local scheme at y and

$$U_y = X_y \times_X U \tag{18.2}$$

for the scheme of étale nearby points of U near y. The k-scheme U_y is regular and geometrically connected. Its fundamental group can be computed by Abhyankar's Lemma as a split extension

$$1 \to \mathrm{Hom}\left(\mathcal{O}^*_{U_y}/\mathcal{O}^*_{X_y}, \hat{\mathbb{Z}}(1)\right) \to \pi_1(U_y) \to \mathrm{Gal}_k \to 1. \tag{18.3}$$

A splitting is constructed for any k-rational point of $\mathrm{N}^0_{Y|X,y}$ below.

Definition 248. The *packet of cuspidal sections* at $y \in Y(k)$ of $\pi_1(U/k)$ is defined as the image of the natural map

$$\mathscr{S}_{\pi_1(U_y/k)} \to \mathscr{S}_{\pi_1(U/k)}.$$

A section of the packet is called a *cuspidal section*. The union of all cuspidal sections will be denote by

$$\mathscr{S}^{\mathrm{cusp}}_{\pi_1(U/k)} \subseteq \mathscr{S}_{\pi_1}(U/k).$$

Since $\pi_1(X_y) = \mathrm{Gal}_k$, there is only one section of $\pi_1(X_y/k)$ and the natural commutative diagram

$$
\begin{array}{ccc}
\mathscr{S}_{\pi(U_y/k)} & \longrightarrow & \mathscr{S}_{\pi(U/k)} \\
\downarrow & & \downarrow \\
\mathscr{S}_{\pi(X_y/k)} & \longrightarrow & \mathscr{S}_{\pi(X/k)}
\end{array}
\tag{18.4}
$$

shows that the packet of cuspidal sections at y all map to $s_y \in \mathscr{S}_{\pi_1(X/k)}$. Deligne's construction of tangential base points in [De89] makes use of an equivalence of categories

$$\exp^*_y \ : \ \mathsf{Rev}(U_y) \xrightarrow{\sim} \mathsf{Rev}((T_y\,X)^0)$$

which to a finite étale cover $V \to U_y$ with normalization $W \to X_y$ associates the finite étale cover of

$$(T_y\,X)^0 = \mathrm{Spec}(k[(\mathcal{O}^*_{U_y}/\mathcal{O}^*_{X_y})^{\mathrm{gp}}])$$

given by

$$\coprod_{w \mapsto y} \mathrm{Spec}(\kappa(w)[(\mathcal{O}^*_{V_w}/\mathcal{O}^*_{W,w})^{\mathrm{gp}}]) \to \mathrm{Spec}(k[(\mathcal{O}^*_{U_y}/\mathcal{O}^*_{X_y})^{\mathrm{gp}}]).$$

Here P^{gp} denotes the Grothendieck group associated to a monoid P. It is easy to check by introducing coordinates and by Abhyankar's Lemma that the functor $\exp^*_y$ is indeed an equivalence. We conclude that there is a natural identification

$$\pi_1(U_y/k) = \pi_1((T_y\,X)^0/k)$$

up to conjugation by $\pi_1(\overline{U}_y)$. We can expand (18.4) to a diagram

$$
\begin{array}{ccccccc}
(T_y\,X)^0(k) & \xrightarrow{\ \kappa\ } & \mathscr{S}_{\pi_1((T_y\,X)^0/k)} & \xrightarrow{\ \exp_y\ } & \mathscr{S}_{\pi(U_y/k)} & \longrightarrow & \mathscr{S}_{\pi(U/k)} \\
\ \downarrow{\scriptstyle\mathrm{pr}} & & \ \downarrow{\scriptstyle\mathrm{pr}_*} & & \downarrow & & \downarrow \\
\mathrm{N}^0_{Y|X,y}(k) & \xrightarrow{\ \kappa\ } & \mathscr{S}_{\pi_1(\mathrm{N}^0_{Y|X,y}/k)} & & \mathscr{S}_{\pi(X_y/k)} & \longrightarrow & \mathscr{S}_{\pi(X/k)}
\end{array}
$$

$$\tag{18.5}$$

in which the maps $\exp_y$ and pr_* are bijections.

Proposition 249. *(1) There is a natural map induced by diagram* (18.5)

$$N^0_{Y|X,y}(k) \to \mathrm{im}\left(\mathscr{S}_{\pi_1(U_y/k)} \to \mathscr{S}_{\pi_1(U/k)} \right)$$

which assigns a cuspidal section s_v of the packet at y to a k-rational normal tangent vector v at $y \in Y = X \setminus U$ that is not tangential to any component of Y passing through y.

(2) The map $v \mapsto s_v$ in (1) is compatible with the natural free transitive action of $\prod_{\alpha \in A_y} k^*$ on $N^0_{Y|X,y}(k)$ and the free transitive action of $\prod_{\alpha \in A_y} \widehat{k^*}$ on $\mathscr{S}_{\pi_1(U_y/k)}$ induced by (18.3).

Proof. We recall that $\widehat{k^*} = \mathrm{H}^1\left(k, \hat{\mathbb{Z}}(1)\right)$, see Sect. 5.2. The rest follows as above by introducing coordinates from the model case

$$U = \mathbb{G}_m \subset X = \mathbb{A}^1_k$$

with $y = 0$, and from the compatibility of π_1 with products.   □

Cardinality of a packet of cuspidal sections. The size of a packet of cuspidal sections was examined in [EsHa08] §6 by tannakian methods, and in [Sx08] §4 by profinite methods. The following proposition summarizes the result since for an algebraic number field or p-adic local field the group $\widehat{k^*}$ is of uncountable cardinality equal to the continuum, see [Sx08] §2.

Proposition 250. *Let k be an algebraic number field or a finite extension of $\mathbb{Q}_p$. Let X/k be a smooth projective geometrically connected curve and let $U \subset X$ be a dense open with complement $Y \hookrightarrow X$, which is a hyperbolic curve. The natural map*

$$U(k) \sqcup \bigsqcup_{y \in Y(k)} \mathscr{S}_{\pi_1(U_y/k)} \to \mathscr{S}_{\pi_1(U/k)}$$

induced by the profinite Kummer map is injective. In particular, the packet of cuspidal sections at a k-rational point y of the boundary Y is of uncountable cardinality equal to the continuum.   □

18.2   The Characterisation of Cuspidal Sections for Curves

The characterisation of cuspidal sections among all sections was achieved by Nakamura in [Na90b] §3, see also [Sx08] §5 for an attempt of a generalization.

As before, U/k will be a hyperbolic curve with smooth projective completion X and boundary $Y = X \setminus U$. We consider U equipped with a base point $\bar{\eta} \in U$ constructed from an algebraic closure $\bar{K}$ of its function field K. Following tradition we call the closed points in Y the *cusps* of U. For every prolongation $\tilde{y}|y$ of a cusp $y \in Y$ considered as a place of K to the maximal subextension $\tilde{K}/K$ unramified over U, we find a *decomposition group*

$$D_{\tilde{y}|y} = \mathrm{im}\left(\pi_1(U_y) \to \pi_1(U, \bar{\eta})\right) \qquad (18.6)$$

and an *inertia group*

$$I_{\tilde{y}|y} = D_{\tilde{y}|y} \cap \pi_1(\overline{U}, \bar{\eta}). \qquad (18.7)$$

Here the choice of $\tilde{y}$ determines a path from the base point of U_y to $\bar{\eta}$.

Lemma 251. *(1) The inertia groups to distinct prolongations intersect trivially.*
(2) $D_{\tilde{y}|y}$ is the normalizer of $I_{\tilde{y}|y}$ in $\pi_1(U, \bar{\eta})$.

Proof. [Na90b] 2.3.2 and [Na90b] 2.4. □

Definition 252. (1) The *anabelian weight filtration* of U is the subset

$$W_{-2}(U) = \bigcup_{\tilde{y}} I_{\tilde{y}|y} \subseteq \pi_1(\overline{U}, \bar{\eta})$$

of all inertia elements. The set $W_{-2}(U)$ is profinite, hence compact, and preserved under conjugation by $\pi_1(U, \bar{\eta})$.
(2) A pro-cyclic closed subgroup $I \subset \pi_1(\overline{U}, \bar{\eta})$ is *essentially cyclotomically normalized* if there is a subgroup N of the normalizer $N_{\pi_1(U,\bar{\eta})}(I)$ of I in $\pi_1(U, \bar{\eta})$ which projects to an open subgroup $\mathrm{Gal}_{k'} \subseteq \mathrm{Gal}_k$ and such that the induced action of N on I is via the cyclotomic character of $\mathrm{Gal}_{k'}$.
(3) A pro-cyclic closed subgroup $I \subset \pi_1(\overline{U}, \bar{\eta})$ is *cyclotomically normalized* if it is essentially cyclotomically normalized with $N = N_{\pi_1(U,\bar{\eta})}(I)$.

Theorem 253 (Nakamura, sec [Na90b] Theorem 1.2). *Let k be a number field, possibly of infinite degree over $\mathbb{Q}$, such that $k\mathbb{Q}^{\mathrm{ab}}/\mathbb{Q}^{\mathrm{ab}}$ is finite. Let U/k be a smooth geometrically connected hyperbolic curve with smooth projective completion X, boundary $Y = X \setminus U$ and base point $\bar{\eta}$ as above. Then the following are equivalent for a nontrivial closed subgroup I in $\pi_1(\overline{U}, \bar{\eta})$.*

(a) There is $\tilde{y}$ with $I \cap I_{\tilde{y}|y} \neq 1$.
(b) I is contained in an inertia subgroup $I_{\tilde{y}|y}$.
(c) I is pro-cyclic and cyclotomically normalized.
(d) I is pro-cyclic and essentially cyclotomically normalized.

Proof. If property (a) holds, then for every $\gamma \in I$ we have

$$1 \neq I \cap I_{\tilde{y}|y} \subseteq I_{\tilde{y}|y} \cap \gamma(I_{\tilde{y}|y})\gamma^{-1} = I_{\tilde{y}|y} \cap I_{\gamma(\tilde{y})|y}$$

and by Lemma 251 thus

$$\gamma \in N_{\pi_1(U,\bar{\eta})}(I_{\tilde{y}|y}) \cap \pi_1(\overline{U}, \bar{\eta}) = I_{\tilde{y}|y},$$

so that $I \subseteq I_{\tilde{y}|y}$ and (b) holds.

Let us now assume property (b). As $I_{\tilde{y}|y}$ is pro-cyclic, the group I is also pro-cyclic. If $\gamma \in \pi_1(U)$ normalizes I then $I_{\gamma.\tilde{y}/\gamma.y} = \gamma I_{\tilde{y}|y} \gamma^{-1}$ intersects $I_{\tilde{y}|y}$ nontrivially in I. Hence $\gamma.\tilde{y} = \tilde{y}$ by Lemma 251 and γ lies in

$$\mathrm{D}_{\tilde{y}|y} = N_{\pi_1(U)}(\mathrm{I}_{\tilde{y}|y}).$$

As $\mathrm{I}_{\tilde{y}|y}$ is cyclotomically normalized the same follows for I and (c) holds.

Property (c) trivially implies (d), hence it remains to deduce (a) from property (d). We may replace I by $\mathrm{I} \otimes \mathbb{Z}_\ell \cong \mathbb{Z}_\ell$ for some prime ℓ, and, moreover, argue by contradiction assuming $\mathrm{I} \cap W_{-2}(U) = 1$. We choose a non-empty compact set $C \subset \mathrm{I}$ avoiding 1. The Hausdorff property yields a normal subgroup N of $\pi_1(\overline{U}, \bar{\eta})$ such that CN/N has empty intersection with $W_{-2}(U)N/N$ in $\pi_1(\overline{U}, \bar{\eta})/N$.

The group $N\,\mathrm{I}/N$ is cyclic of order a power of ℓ and the subgroup generated by $W_{-2}(U) \cap N\,\mathrm{I}$ avoids CN/N. Thus $(W_{-2}(U) \cap N\,\mathrm{I})/N$ is contained in the maximal proper subgroup $(\mathrm{I})^\ell N/N \subseteq \mathrm{I}\,N/N$, and the intermediate ℓ-cyclic cover $U'' \to U'$ of hyperbolic curves over $\bar{k}$ corresponding to $N(\mathrm{I})^\ell \subset N\,\mathrm{I}$ is unramified over the corresponding boundary $X' - U'$. Here X' is the normalization of X in U'. For some finite field extension k'/k, such that U' has a model over k' and such that $\mathrm{Gal}_{k'}$ cyclotomically normalizes I, we obtain a nontrivial Galois invariant map

$$\mathbb{Z}_\ell(1) \cong \mathrm{I} \subset \pi_1(U', \bar{\eta}) \twoheadrightarrow \pi_1(X', \bar{\eta}) \to \mathrm{T}_\ell \operatorname{Pic}^0_{X'}.$$

This contradicts Theorem 1 of Ribet's appendix to [KaLa81].   □

Remark 254. We deviate from the original argument of Nakamura only in the last step by exploiting Ribet's theorem instead of a weight argument.

Corollary 255 (Nakamura). *With notation and assumptions as in Theorem 253, a section $s \in \mathscr{S}_{\pi_1(U/k)}$ is cuspidal if and only if the image $s(\mathrm{Gal}_k)$ cyclotomically normalizes a pro-cyclic subgroup of $\pi_1(\overline{U}, \bar{\eta})$.*

Proof. A cuspidal section at the cusp y cyclotomically normalizes $\mathrm{I}_{\tilde{y}|y}$ for some choice of $\tilde{y}$ depending on various base points and paths.

Conversely, let the image of the section s cyclotomically normalize the pro-cyclic subgroup I. Then by Theorem 253 the group I must be contained in some inertia subgroup $\mathrm{I}_{\tilde{y}|y}$. As above, we see that $s(\mathrm{Gal}_k)$ also normalizes $\mathrm{I}_{\tilde{y}|y}$. Thus s factors over the closed subgroup $\pi_1(U_y) \hookrightarrow \pi_1(U, \bar{\eta})$ and so s is cuspidal.   □

18.3   Grothendieck's Letter

Grothendieck in his letter [Gr83] proposes a different characterisation of the set of cuspidal sections. The improvement of Theorem 253 due to Ribet's theorem shows that both descriptions of $\mathscr{S}^{\mathrm{cusp}}_{\pi_1(U/k)}$ actually agree. For an algebraic number field k we denote by Gal_k^0 the kernel of the cyclotomic character

$$\chi : \mathrm{Gal}_k \to \hat{\mathbb{Z}}^*.$$

Proposition 256. *Let k be an algebraic number field and let U/k be a smooth, hyperbolic curve. A section s of $\pi_1(U/k)$ satisfies exactly one of the following properties.*

(a) The section s is cuspidal.

(b) The action of Gal_k^0 via s and conjugation on $\pi_1(\overline{U}, \bar{\eta})$ has only the trivial fixed point, i.e., the centraliser $Z_{\pi_1(\overline{U})}(s(\mathrm{Gal}_k^0))$ consists only of the trivial group.

Proof. If s is cuspidal, say s belongs to the packet at the cusp y, then $s(\mathrm{Gal}_k)$ cyclotomically normalizes the inertia group $I_{\bar{y}|y}$ for a suitable choice of $\bar{y}$. Hence $s(\mathrm{Gal}_k^0)$ centralises $I_{\bar{y}|y}$. So (a) and (b) cannot hold simultaneously.

If property (b) fails, then there is a pro-cyclic group $\mathbf{1} \neq I$ in $Z_{\pi_1(\overline{U})}(s(\mathrm{Gal}_k^0))$. By Theorem 253 applied to the group I and to the restriction of s to Gal_k^0, we find a cusp y with a prolongation $\bar{y}$ such that $I \subset I_{\bar{y}|y}$. Since $s(\mathrm{Gal}_k^0)$ centralises I it follows from the proof of Theorem 253 that $s(\mathrm{Gal}_k^0) \subseteq D_{\bar{y}|y}$. Moreover, for $\sigma \in \mathrm{Gal}_k$, the group $s(\mathrm{Gal}_k^0)$ is normalized by $s(\sigma)$ and thus

$$s(\mathrm{Gal}_k^0) \subseteq s(\sigma)(D_{\bar{y}|y})s(\sigma)^{-1} \cap D_{\bar{y}|y} = D_{\sigma(\bar{y})|y} \cap D_{\bar{y}|y}.$$

By Theorem 247 (2) applied to a cofinal set of neighbourhoods of s, which essentially relies again on [KaLa81], this is only possible if $\sigma(\bar{y}) = \bar{y}$, whence $\sigma \in D_{\bar{y}|y}$ and thus even $s(\mathrm{Gal}_k) \subseteq D_{\bar{y}|y}$. It follows that y is in fact k-rational and s belongs to the cuspidal packet at y. □

Grothendieck states the section conjecture more precisely as follows, see [Gr83] PS and Conjecture 6 in the introduction of these Lecture Notes.

Conjecture 257. *Let k be a field that is finitely generated over $\mathbb{Q}$. For a smooth geometrically connected hyperbolic curve U/k with geometric point $\bar{u} \in U$, the profinite Kummer map induces a bijection*

$$\kappa : U(k) \to \left\{ s \in \mathscr{S}_{\pi_1(U/k)} ; \ Z_{\pi_1(\overline{U}, \bar{u})}(s(\mathrm{Gal}_k^0)) = \mathbf{1} \right\}.$$

Proposition 256 shows that Conjecture 257 is in fact well-posed because the image of the profinite Kummer map satisfies the centraliser condition, and that the complement in the set of all sections is $\mathscr{S}_{\pi_1(U/k)}^{\mathrm{cusp}}$.

18.4 The Birational Section Conjecture

Let X be a smooth geometrically connected variety over a field k. The absolute Galois group of the function field $k(X)$ of X sits in a natural short exact sequence

$$1 \to \mathrm{Gal}_{\bar{k}(\overline{X})} \to \mathrm{Gal}_{k(X)} \to \mathrm{Gal}_k \to 1, \tag{18.8}$$

where $\bar{k}(\overline{X})$ is the function field of $\overline{X} = X \times_k \bar{k}$. Strictly speaking, the extension (18.8) requires the choice of a geometric point $\bar{\eta} \in \overline{X}$ above the geometric point of $\overline{X}$, whence (18.8) is nothing but the extension $\pi_1(\mathrm{Spec}(k(X))/k, \bar{\eta})$ that we abbreviate as $\pi_1(k(X)/k)$. We have

$$\mathrm{Gal}_{k(X)} = \varprojlim_U \pi_1(U, \bar{\eta}), \tag{18.9}$$

where U ranges over all dense open subsets $U \subset X$.

Definition 258. A *birational sections* for X/k is a section of (18.8). The space of $\mathrm{Gal}_{\bar{k}(\overline{X})}$-conjugacy classes of birational sections is $\mathscr{S}_{\pi_1(k(X)/k)}$.

Lemma 259. *The natural maps* $j_U : \mathrm{Spec}(k(X)) \to U$ *induce a bijection*

$$j_* : \mathscr{S}_{\pi_1(k(X)/k)} \xrightarrow{\sim} \varprojlim_U \mathscr{S}_{\pi_1(U/k)}.$$

Proof. If two classes of sections $[s], [t] \in \mathscr{S}_{\pi_1(k(X)/k)}$ with representatives s, t agree in $\mathscr{S}_{\pi_1(U/k)}$ for every U, then the sets

$$C_U = \{\gamma \in \pi_1(\overline{U}, \bar{\eta}) ; \gamma(-)\gamma^{-1} \circ j_{U,*}(s) = j_{U,*}(t)\}$$

form a projective system of non-empty compact sets. By [Bo98] I §9.6 Proposition 8, the projective limit of the C_U is non-empty and any γ in $\varprojlim_U C_U \subset \mathrm{Gal}_{k(X)}$ yields

$$\gamma(-)\gamma^{-1} \circ s = t.$$

For surjectivity of j_* we consider a collection of compatible conjugacy classes

$$([s_U])_U \in \varprojlim_U \mathscr{S}_{\pi_1(U/k)}.$$

The conjugacy classes

$$[s_U] = \{t : \mathrm{Gal}_k \to \pi_1(U) ; t \in [s_U]\}$$

are compact and thus form a non-empty projective system, again by [Bo98] I §9.6 Proposition 8. Any

$$(t_U)_U \in \varprojlim_U [s_U]$$

yields a section

$$s = \varprojlim_U t_U : \mathrm{Gal}_k \to \varprojlim_U \pi_1(U) = \mathrm{Gal}_{k(X)}$$

of $\pi_1(k(X)/k)$ with $[j_{U,*}(s)] = [s_U]$, showing surjectivity of j_*. $\square$

The birational profinite Kummer map. Let now X/k be a smooth projective geometrically connected curve. Let $TX^0 \subset TX$ be the complement of the zero section in the tangent bundle of X. The construction from Sect. 18.1 is compatible under open inclusions and moreover lifts to birational sections to yield a natural diagram

$$
\begin{array}{ccc}
TX^0(k) & \xrightarrow{\kappa_{\mathrm{bir}}} & \mathscr{S}_{\pi_1(k(X)/k)} \\
\Big\downarrow{\mathrm{pr}} & & \Big\downarrow{j_*} \\
X(k) & \xrightarrow{\kappa} & \mathscr{S}_{\pi_1(X/k)}
\end{array}
\tag{18.10}
$$

Conjecture 260 (Birational version of the section conjecture). *Let k be a field that is finitely generated over $\mathbb{Q}$. For a smooth projective geometrically connected curve X/k, any section of $\pi_1(k(X)/k)$ is cuspidal.*

Remark 261. (1) Conjecture 260 says that there is a lift of j_* in (18.10)

$$
\tilde{j}_* : \mathscr{S}_{\pi_1(k(X)/k)} \to X(k)
$$

such that above $a \in X(k)$ the birational profinite Kummer map reads

$$
k^* \cong \mathrm{pr}^{-1}(a) = (T_a X)^0(k) \to \tilde{j}_*^{-1}(a) \cong H^1(k, \hat{\mathbb{Z}}(1)) = \widehat{k^*}
$$

with a compatible action by $k^* \to \widehat{k^*}$.

(2) Conjecture 260 applies to all curves, since birationally all curves are hyperbolic.

(3) A geometrically abelianized version of Conjecture 260 was discussed by Esnault and Wittenberg in [EsWi10] with respect to abelian birational sections in relation to k-rational zero cycles of degree 1.

Known results. Koenigsmann was able to prove the analogue of the birational section conjecture for smooth projective curves over finite extensions $k/\mathbb{Q}_p$. Koenigsmann's proof exploits model theory of p-adic closed fields in an essential way.

Theorem 262 (Koenigsmann [Ko05] Proposition 2.4 (2)). *Let $k/\mathbb{Q}_p$ be a finite extension and let X/k be a smooth projective geometrically connected curve. Then every section of $\pi_1(k(X)/k)$ is cuspidal.*

A higher dimensional analogue proved along the lines of Koenigsmann's theorem is as follows.

Theorem 263 ([Sx12a] Theorem 6). *Let X/k be a geometrically irreducible, normal, proper variety over a finite extension $k/\mathbb{Q}_p$. Let K be the function field* *of X. Then every section of the natural projection*

$$\mathrm{res}_{K/k} : \mathrm{Gal}_K \to \mathrm{Gal}_k$$

has image in the decomposition subgroup $D_{\bar{v}} \subset \mathrm{Gal}_K$ for a unique k-valuation $\bar{v}$ of $\overline{K}$ with residue field of $v = \bar{v}|_K$ equal to k.

In particular, the space $\mathscr{S}_{\pi_1(k(X)/k)}$ decomposes in a disjoint union of non-empty packets associated to each k-valuation v of K with residue field k.

Koenigsmann's theorem was given an algebraic proof by Pop as an easy corollary of the following result. For a group G, we set

$$G' = [G,G]G^p = \ker(G \twoheadrightarrow G^{\mathrm{ab}} \otimes \mathbb{F}_p)$$

and $G'' = (G')'$.

Theorem 264 (Pop [Po10]). *Let $k/\mathbb{Q}_p$ be a finite extension with $\mu_p \subset k$ and let X/k be a smooth projective geometrically connected curve. A section s of*

$$\mathrm{pr}_* : \mathrm{Gal}^{\mathrm{ab}}_{k(X)} \otimes \mathbb{F}_p \to \mathrm{Gal}^{\mathrm{ab}}_k \otimes \mathbb{F}_p$$

is induced by a cuspidal section of $\mathrm{pr}_* : \mathrm{Gal}_{k(X)} \to \mathrm{Gal}_k$ if and only if s lifts to a section of the maximal $\mathbb{Z}/p\mathbb{Z}$ elementary abelian meta-abelian quotients

$$\mathrm{pr}_* : \mathrm{Gal}_{k(X)}/(\mathrm{Gal}_{k(X)})'' \to \mathrm{Gal}_k/(\mathrm{Gal}_k)''.$$

It follows from Theorem 264 that every section of $\pi_1(k(X)/k)$ is congruent to a cuspidal section modulo $[\mathrm{Gal}_{k(X)}, \mathrm{Gal}_{k(X)}]\,\mathrm{Gal}^p_{k(X)}$. Applying this result for all neighbourhoods of a section and using the usual compactness argument we find that every birational section is cuspidal. An algebraic proof of Theorem 262 in case k does not contain μ_p is achieved by a descent form $k(\mu_p)$ to k using Corollary 107.

A first affirmative result in the case of curves over an algebraic number field was obtained by Stoll [St06] §9, see [HaSx12] Theorem 3.3 for a complete account of the following theorem.

Theorem 265 (Stoll, Harari–S.). *Let k be an algebraic number field and let X/k be a smooth projective geometrically connected curve such that there is a nonconstant map $X \to A$ to an abelian variety A/k with finitely many k-rational points and finite Tate–Shafarevich group $\mathrm{III}(A/k)$. Then every section of $\pi_1(k(X)/k)$ is cuspidal: the birational section conjecture is true for such curves X/k.*

For more results similar to Theorem 265 we refer to [Sx12b] Corollary 14 which in particular asks k to be $\mathbb{Q}$ or imaginary quadratic and treats adelic sections, or [Sx12b] Corollary 15 which works for totally real algebraic number fields k and treats birationally adelic sections.

18.5   Birationally Liftable Sections

Let X/k be a smooth, projective curve. It has come to the attention of various authors that a useful version of the section conjecture will restrict the classes of sections of $\pi_1(X/k)$ by imposing conditions coming from group theory, and that are satisfied by all Diophantine sections.

A commonly investigated extra condition is that of being *birationally liftable*, which occurs in the literature under various names and modifications.

Definition 266. A *birationally liftable* section s of $\pi_1(X/k)$ is a section that belongs to the subset

$$\mathscr{S}^{\mathrm{bir}}_{\pi_1(X/k)} = \bigcap_{j:U\subseteq X} \mathrm{im}\left(j_* : \mathscr{S}_{\pi_1(U/k)} \to \mathscr{S}_{\pi_1(X/k)} \right)$$

where U ranges over all dense open subsets of X.

The *cuspidalization problem* whether every section of $\pi_1(X/k)$ is actually birationally liftable is a hard one. Therefore truncated versions of the cuspidalization problem are being considered. For example, in [BoEm12] Borne and Emsalem consider only lifts to the geometrically maximal abelian pro-ℓ extensions. In [Sa10] Saïdi considers a modification, where sections are only required to lift to the maximal cuspidally central intermediate quotient

$$\pi_1(U) \twoheadrightarrow \pi_1^{\mathrm{cc}}(U) \twoheadrightarrow \pi_1(X)$$

for every dense open $U \subset X$. These sections are termed good in [Sa10], see also [Sa12] for a survey. It is not difficult to see that the cohomological obstruction killed by this lifting property is the class

$$c_1(\mathcal{O}_X(D)) \cup \mathrm{cl}_s \in \mathrm{H}^2\left(k, \hat{\mathbb{Z}}(1)\right)$$

with D any divisor with support in $X \setminus U$. If we now assume that k is a finite extension of $\mathbb{Q}_p$, then for such sections it follows by Tate–Lichtenbaum duality that the cycle class cl_s of the section s belongs to the image of

$$c_1 : \mathrm{Pic}(X) \otimes \hat{\mathbb{Z}} \to \mathrm{H}^2\left(k, \hat{\mathbb{Z}}(1)\right)$$

and X must have a k-rational zero cycle of degree 1, so $\mathrm{index}(X) = 1$, see [Sa10].

The role played in the section conjecture by the property of being birationally liftable is explained in the following proposition.

Proposition 267. *Let k be a field that is finitely generated over $\mathbb{Q}$. For a smooth projective geometrically connected curve X/k of genus ≥ 2, we consider the following three assertions.*

(i)   *Every section of $\pi_1(X/k)$ is birationally liftable.*
(ii) *Every birationally liftable section of $\pi_1(X/k)$ lifts to a section of $\pi_1(k(X)/k)$.*
(iii) *The birational section conjecture holds for X/k.*

Then (i)–(iii) for X/k imply the section conjecture for X/k, which conversely implies (i) and (ii). Moreover, if the section conjecture holds for all smooth projective geometrically connected curves over k of genus ≥ 2, then also assertion (iii) follows.

Proof. By Lemma 259, the first map in

$$j_* : \mathscr{S}_{\pi_1(k(X)/k)} \xrightarrow{\sim} \varprojlim_U \mathscr{S}_{\pi_1(U/k)} \to \mathscr{S}^{\mathrm{bir}}_{\pi_1(X/k)} \subseteq \mathscr{S}_{\pi_1(X/k)}$$

is bijective, while the composition j_* is surjective if and only if (i) and (ii) hold. If every section of $\pi_1(X/k)$ comes from a rational point then j_* is surjective and (i) and (ii) follow. If, moreover, the section conjecture holds for all (branched) neighbourhoods $X' \to X$ of a section s of $\pi_1(k(X)/k)$, then the limit argument shows that s takes values in a decomposition group of a k-rational place of k. This proves the converse direction.

Let us assume assertions (i)–(iii) hold for X/k and then show that the section conjecture holds for X/k as well. By the above j_* is surjective, so that any section $s : \mathrm{Gal}_k \to \pi_1(X)$ lifts to a section of $\pi_1(k(X)/k)$, whence by (iii) the lift takes values in a decomposition group of a k-rational point $a \in X(k)$. By functoriality we conclude that $s = s_a$, and the proof is complete.   □

Remark 268. (1) Unfortunately, no nontrivial application of Proposition 267 is known. In the factorization

$$\kappa : X(k) \to j_*\big(\mathscr{S}_{\pi_1(k(X)/k)}\big) \subseteq \mathscr{S}^{\mathrm{bir}}_{\pi_1(X/k)} \subseteq \mathscr{S}_{\pi_1(X/k)}$$

we know in a few cases that the first map is bijective, see Sect. 18.4, but unfortunately the only case where the second or third inclusion is provably a bijection is in the case that has no point, i.e. if $\pi_1(X/k)$ does not split.

(2) The naive limit argument for assertion (ii) of Proposition 267 does not work. The space $\mathscr{S}_{\pi_1(U/k)}$ for an open $U \subset X$ is in general expected to be non-compact, in contrast to the compactness in the projective case, see Proposition 97. Due to Proposition 96 circumstances are different if k is a finite extension of $\mathbb{Q}_p$, or more generally a field k with topologically finitely generated Gal_k. Then $\mathscr{S}_{\pi_1(U/k)}$ is compact and (ii) always holds.

We end by stating a theorem on the birational section conjecture that imposes an additional group theoretic condition, hence may be considered conditional, and works over the base field $k = \mathbb{Q}$. See also [Ho12] for related results.

Theorem 269 ([Sx12b] Theorem C). *Let $X/\mathbb{Q}$ be a hyperbolic curve. A birationally liftable section*

$$s : \mathrm{Gal}_{\mathbb{Q}} \to \pi_1(X)$$

comes from a rational point $a \in X(k)$ *or is cuspidal, if and only if for every open* $U \subseteq X$ *with geometric point* $\bar{u} \in U$, *a lift* $\bar{s} : \mathrm{Gal}_{\mathbb{Q}} \to \pi_1(U, \bar{u})$, *and every family* E/U *of elliptic curves with geometric fibre* $E_{\bar{u}}$ *over* $\bar{u}$ *the associated family of 2-dimensional representations*

$$\rho_{\bar{s}, E/U, \ell} = \rho_{E/U, \ell} \circ \bar{s} : \mathrm{Gal}_{\mathbb{Q}} \to \mathrm{GL}(\mathrm{T}_\ell\, E_{\bar{u}})$$

has one of the following properties:

(a) *There exists a finite set of places* S *independent of* ℓ *such that* $\rho_{\bar{s}, E/U, \ell}$ *is unramified outside* ℓ *and the places in* S.
(b) *There is a character* $\delta : \mathrm{Gal}_{\mathbb{Q}} \to \{\pm 1\}$ *such that for all* ℓ *we have an exact sequence*

$$0 \to \delta\varepsilon \to \rho_{\bar{s}, E/U, \ell} \to \delta \to 0$$

where ε *is the* ℓ-*adic cyclotomic character.*

List of Symbols

The items below are listed by order of appearance. The list is not meant to be exhaustive, but symbols that occur repeatedly and which are potentially nonstandard have been enclosed in the list.

$\bar{k}$	The algebraic closure of the field k, see (2.7)
Gal_k	The absolute Galois group of the field k, see (2.8)
$\overline{X}$	The base change $\overline{X} = X \times_k \bar{k}$ of X to $\bar{k}$, see (2.9)
$\pi_1(X, \bar{x})$	The étale fundamental group of X with base point $\bar{x}$, see Definition 16
$\pi_1(\overline{X}, \bar{x})$	The geometric fundamental group of X, see page xiii
$\pi_1(X/k)$	The fundamental exact sequence associated to the geometrically connected X/k, see Definition 19
s_a	The Diophantine section associated to the rational point a, see Definition 4
$[s]$	The conjugacy class of the section s, see Definition 7
$\mathscr{S}_{\pi_1(X/k)}$	The space of conjugacy classes of sections of $\pi_1(X/k)$; also denoted by $\mathscr{S}_{\pi_1(X/k,\bar{a})}$ when the base point $\bar{a} \in \overline{X}$ is emphasized, see page xiv and Definition 23
κ	The (profinite) Kummer map, see Definition 1
Gal_k^0	The kernel of the cyclotomic character $\mathrm{Gal}_k \to \hat{\mathbb{Z}}^*$, see Conjecture 6
$\mathscr{S}_{\pi_1(X/k)}(k')$	The space of conjugacy classes of sections of $\pi_1(X/k)$ defined over $\mathrm{Gal}_{k'}$ for an extension k'/k, see Definition 27
$\mathscr{S}_{\pi \to \Gamma}$	The space of sections of $1 \to N \to \pi \to \Gamma \to 1$ up to N-conjugacy, see Definition 7
$\mathscr{S}_{N \rtimes \Gamma}$	The pointed space of sections of $N \rtimes \Gamma \twoheadrightarrow \Gamma$ up to conjugation by N, with the canonical splitting as the special element, see Definition 7
$\mathrm{Tors}_\Gamma(N)$	The pointed set of Γ-equivariant right N-torsors, see page 4
$\delta(t, s)$	The difference cocycle of two sections s and t, see (1.2)

J. Stix, *Rational Points and Arithmetic of Fundamental Groups*, Lecture Notes in Mathematics 2054, DOI 10.1007/978-3-642-30674-7,
© Springer-Verlag Berlin Heidelberg 2013

$\mathrm{Ext}(\Gamma, G')$	The set of extensions $1 \to G' \to E \to \Gamma \to 1$ up to isomorphy, see Definition 9
$[\mathrm{Ext}(\Gamma, G')]_G$	The orbit space of $\mathrm{Ext}(\Gamma, G')$ by the G-action by pushout with conjugation where $G' \trianglelefteq G$ is a normal subgroup, see Definition 9
$Z_{\overline{\pi}}(s_0)$	The centraliser of (the image of) a section s_0 in $\overline{\pi}$, see Definition 11
$\mathsf{Rev}(X)$	The Galois category of finite étale covers of X, see (2.1)
$Y[\bar{a}]$	The fibre in $\bar{a}$ of the object Y, see (2.2)
$P_{\bar{a}}$	The path space pro-representing the fibre functor $\bar{a}$, see (2.3)
$\mathsf{Pro{-}Rev}(X)$	The pro-category associated to $\mathsf{Rev}(X)$, see (2.3)
G^{opp}	The opposite group to G with the same elements but composition reversed, see page 14
$\Pi_1(X)$	The fundamental groupoid of fibre functors of $\mathsf{Rev}(X)$, see Definition 16
$\pi_1(X; \bar{a}, \bar{b})$	The set of étale paths from $\bar{a}$ to $\bar{b}$, see Definition 16
k^{sep}	The separable closure of the field k, see (2.7)
$\pi_1(X/k, \bar{a})$	The fundamental exact sequence associated to the geometrically connected X/k with base point $\bar{a} \in \overline{X}$ emphasized, see Definition 19
Γ^{ab}	The abelianization of the profinite group Γ, see (3.1)
$\pi_1^{\mathrm{ab}}(X/k)$	The maximal geometrically abelian quotient extension of the fundamental group extension $\pi_1(X/k)$, see Definition 26
s^{ab}	The abelianization of the section s, see Definition 26
Alb_X	The Albanese variety of X/k, see (3.2) and (7.1)
$\alpha_X : X \to \mathrm{Alb}_X^1$	The Albanese torsor map of X/k, see (3.3) and (7.1)
$\mathrm{res}_{K/k}$	The restriction map $\mathrm{Gal}_K \to \mathrm{Gal}_k$ for a field extension K/k, see (3.4)
$s_K = s \otimes K$	The base change of the section s with respect to the field extension K/k, see Definition 27
π_s	The anabelian fibre in a section s with respect to a fibration, see Definition 30
$h^{-1}(s)$	The anabelian fibre in a sections s along a finite étale map $h : Y \to X$, see Definition 32
$\underline{M}$	The étale sheaf of sets on $\mathrm{Spec}(k)_{\mathrm{ét}}$ associated to a finite Gal_k-set M, see Definition 32
$\mathrm{Ext}\,[G]$	The category of (continuous) extensions $1 \to N \to E \to G \to 1$ of a pro-finite group G, see Definition 37
$\mathrm{Ind}_H^G(-)$	The non-abelian induction, see page 34
$R_{L|K}\,X$	The Weil restriction of scalars of the variety X/L, see (3.7)
$\mathrm{Sub}(\pi)$	The space of closed subgroups of π, see Section 4.1
$(\mathbb{N}, <)$	The set $\mathbb{N}$ as an ordered set with respect to $<$, see page 38
$Q_n(\overline{\pi})$	The characteristic quotient of $\overline{\pi}$ related to finite quotients of order $\le n$, see (4.1)

$Q_n(\pi_1(X/k))$	The characteristic quotient extension of $\pi_1(X/k)$ related to finite quotients of $\pi_1(\overline{X})$ of order $\leq n$, see (4.2)
X_s	The decomposition tower of a section s of $\pi_1(X/k)$, see Definition 51
κ_f	The Kummer torsor associated to a unit f, see (5.1)
ev_s	The evaluation map of units in sections, see Definition 56 and Definition 57
$\mathbb{N}'$	The partially ordered set of all $n \in \mathbb{N}$ prime to the characteristic of the base field, ordered by divisibility, see (5.2)
$\widehat{k^*}$	The pro-$\mathbb{N}$ completion $\varprojlim_{n \in \mathbb{N}'} k^*/(k^*)^n$ of k^*, see (5.3)
$\widehat{\mathbb{G}}_{\mathrm{m},k}$	The pro-$\mathbb{N}$ completed sheaf $\mathbb{G}_\mathrm{m}$, see (5.4)
$\mathrm{Ab}_{\mathrm{ML}}^{\mathbb{N}'}$	The category of $\mathbb{N}'$-systems of abelian groups localised at Mittag–Leffler-zero objects, see page 48
$T'(M)$	The prime to p Tate module of an abelian group M, see (5.5)
$\mathrm{cl}_s^{\mathrm{graph}}$	The cycle class of a section as defined via graphs, see (6.1)
cl_Z	The cohomological cycle class associated to a cycle Z, see Section 6.1
$\mathrm{cl}_s^{\mathrm{norm}}$	The cycle class of a section as defined by norm compatibility, see (6.3)
$\mathrm{cl}_s^{\mathrm{group}}$	The cycle class of a section as defined by a group extension, see (6.4)
$\mathrm{cl}_s^{\mathrm{dual}}$	The cycle class of a section as defined by duality, see (6.6)
cl_s	The cycle class of a section s, see Definition 62
$\mathrm{W}_{-1}\,\pi_1^{\mathrm{nh}}(\overline{U})$	The weight -1 quotient of $\pi_1^{\mathrm{ab}}(\overline{U})$, see Definition 68
$\mathrm{W}_{-1}\,\pi_1^{\mathrm{ab}}(U/k)$	The weight -1 quotient extension of $\pi_1(X/k)$, see Definition 68
NS_X	The Néron–Severi group scheme of X/k, see (7.2)
G^D	The Cartier dual of a finite flat group scheme G, see page 70
$G^{\mathrm{\acute{e}t}}$	The maximal étale quotient group scheme of a finite flat group scheme G, see page 70
δ_{kum}	The boundary map in the Galois cohomology of the Kummer sequence, see Corollary 71
$\pi_1^{\mathrm{pro}\text{-}p}(X/k)$	The maximal geometrically pro-p quotient extension of the fundamental group extension $\pi_1(X/k)$, see (7.7)
$\pi_1^{\mathrm{pro}\text{-}p}(\overline{X})$	The maximal pro-p quotient of $\pi_1(\overline{X})$, see (7.7)
$\pi_1^{(\mathrm{pro}\text{-}p)}(X)$	The maximal geometrically pro-p quotient of the fundamental group $\pi_1(X)$, see (7.7)
$\rho_{X/k}$	The outer (pro-p) Galois representation associated to X/k, see (7.8)
sp	The specialisation map of fundamental groups, see (8.1)
$R_{\mathscr{X}/S}$	The cokernel of the geometric specialisation map, see (8.3)
$\mathrm{ram}(s)$	The ramification of a section, see Definition 82
$\mathscr{S}_{\pi_1(X/k)}^{\mathrm{nr}}$	The space of unramified sections, see Definition 83
sp_σ	The σ-specialisation map, see (8.6)

$\mathrm{ram}_s^{\mathrm{ab},\ell}$	The pro-ℓ abelianized ramification of a section, see (8.7)
k_v	The completion of an algebraic number field at the place v, see page 107
$\mathrm{index}(X)$	The index of the variety X, see Definition 113
$\mathrm{period}(X)$	The period of the variety X, see Definition 113
$\mathrm{Br}(X/k)$	The relative Brauer group, Definition 113
BS_A	The Brauer–Severi variety associated to the Azumaya algebra A, see (10.1)
$\psi_p(s)$	a $\mathbb{Q}_p$-linear form on the Lie algebra $\mathrm{Lie}(\mathrm{Pic}_X^0)$ induced by a section s, see (10.6)
$\mathrm{Br}^0(X/k)$	Pic^0-part of relative Brauer group, Definition 121
$\mathbb{A}_k$	The adele ring of a number field k, see page 120
$X(\mathbb{A}_k)_\bullet$	The space of modified adelic points of X, see (11.1)
$X(\mathbb{A}_k)_\bullet^{\mathrm{Br}}$	The Brauer kernel, see (11.3)
$\mathscr{S}_{\pi_1(X/k)}(\mathbb{A}_k)$	The space of adelic sections, see Definition 123
$\mathscr{S}_{\pi_1(X/k)}(\mathbb{A}_k)^{\mathrm{Br}}$	The Brauer kernel, see Theorem 127
$X(\mathbb{A}_k)_\bullet^h$	The descent obstruction set induced by a torsor h, see (11.6)
$X(\mathbb{A}_k)_\bullet^{\mathrm{descent}}$	The descent obstruction set, see (11.7)
$\mathscr{S}_{\pi_1(X/k)}(\mathbb{A}_k)^\varphi$	The descent obstruction posed by the torsor φ on adelic sections, see Definition 137
$\mathscr{S}_{\pi_1(X/k)}(\mathbb{A}_k)^{\mathrm{f\text{-}descent}}$	The finite descent obstruction set on adelic sections, see Theorem 138
$\mathscr{S}_{\pi_1(X/k)}(\mathbb{A}_k)^{\mathrm{cf\text{-}descent}}$	The constant finite descent obstruction set on adelic sections, see Definition 143
$\mathscr{S}_{\pi_1(X/k)}(\mathbb{A}_k)^{\mathrm{\acute{e}t\text{-}Br}}$	The étale Brauer–Manin obstruction to adelic sections, see Definition 149
$\mathrm{H}^2(k,(\overline{G},\rho))$	The non-abelian second cohomology set with coefficients in the k-kernel, see page 147
$\mathrm{H}_{\mathrm{nt}}^2(k,(\overline{G},\rho))$	The subset of $\mathrm{H}^2(k,(\overline{G},\rho))$ of neutral classes, see page 148
$\mathrm{H}^1(k,\mathscr{G})$	The non-abelian first cohomology set with coefficients in a k-gerbe $\mathscr{G}$, see (12.1)
$\mathrm{H}_{\mathrm{c}}^2(k,(\overline{G},\rho))$	The non-abelian compactly supported second cohomology with coefficients in a k-kernel, see Definition 164
$\mathrm{H}_{\mathrm{c}}^1(k,\mathscr{G})$	The non-abelian compactly supported second cohomology with coefficients in a k-gerbe, see Definition 167
$\mathrm{Div}(M)$	The maximal divisible subgroup of M, see (13.4)
$\mathrm{div}(M)$	The subgroup of divisible elements of an abelian group M, see (13.5)
G^{sc}	The universal finite étale (isogeny) cover of the semisimple algebraic group G by a simply connected semisimple algebraic group, see (13.7)
G_{lin}	The maximal connected linear algebraic subgroup of a connected algebraic group, see (13.8)
$R(G)$	The radical $R(G)$ of G_{lin}, see (13.8)
$R_{\mathrm{u}}(G)$	The unipotent radical $R_{\mathrm{u}}(G)$ of G_{lin}, see (13.8)

$\mathrm{Div}^0_{\overline{X},\overline{Y}}$	The group of divisors of degree 0 on $\overline{X}$ supported on $\overline{Y}$, see page 170
$X^{(d)}$	The dth symmetric product of X, see page 171
$\pi_1^{\mathrm{nilp}}(\overline{X})$	The maximal pro-nilpotent quotient of $\pi_1(\overline{X})$, see (14.1)
$\pi_1^{\mathrm{nilp}}(X/k)$	The maximal geometrically pro-nilpotent quotient extension of $\pi_1(X/k)$, see (14.2)
κ_{nilp}	The nilpotent Kummer map, see (14.3) and Definition 196
κ_p	The pro-p Kummer map, see (14.3)
$C_\bullet\Gamma$	The descending central filtration on the profinite group Γ, see Definition 194
$C_{\geq -n}\big(\pi_1(X/k)\big)$	The geometrically n-step nilpotent quotient extension of the fundamental group extension $\pi_1(X/k)$, see Definition 195
$C_{\geq -n}\big(\pi_1^{\mathrm{pro}\text{-}\ell}(X/k)\big)$	The geometrically n-step pro-ℓ nilpotent quotient extension of $\pi_1(X/k)$, see Definition 195
κ^{ab}	The abelian Kummer map, see Definition 196
κ_n	The n-step nilpotent Kummer map, see Definition 196
κ_ℓ	The pro-ℓ Kummer map, see Definition 196
$\kappa_{\ell,n}$	The n-step nilpotent pro-ℓ Kummer map, see Definition 196
$\overline{\pi}$	an abbreviation for $\pi_1(\overline{X})$, see (14.4)
$\overline{\pi}^\ell$	an abbreviation for $\pi_1^{\mathrm{pro}\text{-}\ell}(\overline{X})$, see (14.4)
$\mathrm{Lie}(\Gamma)$	The graded $\mathbb{Z}_\ell$-Lie algebra associated to the descending central filtration on a pro-ℓ group Γ, see (14.6)
$\mathfrak{p}$	The graded $\mathbb{Z}_\ell$-Lie algebra associated to $\pi_1^{\mathrm{pro}\text{-}\ell}(\overline{X})$ for a smooth projective curve X/k, see (14.7)
$\mathfrak{p}_n$	The $-n$th graded piece of $\mathfrak{p}$, see page 180
$\mathfrak{p}_K$	The scalar extension to a K-Lie algebra, see (14.8)
$[\mathfrak{p}]$	The Poincaré series of $\mathfrak{p}_K$, see (14.9)
h	The size of the Picard group over a finite field, see 15.2
$\kappa_{\mathbb{R}}$	The version of the Kummer map over $\mathbb{R}$, see (16.1)
k^{cyc}	The maximal cyclotomic extensions $k(\mu_\infty)$ of an algebraic number field k, see page 217
X_y	The étale local scheme $\mathrm{Spec}(\mathcal{O}^{\mathrm{h}}_{X,y})$ of the henselisation of the local ring for a point $y \in X$, see (18.1)
U_y	The scheme of étale nearby points $X_y \times_X U$, see (18.2)
$\mathscr{S}^{\mathrm{cusp}}_{\pi_1(U/k)}$	The space of cuspidal sections, see Definition 248
s_v	The tangential section associated to a tangent vector v, see Proposition 249
$\mathrm{D}_{\bar{y}\|y}$	The decomposition subgroup, see (18.6)
$\mathrm{I}_{\bar{y}\|y}$	The inertia subgroup, see (18.7)
$W_{-2}(U)$	The anabelian weight filtration of U, see Definition 252
$\mathscr{S}_{\pi_1(k(X)/k)}$	The space of birational sections for X/k, see Definition 258
κ_{bir}	The birational Kummer map, see (18.10)
$\mathscr{S}^{\mathrm{bir}}_{\pi_1(X/k)}$	The space of birationally liftable sections, see Definition 266

References

[Ar71]   Arakelov, S.J.: Families of algebraic curves with fixed degeneracies. Izv. Akad. Nauk
          SSSR Ser. Mat. **35**, 1269–1293 (1971)
[AN95]   Asada, M., Nakamura, H.: On graded quotient modules of mapping class groups of
          surfaces. Israel J. Math. **90**, 93–113 (1995)
[Ax68]   Ax, J.: The elementary theory of finite fields. Ann. Math. **88**, 239–271 (1968)
[BaPa95] Bayer-Fluckiger, E., Parimala, R.: Galois cohomology of the classical groups over fields
          of cohomological dimension ≤ 2. Invent. Math. **122**, 195–229 (1995)
[Be84]   Benson, D.J.: Lambda and psi operations on Green rings. J. Algebra **87**, 360–367
          (1984)
[Bl79]   Bloch, S.: Torsion algebraic cycles and a theorem of Roitman. Compos. Math. **39**, 107–
          127 (1979)
[BLR90]   Bosch, S., Lütkebohmert, W., Raynaud, M.: Néron Models. Ergebnisse der Mathematik
          und ihre Grenzgebiete, vol. 21, x + 325 pp. Springer, Berlin (1990)
[Bo98]   Bourbaki, N.: Commutative algebra. Chapters 1–7, reprint. Elements of Mathematics,
          xxiv + 625 pp. Springer, New York (1998)
[BoEm12] Borne, N., Emsalem, M.: Un critère d'épointage des sections ℓ-adiques, preprint. arXiv:
          1201.4589v1 [math.NT] (January 2012)
[BoKa72] Bousfield, A.K., Kan, D.M.: Homotopy limits, completions and localizations. Lecture
          Notes in Mathematics, vol. 304, vi + 348 pp. Springer, Berlin (1972)
[BLM84]   Bremner, A., Lewis, D.J., Morton, P.: Some varieties with points only in a field
          extension. Arch. Math. **43**, 344–350 (1984)
[Br44]   Brandt, A.J.: The free Lie ring and Lie representations of the full linear group. Trans.
          Am. Math. Soc. **56**, 528–536 (1944)
[BrSz12] Brion, M., Szamuely, T.: Prime-to-p étale covers of algebraic groups, preprint. arXiv:
          1109.2802v3 [math.AG] (March 2012)
[By03]   Bryant, R.M.: Free Lie algebras and Adams operations. J. Lond. Math. Soc. (2) **68**,
          355–370 (2003)
[Ca91]   Carlsson, G.: Equivariant stable homotopy and Sullivan's conjecture. Invent. Math. **103**,
          497–525 (1991)
[Ch89]   Chernousov, V.: The Hasse principle for groups of type E_8. Dokl. Akad. Nauk. SSSR
          306, 1059–1063 (1989) (trans: Math. USSR-Izv. **34**, 409–423 (1990))
[CiSx11] Çiperiani, M., Stix, J.: Weil–Châtelet divisible elements of Tate–Shafarevich groups.
          arXiv: 1106.4255v1 [math.NT] (March 2011)
[CiSx12a] Çiperiani, M., Stix, J.: Weil–Châtelet divisible elements in Tate–Shafarevich groups I:
          The Bashmakov problem for elliptic curves over $\mathbb{Q}$, to appear in *Compositio Math.*

[CiSx12b] Çiperiani, M., Stix, J.: Galois sections for abelian varieties over number fields, preprint, 2012

[Co90]   Coleman, R.: Manin's proof of the Mordell conjecture over function fields. Enseign. Math. **36**, 393–427 (1990)

[Cx79]   Cox, D.A.: The étale homotopy type of varieties over $\mathbb{R}$. Proc. Am. Math. Soc. **76**, 17–22 (1979)

[De01]   Debarre, O.: Higher-Dimensional Algebraic Geometry, Universitext, xiv + 233 pp. Springer, Berlin (2001)

[De89]   Deligne, P.: Le groupe fondamental de la droite projective moins trois points. In: Galois groups over $\mathbb{Q}$ (Berkeley, CA, 1987). Mathematical Sciences Research Institute Publications, vol. 16, pp. 79–297. Springer, Berlin (1989)

[De09]   Demarche, C.: Obstruction de descente et obstruction de Brauer–Manin étale. Algebra Number Theor. **3**, 237–254 (2009)

[DMN89]   Dwyer, W., Miller, H., Neisendorfer, J.: Fibrewise completion and unstable Adams spectral sequences. Israel J. Math. **66**, 160–178 (1989)

[EMcL47] Eilenberg, S., MacLane, S.: Cohomology theory in abstract groups II. Group extensions with a non-abelian kernel. Ann. Math. **48**, 326–341 (1947)

[ErSch06] Eriksson, D., Scharaschkin, V.: On the Brauer–Manin obstruction for zero-cycles on curves. Acta Arith. **135**, 99–110 (2008)

[EsHa08] Esnault, H., Hai, Ph.H.: Packets in Grothendieck's section conjecture. Adv. Math. **218**, 395–416 (2008)

[EsWi09] Esnault, H., Wittenberg, O.: Remarks on the pronilpotent completion of the fundamental group. Mosc. Math. J. **9**, 451–467 (2009)

[EsWi10] Esnault, H., Wittenberg, O.: On abelian birational sections in characteristic 0. J. Am. Math. Soc. **23**, 713–724 (2010)

[Fa83]   Faltings, G.: Endlichkeitssätze für abelsche Varietäten über Zahlkörpern. Invent. Math. **73**, 349–366 (1983)

[Fa98]   Faltings, G.: Curves and their fundamental groups [following Grothendieck, Tamagawa and Mochizuki]. Séminaire Bourbaki, vol. 840, (1997/1998). Astérisque **252**, 131–150 (1998)

[FrJa08] Fried, M.D., Jarden, M.: Field Arithmetic, 3rd edn, revised by M. Jarden. Ergebnisse der Mathematik und ihrer Grenzgebiete, vol. 11, xxiv + 792 pp. Springer, Berlin (2008)

[Gi10]   Gille, Ph.: Serre's conjecture II: A survey. In: Garibaldi, S., Sujatha, R., Suresh, V. (eds.) Quadratic Forms, Linear Algebraic Groups, and Cohomology. Developments in Mathematics, vol. 18, pp. 41–56. Springer, Berlin (2010)

[GiSz06] Gille, Ph., Szamuely, T.: Central simple algebras and Galois cohomology. In: Cambridge Studies in Advanced Mathematics, vol. 101, xii + 343 pp. Cambridge University Press, Cambridge (2006)

[GrHa81] Gross, B.H., Harris, J.: Real algebraic curves. Ann. Sci. École Norm. Sup. (4) **14**, 157–182 (1981)

[GrRo78] Gross, B.H., Rohrlich, D.E.: Some results on the Mordell–Weil group of the Jacobian of the Fermat curve. Invent. Math. **44**, 201–224 (1978)

[Gr83]   Grothendieck, A.: Brief an Faltings (27/06/1983). In: Schneps, L., Lochak, P. (eds.) Geometric Galois Action 1. LMS Lecture Notes, vol. 242, pp. 49–58. Cambridge (1997)

[Ha11a]   Hain, R.: Remarks on Non-Abelian Cohomology of Proalgebraic Groups. J. Algebraic Geom. arXiv: 1009.3662v2 [math.AG] (2011)

[Ha11b]   Hain, R.: Rational points of universal curves. J. Am. Math. Soc. **2**, 709–769 (2011)

[Ha59]   Hall, M. Jr.: The Theory of Groups, xiii + 434 pp. Macmillan, New York (1959)

[HaSk02] Harari, D., Skorobogatov, A.N.: Non-abelian cohomology and rational points. Compos. Math. **130**, 241–273 (2002)

[HaSx12] Harari, D., Stix, J.: Descent obstruction and fundamental exact sequence. In: Stix, J. (ed.) The Arithmetic of Fundamental Groups, PIA 2010. Contributions in Mathematical and Computational Science, vol. 2, pp. 147–166. Springer, Berlin (2012)

[HaSz05] Harari, D., Szamuely, T.: Arithmetic duality theorems for 1-motives. J. Reine Angew. Math. **578**, 93–128 (2005)

[HaSz09] Harari, D., Szamuely, T.: Galois sections for abelianized fundamental groups, with an appendix by E.V. Flynn. Math. Annalen **344**, 779–800 (2009)

[Ha65-75] Harder, G.: Über die Galoiskohomologie halbeinfacher Matrizengruppen I. Math. Zeit. **90**, 404–428 (1965) Part II; Math. Zeit. **92**, 396–415 (1966) Part III; J. Reine Angew. Math. **274/5**, 125–138 (1975)

[HaSch10] Harpaz, Y., Schlank, T.M.: Homotopy obstructions to rational points, preprint. arXiv: 1110.0164v1 [math.AG] (October 2011)

[Hi1902] Hilbert, D.: Mathematical problems. Bull. Am. Math. Soc. **8**, 437–479 (1902)

[Ho09]   Hoshi, Y.: Absolute anabelian cuspidalizations of configuration spaces of proper hyperbolic curves over finite fields. Publ. Res. Inst. Math. Sci. **45**, 661–744 (2009)

[Ho10]   Hoshi, Y.: Existence of nongeometric pro-p Galois sections of hyperbolic curves. Publ. RIMS Kyoto Univ. **46**, 829–848 (2010)

[Ho12]   Hoshi, Y.: Conditional results on the birational section conjecture over small number fields, preprint, RIMS-1742 (February 2012)

[Jo1872] Jordan, C.: Recherches sur les substitutions, J. Liouville **17**, 351–367 (1872)

[Ja88]   Jannsen, U.: Continuous étale cohomology. Math. Annalen **280**, 207–245 (1988)

[KaLa81] Katz, N., Lang, S.: Finiteness theorems in geometric classfield theory, with an appendix by K.A. Ribet. Enseign. Math. (2) **27**, 285–319 (1981)

[Ki05]   Kim, M.: The motivic fundamental group of $\mathbb{P}^1 \setminus \{0, 1, \infty\}$ and the theorem of Siegel. Invent. Math. **161**, 629–656 (2005)

[Ki12]   Kim, M.: Remark on fundamental groups and effective Diophantine methods for hyperbolic curves. In: Goldfeld, D., Jorgenson, J., Jones, P., Ramakrishnan, D., Ribet, K., Tate, J. (eds.) Number Theory, Analysis and Geometry, in Memory of Serge Lang, pp. 355–368. Springer, Berlin (2012)

[KiTa08] Kim, M., Tamagawa, A.: The ℓ-component of the unipotent Albanese map. Math. Ann. **340**, 223–235 (2008)

[Ki09]   Kings, G.: A note on polylogarithms on curves and abelian schemes. Math. Zeit. **262**, 527–537 (2009)

[Kn65]   Kneser, M.: Galoiskohomologie halbeinfacher algebraischer Gruppen über p-adischen Körpern I. Math. Zeit. **88**, 40–47 (1965) Part II; Math. Zeit. **89**, 250–272 (1965)

[Kn69]   Kneser, M.: Lectures on Galois Cohomology of Classical Groups, ii + 158 pp. Tata Institute, Bombay (1969)

[Ko05]   Koenigsmann, J.: On the 'section conjecture' in anabelian geometry. J. Reine Angew. Math. **588**, 221–235 (2005)

[Ko95]   Kollár, J.: Shafarevich Maps and Automorphic Forms, M.B. Porter Lectures, x + 201 pp. Princeton University Press, Princeton (1995)

[KPR86]   Kuhlmann, F.-V., Pank, M., Roquette, P.: Immediate and purely wild extensions of valued fields. Manuscripta Math. **55**, 39–67 (1986)

[La66]   Labute, J.: Demuškin groups of rank $\aleph_0$. Bull. Soc. Math. France **94**, 211–244 (1966)

[La67]   Labute, J.: Algèbres de Lie et pro-p-groupes définis par une seule relation. Invent. Math. **4**, 142–158 (1967)

[La70]   Labute, J.: On the descending central series of groups with a single defining relation. J. Algebra **14**, 16–23 (1970)

[La94]   Labute, J.: Groups and lie algebras: The Magnus theory. In: The Mathematical Legacy of Wilhelm Magnus: Groups, Geometry and Special Functions. Contemporary Mathematics, vol. 169, pp. 397–406. American Mathematical Society, Providence (1994)

[LMD90]   Lachaud, G., Martin-Deschamps, M.: Nombre de points des jacobiennes sur un corps fini. Acta Arithmetica **56**, 329–340 (1990)

[LS57]   Lang, S., Serre, J.-P.: Sur les revêtements non ramifiés des variétés algébriques. Am. J. Math. **79**, 319–330 (1957)

[La92]   Lannes, J.: Sur les espaces fonctionnels dont la source est le classifiant d'un p-groupe
          abélian élementaire. Publ. Math. IHES **75**, 135–244 (1992)
[Li69]   Lichtenbaum, S.: Duality theorems for curves over p-adic fields. Invent. Math. **7**, 120–
          136 (1969)
[Li40]   Lind, C.-E.: Untersuchungen über die rationalen Punkte der ebenen kubischen Kurven
          vom Geschlecht Eins. Thesis, 97 pp. University of Uppsala (1940)
[McR71]   MacRae, R.E.: On unique factorization in certain rings of algebraic functions. J.
          Algebra **17**, 243–261 (1971)
[Ma71]   Manin, Y.I.: Le groupe de Brauer-Grothendieck en géométrie diophantienne. Actes
          du Congrès International des Mathématiciens (Nice, 1970), Tome 1, pp. 401–411.
          Gauthier-Villars, Paris (1971)
[Ma55]   Mattuck, A.: Abelian varieties over p-adic ground fields. Ann. Math. (2) **62**, 92–119
          (1955)
[MeTa85] Melnikov, O.V., Tavgen, O.I.: The absolute Galois group of a Henselian field, (Russian).
          Dokl. Akad. Nauk BSSR **29**, 581–583, 667 (1985)
[Mi72]   Miyanishi, M.: On the algebraic fundamental group of an algebraic group. J. Math.
          Kyoto Univ. **12**, 351–367 (1972)
[Mi84]   Miller, H.: The Sullivan conjecture on maps from classifying spaces. Ann. Math. (2)
          120, 39–87 (1984)
[Mi80]   Milne, J.S.: Étale cohomology. Princeton Mathematical Series, vol. 33, xiii + 323 pp.
          Princeton University Press, Princeton (1980)
[Mi82]   Milne, J.S.: Zero cycles on algebraic varieties in nonzero characteristic: Rojtman's
          Theorem. Compos. Math. **47**, 271–287 (1982)
[Mi86]   Milne, J.S.: Abelian Varieties. In: Cornell, G., Silverman, J.H. (eds.) Arithmetic
          Geometry, xvi + 353 pp. Springer, Berlin (1986)
[Mi06]   Milne, J.S.: Arithmetic Duality Theorems, 2nd edn, viii + 339 pp. BookSurge, LLC,
          Charleston (2006). ISBN: 1-4196-4274-X
[Mo99]   Mochizuki, Sh.: The local pro-p anabelian geometry of curves. Invent. Math. **138**(2),
          319–423 (1999)
[Mo03]   Mochizuki, Sh.: Topics surrounding the anabelian geometry of hyperbolic curves. In:
          Galois Groups and Fundamental Groups. Mathematical Sciences Research Institute
          Publications, vol. 41, pp. 119–165. Cambridge University Press, Cambridge (2003)
[Mo07]   Mochizuki, Sh.: Absolute anabelian cuspidalizations of proper hyperbolic curves. J.
          Math. Kyoto Univ. **47**, 451–539 (2007)
[Na90a]   Nakamura, H.: Rigidity of the arithmetic fundamental group of a punctured projective
          line. J. Reine Angew. Math. **405**, 117–130 (1990)
[Na90b]   Nakamura, H.: Galois rigidity of the étale fundamental groups of punctured projective
          lines. J. Reine Angew. Math. **411**, 205–216 (1990)
[Na91]   Nakamura, H.: On Galois automorphisms of the fundamental group of the projective
          line minus three points. Math. Zeit. **206**, 617–622 (1991)
[Ne92]   Neukirch, J.: Algebraische Zahlentheorie, xiii + 595 pp. Springer, Berlin (1992)
[NSW08]   Neukirch, J., Schmidt, A., Wingberg, K.: Cohomology of number fields, 2nd edn.
          Grundlehren der Mathematischen Wissenschaften, vol. 323, xvi + 825 pp. Springer,
          Berlin (2008)
[Pa10a]   Pál, A.: Diophantine decidability for curves and Grothendieck's section conjecture.
          arXiv: 1001.4969v2 [math.NT] (January 2010)
[Pa10b]   Pál, A.: Homotopy sections and rational points on algebraic varieties. arXiv: 1002.
          1731v2 [math.NT] (March 2010)
[Pa11]   Pál, A.: The real section conjecture and Smith's fixed point theorem for pro-spaces. J.
          Lond. Math. Soc. **83**, 353–367 (2011)
[Pa90]   Parshin, A.N.: Finiteness theorems and hyperbolic manifolds. In: The Grothendieck
          Festschrift, vol. III. Progress in Mathematics, vol. 88, pp. 163–178. Birkhäuser, Boston
          (1990)

[PlRa94]   Platonov, V.P., Rapinchuk, A.: Algebraic groups and number theory. Pure and Applied Mathematics, vol. 139, xii + 614 pp. Academic, New York (1994)

[Po06]   Poonen, B.: Heuristics for the Brauer-Manin obstruction for curves. Exp. Math. **15**, 415–420 (2006)

[Po08]   Poonen, B.: Insufficiency of the Brauer-Manin obstruction applied to étale covers. Ann. Math. (2) **171**, 2157–2169 (2010)

[PoSt99]   Poonen, B., Stoll, M.: The Cassels-Tate pairing on polarized abelian varieties. Ann. Math. (2) **150**, 1109–1149 (1999)

[PoVo10]   Poonen, B., Voloch, J.F.: The Brauer-Manin obstruction for subvarieties of abelian varieties over function fields. Ann. Math. (2) **171**, 511–532 (2010)

[Po10]   Pop, F.: On the birational p-adic section conjecture. Compos. Math. **146**, 621–637 (2010)

[PoSx11]   Pop, F., Stix, J.: Arithmetic in the fundamental group of a p-adic curve: on the p-adic section conjecture for curves. arXiv: 1111.1354v1 [math.AG] (November 2011)

[Qu08]   Quick, G.: Profinite homotopy theory. Doc. Math. **13**, 585–612 (2008)

[Qu10]   Quick, G.: Continuous group actions on profinite spaces. J. Pure Appl. Algebra **215**, 1024–1039 (2011)

[Ra11a]   Rastegar, A.: Deformation of outer representations of Galois group. Iran. J. Math. Sci. Inform. **6**(1), 33–52, 101 (2011)

[Ra11b]   Rastegar, A.: Deformation of outer representations of Galois group II. Iran. J. Math. Sci. Inform. **6**(2), 33–41 (2011)

[Ra70]   Raynaud, M.: Spécialisation du foncteur de Picard. Publ. IHES **38**, 27–76 (1970)

[Re42]   Reichardt, H.: Einige im Kleinen überall lösbare, im Grossen unlösbare diophantische Gleichungen. J. Reine Angew. Math. **184**, 12–18 (1942)

[Ru12]   Rungtanapirom, N.: Godeaux–Serre Varieties with Prescribed Arithmetic Fundamental Group, vi + 42 pp. Diplomarbeit, Heidelberg (2011)

[S$^+$08]   Stein, W.A., et al.: Sage Mathematics Software (Version 3.1.4), The Sage Development Team. http://www.sagemath.org (2008)

[Sa10]   Saïdi, M.: Good sections of arithmetic fundamental groups. arXiv: 1010.1313v1 [math.AG] (October 2010)

[Sa12]   Saïdi, M.: Around the Grothendieck anabelian section conjecture. In: Coates, J., Kim, M., Pop, F., Saïdi, M., Schneider, P. (eds.) Non-abelian Fundamental Groups and Iwasawa Theory. Cambridge University Press, Cambridge (2012)

[Sch12]   Schmidt, A.: Motivic aspects of anabelian geometry, in: Galois-Teichmüller Theory and Arithmetic Geometry, Proceedings for a conference in Kyoto (October 2010), H. Nakamura, F. Pop, L. Schneps, A. Tamagawa eds., Advanced Studies in Pure Mathematics **63**, 503–517 (2012)

[Sch25]   Schreier, O.: Abstrakte kontinuierliche Gruppen. Abhand. Hamburg **4**, 15–32 (1925)

[Sch94]   Scheiderer, C.: Real and étale cohomology. Lecture Notes in Mathematics, vol. 1588, xxiv + 273 pp. Springer, Berlin (1994)

[Sch98]   Scharaschkin, V.: The Brauer-Manin obstruction for curves, preprint. www.jmilne.org/ math/Students/b.pdf (December 1998)

[Se51]   Selmer, E.S.: The Diophantine equation $ax^3 + by^3 + cz^3 = 0$. Acta Math. **85**, 203–362 (1951)

[Se77]   Serre, J.-P.: Linear representations of finite groups. Graduate Text in Mathematics, vol. 42, x + 188 pp. Springer, Berlin (1977)

[Se79]   Serre, J.-P.: Local Fields. Graduate Text in Mathematics, vol. 67, viii + 260 pp. Springer, New York (1979)

[Se97]   Serre, J.-P.: Galois Cohomology, new edition, x + 210 pp. Springer, New York (1997)

[SGA1]   Grothendieck, A.: Séminaire de Géométrie Algébrique du Bois Marie (SGA 1) 1960–1961: Revêtements étales et groupe fondamental. Documents Mathématiques vol. 3, xviii + 327 pp. Société Mathématique de France (2003)

[Sk99]   Skorobogatov, A.: Beyond the Manin obstruction. Invent. Math. **135**, 399–424 (1999)

[Sk01]   Skorobogatov, A.: Torsors and rational points. In: Cambridge Tracts in Mathematics, vol. 144, viii + 187 pp. Cambridge University Press, Cambridge (2001)

[Sk09]   Skorobogatov, A.N.: Descent obstruction is equivalent to étale Brauer–Manin obstruction. Math. Ann. **344**, 501–510 (2009)

[Sx02]   Stix, J.: Projective anabelian curves in positive characteristic and descent theory for log étale covers, Thesis, Bonner Mathematische Schriften **354**, xviii+118 (2002)

[Sx05]   Stix, J.: A monodromy criterion for extending curves. Intern. Math. Res. Notices **29**, 1787–1802 (2005)

[Sx08]   Stix, J.: On cuspidal sections of algebraic fundamental groups. In: Nakamura, H., Pop, F., Schneps, L., Tamagawa, A. (eds.) Galois-Teichmller Theory and Arithmetic Geometry. Proceedings for a Conferences in Kyoto (October 2010). Advanced Studies in Pure Mathematics, vol. 63, pp. 519–563. arXiv: 0809.0017v1 [math.AG] (2012)

[Sx10a]   Stix, J.: Trading degree for dimension in the section conjecture: The non-abelian Shapiro Lemma. Math. J. Okayama Univ. **52**, 29–43 (2010)

[Sx10b]   Stix, J.: On the period-index problem in light of the section conjecture. Am. J. Math. **132**, 157–180 (2010)

[Sx11]   Stix, J.: The Brauer–Manin obstruction for sections of the fundamental group. J. Pure Appl. Algebra **215**, 1371–1397 (2011)

[Sx12a]   Stix, J.: Birational p-adic Galois sections in higher dimensions. Israel J. Math. arXiv: 1202.2781v1 [math.AG] (February 2012, to appear)

[Sx12b]   Stix, J.: On the birational section conjecture with local conditions, preprint. arXiv: 1203.3236v1 [math.AG] (March 2012)

[St06]   Stoll, M.: Finite descent obstructions and rational points on curves, draft version no. 8. arXiv: 0606465v2 [math.NT] (November 2006)

[St07]   Stoll, M.: Finite descent obstructions and rational points on curves. Algebra Numb. Theor. **1**, 349–391 (2007)

[Su71]   Sullivan, D.: Geometric topology, Part I: Localization, periodicity, and Galois symmetry. Massachusetts Institute of Technology, 432 pp., revised and annotated version, xiii + 284 pp. Cambridge. www.maths.ed.ac.uk/~aar/surgery/gtop.pdf (1971)

[Sz09]   Szamuely, T.: Galois groups and fundamental groups. Cambridge Studies in Advanced Mathematics, vol. 117, x + 270 pp. Cambridge University Press, Cambridge (2009)

[Sz12]   Szamuely, T.: Heidelberg lectures on fundamental groups. In: Stix, J. (ed.) The Arithmetic of Fundamental Groups. PIA 2010. Contributions in Mathematical and Computational Science, vol. 2, pp. 53–74. Springer, New York (2012)

[Sz79]   Szpiro, L.: Sur le théorème de rigidité de Parsin et Arakelov. In: Journées de Géométrie Algébrique de Rennes (Rennes, 1978) vol. II. Astérisque vol. 64, pp. 169–202. Soc. Math. France, Paris (1979)

[Ta97]   Tamagawa, A.: The Grothendieck conjecture for affine curves. Compos. Math. **109**(2), 135–194 (1997)

[Wg09]   Wickelgren, K.: Lower central series obstructions to homotopy sections of curves over number fields. Thesis, 97 pp. Stanford University (2009)

[Wg10]   Wickelgren, K.: 2-nilpotent real section conjecture. arXiv: 1006.0265v1 [math.AG] (June 2010)

[Wg12a]   Wickelgren, K.: On 3-nilpotent obstructions to π_1 sections for $\mathbb{P}_{\mathbb{Q}}^1 - \{0, 1, \infty\}$. In: Stix, J. (ed.) The Arithmetic of Fundamental Groups. PIA 2010. Contributions in Mathematical and Computational Science, vol. 2, pp. 281–328. Springer, Berlin (2012)

[Wg12b]   Wickelgren, K.: n-Nilpotent obstructions to π_1 sections of $\mathbb{P}^1 - \{0, 1, \infty\}$ and Massey products. In: Nakamura, H., Pop, F., Schneps, L., Tamagawa, A. (eds.) Galois-Teichmller Theory and Arithmetic Geometry. Proceedings for a Conferences in Kyoto (October 2010). Advanced Studies in Pure Mathematics, vol. 63, pp. 579–600 (2012)

[Wi34]   Witt, E.: Zerlegung reeller algebraischer Funktionen in Quadrate. Schiefkörper über reellen Funktionenkörpern. J. Reine Angew. Math. **171**, 4–11 (1934)

[Wi08]   Wittenberg, O.: On the Albanese torsors and the elementary obstruction. Math. Annalen **340**, 805–838 (2008)

[Wi12]   Wittenberg, O.: Une remarque sur les courbes de Reichardt–Lind et de Schinzel. In:
         Stix, J. (ed.) The Arithmetic of Fundamental Groups. PIA 2010. Contributions in
         Mathematical and Computational Science, vol. 2, pp. 329–337. Springer, Berlin (2012)
[Wo09]   Wolfrath, S.: Die scheingeometrische étale Fundamentalgruppe. Thesis, 62 pp., Uni-
         versität Regensburg, preprint no. 03/2009

Index

J. Stix, *Rational Points and Arithmetic of Fundamental Groups*, Lecture Notes in Mathematics 2054, DOI 10.1007/978-3-642-30674-7,
© Springer-Verlag Berlin Heidelberg 2013

LECTURE NOTES IN MATHEMATICS   🐎 Springer

Edited by J.-M. Morel, B. Teissier; P.K. Maini

Editorial Policy (for the publication of monographs)

1. Lecture Notes aim to report new developments in all areas of mathematics and their applications - quickly, informally and at a high level. Mathematical texts analysing new developments in modelling and numerical simulation are welcome.

   Monograph manuscripts should be reasonably self-contained and rounded off. Thus they may, and often will, present not only results of the author but also related work by other people. They may be based on specialised lecture courses. Furthermore, the manuscripts should provide sufficient motivation, examples and applications. This clearly distinguishes Lecture Notes from journal articles or technical reports which normally are very concise. Articles intended for a journal but too long to be accepted by most journals, usually do not have this "lecture notes" character. For similar reasons it is unusual for doctoral theses to be accepted for the Lecture Notes series, though habilitation theses may be appropriate.

2. Manuscripts should be submitted either online at www.editorialmanager.com/lnm to Springer's mathematics editorial in Heidelberg, or to one of the series editors. In general, manuscripts will be sent out to 2 external referees for evaluation. If a decision cannot yet be reached on the basis of the first 2 reports, further referees may be contacted: The author will be informed of this. A final decision to publish can be made only on the basis of the complete manuscript, however a refereeing process leading to a preliminary decision can be based on a pre-final or incomplete manuscript. The strict minimum amount of material that will be considered should include a detailed outline describing the planned contents of each chapter, a bibliography and several sample chapters.

   Authors should be aware that incomplete or insufficiently close to final manuscripts almost always result in longer refereeing times and nevertheless unclear referees' recommendations, making further refereeing of a final draft necessary.

   Authors should also be aware that parallel submission of their manuscript to another publisher while under consideration for LNM will in general lead to immediate rejection.

3. Manuscripts should in general be submitted in English. Final manuscripts should contain at least 100 pages of mathematical text and should always include

   - a table of contents;
   - an informative introduction, with adequate motivation and perhaps some historical remarks: it should be accessible to a reader not intimately familiar with the topic treated;
   - a subject index: as a rule this is genuinely helpful for the reader.

   For evaluation purposes, manuscripts may be submitted in print or electronic form (print form is still preferred by most referees), in the latter case preferably as pdf- or zipped psfiles. Lecture Notes volumes are, as a rule, printed digitally from the authors' files. To ensure best results, authors are asked to use the LaTeX2e style files available from Springer's web-server at:

   ftp://ftp.springer.de/pub/tex/latex/svmonot1/ (for monographs) and
   ftp://ftp.springer.de/pub/tex/latex/svmultt1/ (for summer schools/tutorials).

Additional technical instructions, if necessary, are available on request from lnm@springer.com.

4. Careful preparation of the manuscripts will help keep production time short besides ensuring satisfactory appearance of the finished book in print and online. After acceptance of the manuscript authors will be asked to prepare the final LaTeX source files and also the corresponding dvi-, pdf- or zipped ps-file. The LaTeX source files are essential for producing the full-text online version of the book (see http://www.springerlink.com/openurl.asp?genre=journal&issn=0075-8434 for the existing online volumes of LNM). The actual production of a Lecture Notes volume takes approximately 12 weeks.

5. Authors receive a total of 50 free copies of their volume, but no royalties. They are entitled to a discount of 33.3 % on the price of Springer books purchased for their personal use, if ordering directly from Springer.

6. Commitment to publish is made by letter of intent rather than by signing a formal contract. Springer-Verlag secures the copyright for each volume. Authors are free to reuse material contained in their LNM volumes in later publications: a brief written (or e-mail) request for formal permission is sufficient.

Addresses:
Professor J.-M. Morel, CMLA,
École Normale Supérieure de Cachan,
61 Avenue du Président Wilson, 94235 Cachan Cedex, France
E-mail: morel@cmla.ens-cachan.fr

Professor B. Teissier, Institut Mathématique de Jussieu,
UMR 7586 du CNRS, Équipe "Géométrie et Dynamique",
175 rue du Chevaleret
75013 Paris, France
E-mail: teissier@math.jussieu.fr

For the "Mathematical Biosciences Subseries" of LNM:

Professor P. K. Maini, Center for Mathematical Biology,
Mathematical Institute, 24-29 St Giles,
Oxford OX1 3LP, UK
E-mail : maini@maths.ox.ac.uk

Springer, Mathematics Editorial, Tiergartenstr. 17,
69121 Heidelberg, Germany,
Tel.: +49 (6221) 4876-8259

Fax: +49 (6221) 4876-8259
E-mail: lnm@springer.com